FROM QUARKS TO QUASARS

Volume 7 *University of Pittsburgh Series in the Philosophy of Science*

From Quarks

Editor

ROBERT G. COLODNY

ALBERTO COFFA
JOHN A. WINNIE
DAVID B. MALAMENT
MICHAEL R. GARDNER
JOHN STACHEL

to Quasars

Philosophical Problems of Modern Physics

University of Pittsburgh Press

Published by the University of Pittsburgh Press, Pittsburgh, Pa., 15260

Feffer and Simons, Inc., London
Manufactured in the United States of America

Library of Congress Cataloging in Publication Data

Main entry under title:

From quarks to quasars.

(University of Pittsburgh series in the philosophy of science; v. 7)
Bibliography: p.
Includes index.
1. Physics—Philosophy. 2. Quantum theory.
I. Colodny, Robert Garland. II. Coffa, Alberto.
III. Series.
QC6.F76 1986 530'.01 84-29456
ISBN 0-8229-3515-5

Alberto Coffa

In memoriam

Contents

Preface

Each year, the Center for Philosophy of Science at the University of Pittsburgh invites philosophers or scientists to present a public lecture and/or to conduct one or more seminars on some topic in the philosophy of the physical, biological, or human sciences. This series is supported by a grant from the R. K. Mellon Foundation. In recent years, Larry Laudan and I planned these presentations so as to center on one or another of several specific topical foci.

Among the ensuing essays, those which cluster around any one such motif have been gathered to form a separate volume having a high degree of thematic unity. Thus, the present book is addressed to readers concerned with the cardinal contemporary issues in the philosophy of physics.

Its publication affords me the welcome opportunity to express our warm gratitude to Professor Robert G. Colodny. For the past twenty years, he has generously taken time from his normal scholarly work as a historian to do the arduous editing of the first seven volumes of our *Pittsburgh Series*. The appearance of this seventh volume concludes his valued service as Editor.

ADOLF GRÜNBAUM
Chairman, Center for Philosophy of Science
University of Pittsburgh

ROBERT G. COLODNY
University of Pittsburgh

Introduction

> A good physicist uses formalism as a poet uses language.
> —Yu. A. Manin

There are two truisms about our intellectual world that astonish the historian. The first is that only three hundred years separate us who live in the space age from our great ancestors who, in the age of Galileo, Kepler, and Newton, built the foundations of classical, that is, modern science. The second is that in the memory of men still among us, the foundations of that science were completely restructured. Relativity theory and quantum mechanics with elementary particle physics and the new cosmologies, conjointly represent a radical departure not only from old ideas about the physical world but equally revolutionary methods of analyzing the sources of our knowledge. Physics, one might say has again become "Natural Philosophy." This means that at the multifaceted junctures between mathematics and physics and logic and philosophy there arise of necessity, debates, or if one prefers an older terminology, dialogues. These concern not merely the structure of the formalisms but their interpretations; and these not only have as reference points specific domains of physics, but imply modifications for the entire mosaic of modern understanding in such realms of inquiry as astrophysics, biophysics, and the philosophy of science.

If this is a proper statement of our intellectual moment, then it becomes apparent that the authors whose works are brought together in this volume are engaged in a humanistic enterprise. This is so for a double reason: contemporary science has reestablished the centrality of the human mind in its relation to the known and restored the intimacy of *homo sapiens*, the explorer with the cosmos explored.

Only an unworldly optimist would have expected that the transition from one world view to its replacement would have been devoid of intellectual anguish, or that some of the fruits harvested from the tree of knowledge would not be bitter. Both human tragedies are hinted at by one scholar who worked in the very center of the "revolution"—Max Born. His words are repeated here because they put into proper perspective important philosophical *issues* that appear in most of the essays that follow.

"In 1921 I believed—and I shared this belief with most of my contemporary physicists—that science produced an objective knowledge of the world, which is governed by deterministic laws. The scientific method seemed to me superior to other more subjective ways of forming a picture of the world: philosophy, poetry, religion, and I even thought the unambiguous language of science to be a step towards a better understanding between human beings.

"In 1951 I believed in none of these things. The border between object and subject had been blurred; deterministic laws had been replaced by statistical ones, and although physicists understood one another across all national frontiers, they had contributed nothing to a better understanding of nations, but had helped in inventing and applying the most horrible weapons of destruction."

Thirty years later, one cannot announce a new era of fraternity among nations, yet it is also indisputably the case that science, particularly mathematical physics, is a house of many mansions, in which one will find not a trace of nation of origin in any of the rooms. What is Irish about the mathematics of Hamilton or French about the work of Poincaré or Jewish about the field equations of Einstein?

With the many mansions metaphor in mind, it is proper to mention that the philosophers who appear here are not strangers looking in from the outside or casual visitors. They are permanent residents. This is one of the enduring achievements of contemporary scholarship. The division of the university into academic fiefdoms has not prevented the emergence of scholars with a transdisciplinary competence, learned in physics, mathematics, logic, and philosophy.

The readers who follow the arguments that fill these pages will, of course, note that the powerful light of modern mathematics is used to illuminate the more obscure parts of physical hypotheses. It could not be otherwise.

At the dawn of this scientific revolution, Alfred North Whitehead wrote these prophetic words: "We are entering an age of reconstruction, in science, and in political thought. Such ages, if they are to avoid ignorant oscillations between extremes, must seek truth in its ultimate depths. There can be no vision of this depth of truth apart from a philosophy which takes full account of those ultimate abstractions, whose interconnections it is the business of mathematics to explore." In a later passage pertaining to quantum mechanics, he wrote: "We have come back to a version of the doctrine of old Pythagoras from whom mathematics and mathematical physics take their rise. He discovered the importance of dealing with abstractions, and in particular directed attention to numbers as characterizing the periodicities of music. The importance of the abstract idea of periodicity was thus present at the very beginning both of mathematics and philosophy." Solomon Bochner has given a terse explanation for the level of abstraction mandatory for the power of mathematics, particularly of the kind that is used in these essays: "The Greeks were more philosophers and poets than mathematicians. They did not have in their creativeness the instinctive realization that in the actual creation of mathematical subject matter, the complex comes before the simple, the recondite before the obvious, the hidden before the manifest."

There was a long prologue in the development of modern mathematics that is the subject of Alberto Coffa's essay, "From Geometry to Tolerance: Sources of Conventionalism in Nineteenth-Century Geometry." As Herman Weyl put it: "The historical development of the problem of space teaches how difficult it is for human beings entangled in external reality to reach a definite conclusion. A prolonged phase of mathematical development, the great expansion of geometry dating from Euclid to Riemann, the discovery of the physical facts of nature and their underlying laws from the time of Galileo together with the incessant impulses imparted by new empirical data, finally the individual genius of great minds, Newton, Gauss, Riemann, Einstein, all these factors were necessary to set us free from the external, accidental, non-essential characteristics which would otherwise have held us captive."

Coffa's concern is not with the technical problems posed by the emergence of non-Euclidean geometries: Lobachevski, Bolyai, Gauss, and others, but with the structure of the arguments about the nature of geometrical propositions as these arose in philosophy. As Coffa states: "Whoever cast their roles can hardly have done better: the two greatest philosophers and the two greatest mathematicians of their generation met and dueled artfully for their respective disciplines": Frege, Russell, Hilbert, and Poincaré.

The intellectual problems posed beyond the boundaries of mathematics by the new geometries, as Coffa demonstrates, had profound consequences for the development and refinement of epistemology and thus affected all philosophy. And as Ernest Nagel, among others, has observed, these debates prepared the way for the acceptance of relativity theory in the next century. Readers will also note a similarity between these geometrical-logical debates and the controversies concerning the foundations of quantum physics as portrayed by Gardner and Stachel.

John Winnie's essay, "Invariants and Objectivity: A Theory with Applications to Relativity and Geometry," fits into that grand tradition of logico-mathematical investigations that follow from the *Principia Mathematica*. As Winnie stated (in a note to the editor): "First a general theory of sets—theoretical structures—is developed and the consequences of choosing invariance as the criterion of objectivity are examined in detail. Next, applications of that theory to the conventionalism issue in the foundations of geometry and special relativity are provided. Finally, free invariance is presented as an alternative theory of objectivity, and the conditions under which invariance and free invariance coincide are established. An application to Klein's Erlanger Program is forthcoming at once." In a later commentary, he notes, "The objectivity problem also arises frequently in connection with the various cosmological models of general relativity. In highly symmetrical models, the theory of global invariance would apply; for less symmetrical or rigid models, the theory of free invariance (sec. 5) is a possible approach."

David Malament's "Newtonian Gravity, Limits, and the Geometry of Space" sheds new light on the relationship between the Newtonian gravitational theory and Einstein's general theory of relativity. He shows that various features of general relativity, once thought to be uniquely characteristic of it, *do not* distinguish it from Newtonian theory. He thus provides the means with which to make geometric sense of the standard claim that "Newtonian gravitational theory is the classical limit of general relativity." The relationship of this beautiful work to the other essays with a geometrical core in this volume is suggested by Malament's assertion that: "Much has been written about the geometry of space, but almost exclusively in connection to the claims for or against conventionality. There has been little consideration of reasons '*internal* to physics' why space should have one geometric structure rather than another. My remarks concern one such (reason)." It is clear that exploration along the boundary of two theories, particularly where one is historically the replacement of the earlier, must clarify the foundations of both. Futhermore, Malament's insights will mandate a rewriting of many standard accounts of the history of physics.

Quantum mechanics as a more or less complete theory of the subatomic realm is scarcely fifty years old, and so it is not surprising that the shock waves from this shift in the foundations of physics continue to produce fissures in all othodox walls constructed to constrain the transformation of *philosophical* theories which had as their objective the explanation of scientific reasoning or the nature and function of scientific theories. The debates that began in the era of those Titan-like iconoclasts Planck and Einstein continue in this volume. The radical discontinuity between fundamental aspects or tenets of classical physics and the emergent quantum microworld, when reflected in philosophical thought, stimulated, if it did not mold, the ideas that crystallized in the varieties of logical positivism. As the logico-mathematical foundations were probed, the logical foundations of patterns of explanation were also called into question. It is debates at this level that Michael Gardner and John Stachel join herein.

Gardner argues that what he, somewhat scornfully, refers to as the "Received View" is an oversimplification of, and an overgeneralization from, certain quite peculiar features of quantum theory. He defends what is called the *minimal statistical interpretation of quantum mechanics* from numerous competing if not contradictory interpretations.

John Stachel's arguments against the proponents of a new quantum logic have echoes of some of the most passionate debates within the western philosophical tradition. As he puts it: "We are told that the reason we have such trouble in our intercourse with the microworld—the world of the electrons, protons, photons—is that we are using the wrong logic in our attempts to understand them. . . . What I am arguing against, basically, is the idea that there is some sort of simple mirroring relationship between the structure of language and the structure of the extra-linguistic world; so that logical relations must mirror structural relations of 'reality' in some direct, unmediated fashion." For Stachel, this enterprise of seeking a new logic is a version of "the uncritical positivism and equally uncritical idealism which Marx stigmatized in his Critique of Hegel." Furthermore, Stachel contends that this path leads away, dangerously, from the more urgent task of resolving the tensions within the existing theoretical and experimental structure, or between that structure and other unassimulated elements—tensions that might lead to changes in, or even the complete overthrow of that structure.

It is clear that the physics of the microworld is not a completed structure; that not only are there elements being added to this House of Many Mansions, but the foundations are being altered. Even as these notes are being written, the news is coming in of the isolation of the W particle at the CERN laboratories in Europe. Long predicted, this object awaited the building of more powerful machines of high energy physics. This breakthrough reminds us of similar turning points in the extraordinary history of the "new" physics: de Broglie matter waves, Dirac positron, Yukawa meson, and similar episodes. The pattern of the recent history indicates that as experimentalists penetrate even deeper into the fine grain of nature, grander, more general unified theories emerge. The philosophers also confront "possible worlds" of baffling complexity and beauty. Thus there now exists a Penrose world which, in the analysis of Manin, "has turned out to be very useful in the study of Maxwell's equations and their generalizations—the Yang-Mills equations which, we now believe, describe gluon fields connecting the quarks in a nucleon!"

It is no exaggeration to say that the progress of quantum mechanics has absorbed and stimulated the creative energies of the most powerful minds of the twentieth century. Albert Einstein's role in the introduction and development of these theories, particularly in the years 1900–1915 has been chronicled before. However, there have been gaps in the story and misconstruals of the significance of some of the evidence.

John Stachel, a professor of physics, is also editor of the Einstein Project of Princeton University. His research in this archival treasure has now permitted him to shed new light on the evolution of Einstein's philosophy of nature and his philosophy of science with particular reference to the development of emerging quantum theory. In some cases, the alterations of the standard view are matters of nuance. But for a scientist who carried on a dialogue with his God and found him *subtle* but not wicked, nuances may be everything. "Einstein and the Quantum: Fifty Years of Struggle" will astonish and delight the world of scholars.

NOTE

Debates concerning the philosophy of quantum physics have appeared in earlier volumes of this series in the Philosophy of Science: Paul K. Feyerabend, "Problems of Microphysics," in *Frontiers of Science asnd Philosophy* (Pittsburgh: University of Pittsburgh Press, 1962); Hillary Putnam, "A Philosopher Looks at Quantum Mechanics," in *Beyond the Edge of Certainty: Essays in Contemporary Science and Philosophy* (New York: Prentice Hall, 1965); Henry Margenau, "The Philosophical Legacy of Quantum Mechanics," in *Mind and Cosmos* (Pittsburgh: University of Pittsburgh Press, 1966); and essays by Arthur Fine, Gerald Feinberg, David Finkelstein; Clifford Hooker, Bas Van Frasson, and Howard Stein in *Paradigms and Paradoxes: The Philosophical Challenge of the Quantum Domain* (Pittsburgh: University of Pittsburgh Press, 1972).

In preparing the introduction, the editor acknowledges the assistance of the contributors and is indebted to the works of E. Nagel, S. Bochner, H. Weyl, and Yu. I. Manin.

FROM QUARKS TO QUASARS

ALBERTO COFFA
Indiana University

From Geometry to Tolerance: Sources of Conventionalism in Nineteenth-Century Geometry

> Let us for a moment consider . . . any axiom of geometry, for example the following:—through two points in space there always passes one and only one straight line. How is this axiom to be interpreted in the older sense and in the more modern sense?
>
> The older interpretation: —Everyone knows what a straight line is, and what a point is. Whether this knowledge springs from an ability of the human mind or from experience, from some collaboration of the two or from some other source, is not for the mathematician to decide. He leaves the question to the philosopher. Being based upon this knowledge, which precedes all mathematics, the axiom stated above is, like all other axioms, self-evident, that is, it is the expression of a part of this a priori knowledge.
>
> The more modern interpretation: —Geometry treats of entities which are denoted by the words straight line, point, etc. These entities do not take for granted any knowledge or intuition whatever, but they presuppose only the validity of the axioms, such as the one stated above, which are to be taken in a purely formal sense, i.e., as void of all content of intuition or experience. These axioms are free creations of the human mind. . . . The matter of which geometry treats is first defined by the axioms.
>
> —Albert Einstein
> *Geometrie und Erfahrung*

Most students of philosophy have their favorite way of classifying empiricists. My own is this: there are two kinds of empiricists, the strict ones and the loose ones. Strict empiricists believe that all sentences in a system of beliefs—scientific or otherwise—fall into one or another of the following two categories: (1) those whose truth value is determinable only on the basis of empirical facts and (2) those that are truths or falsehoods of logic, or reducible to such via definitions.

The loose empiricist, on the other hand, refuses to acknowledge the exhaustiveness of the strict empiricist's categories. For, as he examines the sentences he is most impressed by, those of the empirical sciences and mathematics, he finds that he must call for Procrustean help much too often in order to force some of them into one or another of the strict empiricist's categories.

Examples of such hard-to-classify sentences range from the sublime to the pedestrian. Among the former, the loose empiricist will include the axioms of geometry, some principles of physics—often invariance laws—the attribution of continuity or discreteness, finitude or infinitude to space and time, and so on. Among the least exciting kind he includes statements such as that physical objects are in space and endure through time and that they are colored through and through and through. All of these, and many others, exhibit in the loose empiricist's eyes a strange dual nature; for even though their syntactic appearance is no different from that of sentences that do convey information, as we look at the way in which these sentences are used by their most competent users, we see that they are treated as if facts were largely irrelevant to the determination of their truth value (if not, perhaps, to their adoption or rejection).

One of the most interesting episodes in the last two hundred years of philosophy, I believe, is the sequence of loosely empiricistic attempts to develop a theory of what these claims are and do. From Kant's theory of the synthetic a priori to current theories of categorical principles (Sellars), framework principles (Putnam), paradigms (Kuhn), research program cores (Lakatos), and protophysics (Lorenzen), there extends a long series of variously enlightening attempts to provide a satisfactory account of their idiosyncratic and elusive nature. It should go without saying that behind this topic hides the largest sleeping giant of modern analytic epistemology, the problem of the a priori. One aim of this essay is to show that in that complex fabric of ideas known as conventionalism, there is a thread, going from Poincaré and Hilbert to Carnap and Wittgenstein, that may be seen as an enlightening attempt to solve that millennial puzzle.

As I see it, this attempt draws its inspiration from that most celebrated of intellectual episodes in mathematics, the decline of Euclideanism in nineteenth-century geometry. The variety of conventionalism with which I deal here developed as a response to a certain conflict between the classical (Aristotelian) conception of science and the appearance of non-Euclidean geometries. As I shall argue in section 1, the mathematicians' original reaction to this philosophical problem was to view it as a problem of interpretation, one which Beltrami and Klein finally solved. But the bounds of admissible geometries soon grew far beyond the domain that their line of thought could justify. As a consequence, increasing numbers of geometers came to the conclusion that a more radical break with the classical understanding of science was required. The nature and the magnitude of the challenge they unwillingly posed to that conception of science becomes clearest in the debates that developed between the most outstanding proponents of these two conflicting standpoints.

Whoever cast their roles could hardly have done better: the two greatest philosophers and the two greatest mathematicians of their generation met and dueled artfully for their respective disciplines. The struggles between Frege, Russell, Hilbert, and Poincaré, to be surveyed in sections 2 and 3, reveal better than any other source I know, the role of the two central themes of this essay: the thesis of semantic atomism and the nature of indefinables. The former was the basic presupposition of the problem under debate; the latter, its kernel.

During the first three decades of our century, the lessons of geometry slowly soaked into philosophy. Then, around 1930, there was a sudden, dramatic change.

Up to the 1920s, positivists had rejected all earlier theories of the a priori, claiming that they were based on a variety of confusions. Faced with a candidate for the role of an a priori claim, positivists either would dismiss it as meaningless nonsense or would try to force it into one or another of the strict empiricist's categories. In all cases they failed to articulate a theory of the nature and function of those statements, thus lending support to the conviction—often justified—that positivists were strict empiricists.

The conjecture I will put forth in sections 4 and 5 is that in the late twenties and thirties this situation changed dramatically. For in that period and, as far as I can tell, for the first time in the history of positivism, its two leaders, Carnap and Wittgenstein, came to acknowledge the significance of the a priori, and then proceeded to develop a radically new theory of it. Instead of limiting themselves either to offering a variety of arguments to dismiss a priori claims as nonsense or to viewing them as truths or falsehoods of logic or of some generalized physics, they acknowledged their idiosyncratic character and the inability of a philosophy to get off the ground unless it begins by recognizing the peculiar and essential role played by them in all forms of knowledge. Both Carnap and Wittgenstein used the word "syntax" in connection with their theories of the a priori. As we shall see, Wittgenstein's philosophical grammar and Carnap's logical syntax are more likely to be understood and appreciated when we regard them as extensions of the conventionalist insights concerning the nature of geometry and of its indefinables.

Wittgenstein's grammar and Carnap's syntax were the sources of a version of conventionalism that is widely believed to have been refuted by Quine's attack against "truth by convention." On the contrary, that attack left the kernel of linguistic conventionalism untouched. The rush to holism and the attendant anticonventionalism of recent decades were partly fueled by widespread misunderstandings of the conventionalist doctrines and of what it took to refute them. As holism became more widely entrenched, the familiar mechanisms of cognitive dissonance failed to encourage a balanced historical appraisal of the standpoint that was being rejected. As a result, our communal picture of conventionalism is, to put it mildly, flawed. There is, however, a simple remedy known among lazy philosophers as "the genetic fallacy"; for the true nature of conventionalism is best displayed in the historical dialectic (*sit venia verbi*) which led to Carnap's principle of tolerance. The story is far too complex to be told here, but perhaps the syncopated, sketchy account that follows may help draw a wedge between conventionalism and the silly doctrine that now bears its name.

1. GEOMETRIES

Until the nineteenth century it was a commonplace among mathematicians and philosophers to define the different branches of mathematics on the basis of their objects, as, for example, the science of number, space, extension, and so forth. What has been called the "Aristotelian" conception of science[1] demanded that every genuine field of knowledge should *begin* by identifying a genus of entities whose truths it was aimed to capture. One could hardly attempt to identify a class of true sentences, it was thought, without first determining what the sentences in question are about or, at the very least, what their primitive terms mean.

Together with this conviction, and from very early on, there had developed in mathematics a sequence of situations in which the tacit enforcement of that semantic dogma had generated the sort of uneasiness which is usually described as philosophical. From the discovery of irrationals to the discovery of points at infinity in projective geometry through the discoveries of the number zero, the negatives, infinitesimals, the square root of minus one, and so many other alleged mathematical entities, one finds the same process recurring again and again. At first, a few mathematicians of genius observe that reasons of a formal or practical nature make it advisable to extend the existing linguistic forms, allowing for new forms of expression that, as far as their syntactic behavior is concerned, are practically indistinguishable from those that the community of mathematicians already considers to be legitimately referential. Mathematicians proceed to notice that it is hard to say what the meanings or referents of these new expressions are; but, in the absence of a philosophy that justifies the use of the new expressions without presupposing that they designate, it is concluded that the new names must have a meaning.[2] The distrustful attitude of early generations towards these alleged semantic objects is revealed by their use, in this connection, of ontologically derogatory adjectives such as "imaginary," "impossible," and the like. The most philosophically inclined will conclude that they are facing a problem of interpretation, by which they usually mean the problem of finding a way to describe the new meanings in the old language. By and large, however, the community of mathematicians wisely chooses not to hold its breath until this problem is solved. Instead, it plunges into the use of the new linguistic forms, inspired less by faith in the semantic propriety of its claims than the advice attributed to d'Alembert: "Allez en avant et la foi vous viendra!"

One of the most remarkable facts about nineteenth-century geometry is that in it, perhaps for the first time in the history of science, that faith did not come. Not only were the points and lines of hyperbolic geometry never to achieve that familiar status which appeared eventually to befall all "impossible" objects; the loss of semantic faith eventually infested their geometric neighbors, the indefinables of Euclid's geometry. Indeed, whereas early in the century almost everybody appeared to be quite certain that geometry was about, say, points and straight lines, and that there couldn't really be any serious question about what those words designated, by the end of the century geometers thought that geometry was about nothing in particular, and that words such as 'point' and 'straight line' either had literally no meaning at all, or else (almost equivalently) had pretty much any meaning you might think of. In spite of the attention received by other aspects of the geometric revolution, not enough has been devoted to the forces that led from one end of this semantic spectrum to the other.[3]

1.1. The Multiplicity of Geometries

When non-Euclidean geometries first appeared on the scene, they were naturally understood as rivals of Euclid, and their primitives were therefore taken to designate the same things as Euclid's, whatever they might be. Presumably, since both Euclid's and Lobachevskii's were *geometries*, they had to be interpreted as attempts to identify the class of truths concerning the very same domain of entities. Their points, straight lines, and so on had to be the same in both theories.

As soon as geometers became convinced that non-Euclidean geometries are consistent, the question that philosophy instructed them to pose was: Which geometry is the true one? And the answer to this question obviously depended on the answer to a previous question: What is geometry really about? What *are* the points and straight lines that geometry deals with? The problem of the topic of geometry became now more than a need to satisfy a vaguely philosophical discomfort; it became a pressing issue on the frontiers of mathematics. The problem was further complicated by the introduction of a new, indeed, unprecedented development.

During the second half of the nineteenth century, through a process still awaiting explanation, the community of geometers reached the conviction that all geometries were here to stay. Given the then current philosophical opinions concerning the nature of science in general and of geometry in particular, this had all the appearance of being the first time that a community of scientists had agreed to accept in a not-merely-provisory way all the members of a set of mutually inconsistent theories about a certain domain. I shall refer to the emerging situation as *the problem of the multiplicity of geometries*. It was now up to philosophers to solve it, that is, to make epistemological sense of the mathematicians' attitude toward geometry.

The challenge was a difficult test for philosophers, a test which (sad to say) they all failed. Frege, for example, the most enlightened and penetrating among them, argued that all non-Euclidean geometries were false and that they should therefore be placed together with astrology and alchemy in the category of the pseudosciences. The geometers returned such advice with thanks and turned to the task of solving their own philosophical problems without outside help.

The first attempt by geometers to solve the problem of the multiplicity of geometries was framed well within the confines of an Aristotelian picture of knowledge. Noting that the choice of a unique, intertheoretic set of meanings for the geometric primitives was inconsistent with the simultaneous acceptance of all geometries, some geometers chose to conceive their problem as one of interpretation.

A problem of interpretation, in the intended sense, normally arises when we happen to have good reasons to want to preserve a certain theory, but we are not quite clear about what the theory says. This semantic difficulty is to be distinguished from a different source of uneasiness toward otherwise satisfactory theories which arises when we know full well what the theory is about and what it says but, for some reason or other, we find it hard to believe some of its claims. Quantum physics offers perhaps a clear instance of the first sort of semantic difficulty; the Copernican theory, one of the second. One should not confuse the "instrumentalist" *proposal* to alter the very clear meaning of certain expressions in a theory in order to preserve its welcome predictive power while removing its offensive implications, with the *recognition* that it is highly unclear what, if anything, certain sentences of a theory assert. A problem of interpretation, as here conceived, does not arise out of unreasonable epistemic demands on the evidential grounds of a theory; it arises from semantic questions concerning what the theory says and what, if anything, it is about.

The point is then that the first serious attempt to solve the problem of the multiplicity of geometries and to find room for all geometries within mathematics, conceived that problem as one of interpretation. Beltrami's solution to it was, as we shall see, entirely classical—but partial. In Klein's more adequate treatment there was already an implicit acknowledgment that not everything was as clear concerning meaning and reference in geometry as one might have hoped, a theme that will recur throughout the following pages. For, instead of taking for granted that the points and lines of hyperbolic geometry are the same allegedly familiar objects that Euclidean geometry studies, Klein came to the conviction that it was more appropriate to treat these geometric entities the way mathematicians had treated the point at infinity or the square root of minus one. A brief survey of the basic elements of Beltrami's and Klein's respective solutions will help us grasp the crisis of semantic intuition in midcentury geometry.

In an article published in *Crelle's Journal* for 1839–1840, Minding had noticed that the trigonometric expressions for geodesic triangles on a surface of negative curvature were formally identical to the trigonometric formulas valid on a sphere, except for the fact that in the former the radius always occurred multiplied by *the square root of minus one.* Two years earlier, and in that same journal, Lobachevskii had observed that the trigonometry of his "imaginary" (i.e., hyperbolic) geometry was the imaginary counterpart of spherical trigonometry in that the formulas in question were identical except for the occurrence of *r times the square root of minus one* in the hyperbolic case wherever the radius *r* of the sphere occurs in the Euclidean case. Beltrami was the first one to put these two facts together[4] and to derive from them a (partial)[5] interpretation of Lobachevskii's geometry. In his "An Attempt to Interpret Non-Euclidean Geometry" (3) and later in his "Fundamental Theory of Spaces of Constant Curvature" (4), he put forth the famous pseudospherical model of hyperbolic geometry.

Beltrami's "Saggio" (3) starts with a clear statement of his goals. His justification of non-Euclidean geometry, such as it may be, is not intended to challenge in any way the legitimacy of Euclidean geometry. "In mathematics," he tells us, "the triumph of new concepts cannot put in jeopardy the already acquired truths" (p. 374). On the contrary, Euclidean geometry remains the undisputed true doctrine of space. Lobachevskii's theory is to be justified by showing that one can identify "a real substratum for that doctrine, instead of admitting the necessity of a new order of entities and concepts" (p. 375). The pseudospherical model is then developed in order to argue that two-dimensional hyperbolic geometry is no more than the (Euclidean) geometry of the pseudosphere. The fear of the new geometry vanishes once we realize that "non-Euclidean planimetry is nothing else than the geometry of a [Euclidean] surface of constant negative curvature" (4, p. 425).

The predominance of Euclidean geometry is further emphasized by Beltrami's attitude toward the three-dimensional case. According to him, whereas the results of hyperbolic planimetry receive "a true and appropriate interpretation because they become constructible on a real surface, those which refer to three dimensions are capable of only an analytical representation" (4, p. 427). The goal of finding a "real substratum" for three-dimensional hyperbolic geometry "cannot be achieved" (3, p. 375).

Klein's intentions were even more clearly reductionistic than Beltrami's, although in his case the reduction was not to Euclidean but to projective geometry. Still in the 1920s, he was complaining about the widespread misinterpretation of his 1871 article as an attempt to provide a Euclidean model for non-Euclidean geometries (66, p. 242). His basic goal, as he had already pointed out in 1871, was to display the "inner essence" of all metric geometries and to offer new foundations for them by reducing them to projective principles.

I wish I could recount here the fascinating story of how geometers came upon the idea that behind the metric structure that was the subject of Euclidean geometry, there is a deeper, underlying, nonmetric domain, desribed by what they called "descriptive" or "projective" geometry. This idea evolved from the tacit appeal to what became one of the great conceptions of nineteenth-century geometry, the notion of invariance under a group of transformations. Practical concerns had led geometers to focus on certain geometric transformations of a kind that they called "projective" because they were basically obtained through processes of projection and section. Considering these transformations, they noticed that although some features of the projected figures, including the specifically metric ones, varied radically under these transformations, other features remained invariant. This led to the idea of fixing attention on the properties invariant under those tranformations and of identifying that fragment of geometry that circumscribed itself to the analysis of such invariants. This was the task of projective geometry.

During the first half of the nineteenth century, projective geometry became the leading branch of geometry, and in the 1850s and 1860s it was widely regarded as a more promising geometric territory than the one discovered by Riemann and Helmholtz.[6] Cayley proposed the program of constructing all geometry from this seemingly small fragment of it, and when he succeeded in reducing Euclidean to projective geometry he felt justified in issuing his famous motto: "descriptive [i.e., projective] geometry is all of geometry."

Cayley's reduction was, in effect, grounded on the fact that if we consider a certain kind of conic in the projective plane and if we look at the projective transformations that leave this conic invariant, it turns out that these projectivities coincide with the similarities in the Euclidean plane. The Euclidean metric structure (up to a scale factor) can thus be said to be contained within the projective structure, since the former can be fully defined in terms of the latter. The specific way in which Cayley showed the reduction to hold involved the definition of a "distance" function, that is, a function that associated a real number with every pair of points and that fulfilled certain standard properties. The function in question was projective, in the sense that its characterization appealed only to projective coordinates[7] (a point on which Russell would later focus) and projective primitives. Cayley established that if the numbers assigned by this function to arbitrary pairs of points are regarded as their distance, then the geometry of the corresponding space is Euclidean.

In 1869 Klein stumbled upon Cayley's memoir, and shortly thereafter, he came across Lobachevskii's papers. In a remarkable stroke of genius, he concluded that there was a deep connection between them. The question Klein posed was this: What if, instead of fixing attention on the specific projective conic that Cayley had considered, we reproduce Cayley's construction for a different conic, for example, for the kind mathematicians call real and nondegenerate? The answer was given in Klein's (66): the group of projectivities leaving a real nondegenerate conic invariant is (isomorphic with) the group of hyperbolic isometries, that is, with the isometries of Lobachevskii's geometry. Klein proceeded to put forth another function from pairs of projective points (internal to the given conic) to reals, which was defined in purely projective terms. If this function is taken to be the distance between points inside the conic, then the geometry of these points can be shown to be Lobachevskii's. He went on to argue that for any projective conic, the group of projectivities leaving it invariant should be seen as the metric group of some geometry. Many decades later Klein would return to this remarkably beautiful idea of his, noting that it had placed the geometer in the position of "the traveler who, standing on top of a mountain, contemplates several valleys at the same time, and who at an earlier time while standing in one of the valleys could barely manage to imagine the course of the others" (65, p. 542).

This sketchy account of Beltrami's and Klein's ideas may suffice to show the sense in which their attempts to solve the problem of the multiplicity of geometries were reductionistic. One might say that they were both attempting to tame the wild languages of the new geometries into the more acceptable—Euclidean or projective—frameworks.

But Klein himself, led by the internal logic of his own discovery, put forth in 1872 a new conception of geometry (68). In Weyl's words, the dictatorial regime of the projective idea in geometry was finally overthrown by Klein's Erlangen program, which proclaimed all groups of transformations as geometrically equal. (It would soon turn out, as usual, that some groups were more equal than others.) As a consequence, whereas by midcentury the question was whether the number of admissible metric geometries was one, or two, or perhaps even three, the increasing acceptance of the Erlangen program as well as other geometric developments (such as the growing recognition of the significance of nonhomogenous Riemannian geometries) soon led to the conviction that there were many more admissible geometries than Beltrami and Klein could account for through their reductionistic methods. Their approach, conceived with the narrow class of linear geometries in mind, had nothing to say about what to do with this unruly and ever-growing class of geometries. A radically new approach was needed.

Under these circumstances, by the end of the nineteenth century, a large number of geometers, with Poincaré, Hilbert, and the Italian school in the lead, took it upon themselves to solve the problem of the accommodation of all geometric systems within geometry by providing a new understanding of the nature of geometric knowledge. The "new geometry" which they put forth was less a bold new idea of a few brilliant minds than the explicit recognition of a fact that had been making its presence felt with growing urgency for quite some time, a fact already displayed in the preceding considerations, and which I must now attempt to characterize more precisely.

1.2. Semantic Uncertainties

Almost from the beginning of geometry there had been what one might call an element of semantic uncertainty that had slowly developed through the centuries and had finally acquired crisis proportions with the advent of non-Euclidean geometries. A quick glance at some of the most revealing episodes will illustrate the point.

One of the most familiar facts about the prehistory of non-Euclidean geometry is the seemingly interminable sequence of attempts to prove the parallel postulate. As usually told, there is an implausibly comic element in the episode, with each member of that sequence rejecting his predecessors' proofs and proceeding to offer another one, soon to be rejected by all of his successors. On the standard account, the problem was that all proofs relied on "tacit assumptions" which, once identified, were seen to be no less questionable than the postulate in need of proof. What we would have, then, is a long list of oversights that could have been avoided had geometers exercized a greater logical care.

In fact, logic had little to do with this episode. From Proclus on, a large proportion of the "tacit assumptions" that would be questioned by later geometers were either explicitly stated or else regarded as too obvious to merit a sentence in the proof. Proclus's proof, for example, assumed (tacitly) that the distance between parallel straight lines does not grow without limit (81, pp. 371—72); Cataldi assumed (explicitly) that straight lines that are not equidistant converge in one direction and diverge in the other; Wallis's celebrated proof assumes (quite explicitly) that for every triangle there is another similar triangle of arbitrary size.[8] The list could be prolonged almost indefinitely.

A standard account of this episode sees in it a growing insight on the nature of the parallel axiom; I can find nothing of the kind. For if one is willing to accept anything at all as self-evident, as no doubt all participants in these proceedings were, what could possibly be more self-evident than the existence of two similar triangles, or any of the other assumptions mentioned above? Few geometers have received more praise from philosophers than Proclus has, on account of his celebrated objection to the self-evidence of the parallel postulate;[9] but I can find no answer to Lambert's complaint that one cannot "derive from the consideration of asymptotes objections to the representation of straight lines, for in an entirely analogous manner one would be led to wonder whether two straight lines could not be placed so that they enclose a space" (Stäckel and Engel 94, p. 155).

These attitudes toward the self-evidence of the parallel postulate and of their sufficient conditions could not reflect a growing realization that there was something wrong with the postulate, for the simple reason that there was, in fact, *nothing* wrong with it that wasn't also wrong with all other postulates. What this sequence of "proofs and refutations" reflects is, merely, that what one geometer regarded as evident, most others regarded as less than obvious; that, as I would prefer to put it, what counted for some as meaning postulates for the geometric primitives, or as truths *ex vi terminorum*, were, for most others, uncertain assumptions. The geometers could hardly fail to perceive the ever sharper distinction between the domain of intersubjective agreement in their discipline, largely restricted to what we would *now* call its purely logical element, and the domain of subjective inclination and widespread disagreement which focused, apparently, on how exactly to read the geometric primitives. This semantic uncertainty surrounding the parallel postulate and its sufficient conditions could hardly fail to infest its neighboring postulates and axioms, as it eventually did in the late nineteenth century.

There is one other symptom of semantic uncertainty in geometry that requires special mention here, since it was also one of the main points of contention in the Frege-Hilbert debate.

Even though Klein's (66) offers no direct reference to the issues of independence and consistency, his next article on the topic already includes a claim that Klein's result gives a definite negative answer to the question whether the parallel postulate is a "mathematical consequence" of the remaining axioms (68, p. 493). The reason given in 1873 is that appealing only to those remaining axioms, one can construct a coherent system ("ein in sich konsequentes Lehrgebäude") that encompasses a system of non-Euclidean geometry as a special case (67, p. 312).

No one is likely to be very surprised by the preceding piece of information. After all, any logic book will tell us that the way to show that a sentence is independent of an axiom system is to give a "meaning" (interpretation) to the nonlogical signs of that system that makes the axioms true and the sentence in question false; and that is what Klein had done. All we need assume in order to understand how Klein came to recognize in his argument an independence proof is that (1) he had something like an axiomatization (however rough) of hyperbolic geometry in mind and that (2) his instinctive knowledge of logic was sufficient to give him the distinction between logical and descriptive (geometric) primitives (the form vs. matter distinction, if you will) *plus* the knowledge that one may freely alter the meanings of descriptive terms in order to establish independence and consistency. As we are often told, the moving force behind independence results such as Klein's is the correct understanding of the essence of an axiomatic system or of the consequence relation or of the idea of interpretation. The common feature of all these accounts is that a logical insight is claimed to be the cause of a geometric insight.[10]

But logic, once again, had little to do with this. Let me first attempt to show that there is some reason to be surprised at the fact that Klein, and then the geometric community, took his result to be an independence, as well as a consistency proof.

Imagine that someone is wondering whether the proposition (not the sentence) *a is red* can be consistently conjoined with the negation of *a is colored;* in other words, whether the information contained in the former is consistent with the denial of the latter. Imagine further that someone offered the following proof that, indeed, it is: Let 'a' designate John, who is tall and thin, let 'red' mean *tall* and let 'colored' mean *fat*. Under this interpretation 'a is red' (the sentence, not the proposition) is true and 'a is colored' is false. Therefore, the proposition *a is red* can be consistently conjoined with *a is not colored;* they *can* both be true.[11]

Clearly this argument is absurd. But how does it differ from Klein's? Formally, they are identical; both of them establish that from certain propositions it is not possible to derive others by purely logical means. But it would be absurd to conclude from this that the negation of one proposition can be conjoined with the other to produce a consistent system of claims.

Klein's proof, of course, establishes that it is impossible to derive a contradiction from the axioms of hyperbolic geometry by means of logical rules of inference. In other words, it establishes that if we ignore the meanings of descriptive terms (such as 'point', 'straight line', and so on) and consider only the information conveyed by the logical words and the syntactic structure of the axioms, then there is no inconsistency to be found in the (semantically mutilated) axioms of hyperbolic geometry.[12]

What is often forgotten is that the problem with which geometers had been coping for centuries was whether what the remaining axioms of Euclidean geometry *say* already contains the information conveyed by the fifth postulate, and *not* whether the postulate is contained in the remaining axioms when you ignore the meanings of all geometric words. Klein's proof clearly provides an answer to the second problem; but the geometric community took it to be an answer to the traditional problem, that is, to the first question. This is what should have surprised us.

There are only two possible reactions to this situation. Either we conclude that the geometric community was guilty of a monumental logical blunder,[13] or we conclude that it was thereby recognizing that, even when taken in their full-blown semantic splendor, the axioms of geometry convey no more information than what we gather when we assume the descriptive, geometrical words to have no (preaxiomatic) meaning. This latter alternative is, to my mind, the only reasonable one. As I understand it, by recognizing in Klein's reasoning a proof of semantic independence, the geometric community was implicitly giving one more sign of its acknowledgment that even though we are all welcome to attach any pictures we want to geometric primitives, this is a strictly personal matter which in no way concerns geometry. One could, for example, associate with the expression 'straight line' a certain mental representation or a light ray or a tight string or a certain kind of motion or (for the singularly endowed) a Platonic straight line. Why not? Hilbert would soon be urging everybody to associate tables and beer-mugs with those words, if it helps or pleases us. As far as geometry is concerned, however, all the intersubjectively manageable meaning that there is in it is conveyed by what would later *become* the "logical structure" of the axioms. Representations can, no doubt, be tremendously useful for heuristic purposes, as Klein so frequently urged. But they can also be a hindrance; and, in any case, geometry is not concerned with them.[14]

My inclination, therefore, is to think that logical insights had little to do with the course of nineteenth-century geometry. This is, after all, what one would have expected; geometers are supposed to know more geometry, *not* more logic than logicians; and vice versa. It is worth recalling that in 1872, when Klein described his proof as, in effect, an independence result, logic was a sorry subject from which he couldn't possibly have learned anything of consequence. At its best it was the emerging doctrine of classes that Boole, Jevons, and Schröder struggled to formulate, with limited success. The ideas of quantification and propositional function and the first careful account of formal inference which they made possible, were still seven years in the future. More revealingly, when those basic logical distinctions were first put forth in Frege's *Begriffsschrift,* they occurred in the context of a philosophy that found nothing but irreparable confusion in the geometers' "logical insights" on independence and consistency. Could the greatest logician of his century have missed a logical point that even third-rate geometers were quick to recognize? I would perfer to believe that the conflict between the logician-philosophers and the geometers was not on a matter of logic but on one of geometry; that neither Frege nor Russell misunderstood "the essence of axiomatics." What makes us think differently is that in the early decades of our century logic evolved, adjusting to the picture of knowledge that has emerged in geometry. The distinctions that first appeared in geometry (and in other branches of mathematics) for reasons of its own were, decades later, embedded in a logic oblivious to its ancestry. Now they are second nature to us and to our logic; small wonder that we miss the shocking aspect of Klein's interpretation of his theorem, or that we tend to misdescribe his insight as logical.

As I interpret it, therefore, the various reactions elicited by the parallel postulate through the centuries were the symptoms of a deep-rooted semantic uncertainty concerning the preaxiomatic meaning of its geometric primitives. That uncertainty eventually turned into neglect when Beltrami's and Klein's reductionistic intentions were ignored and their proofs reinterpreted as independence and consistency results. All that was left for Hilbert to do was to make explicit the implicit recognition that, as far as geometry was concerned, geometric primitives have no preaxiomatic meanings.

Indeed, as the communtiy of geometers turned to formalism,[15] the question that was being decided was this: Do we know anything at all about points, straight lines, and so on, beyond what the geometric axioms tell us? The classical answer had been the one Frege and Russell would soon reiterate in the debates to be presently surveyed: Of course we do; we know the *meaning* of the geometric primitives and, therefore, we know what sorts of things points and straight lines are *before* we even formulate the axioms that state their mutual relations. This preaxiomatic knowledge suffices to determine an infinitude of trivial claims such as, for example, that Frege's pocket watch is not a point. Such items of information are not explicitly included in geometry merely because of their triviality; but they are nonetheless an essential part of geometry, since they determine its topic and, hence, by implication, the class of truths that geometry aims to capture.

Up to the nineteenth century this classical response had not interfered with the goals of geometry, but by the end of that century, as we saw, a conflict had developed. Hilbert's and Poincaré's vision of geometry stressed different aspects of a common trend toward formalism, the former by emphasizing the fact that primitives have no preaxiomatic meaning, the latter by insisting that primitives do have a meaning, after all, but only as much as the axioms give them. The common, shocking assumption was that the axioms of geometry, those traditional paradigms of certain knowledge, do not even express propositions.

For decades professional philosophers had remained largely unmoved by the new developments, watching them from afar or not at all. The few who had any reasonable idea of what had happened adopted an attitude of benign dismay toward the unruly course of events in mathematics. As the trend toward formalism became stronger and more definite, however, some philosophers concluded that the noble science of geometry was taking too harsh a beating from its practitioners. Perhaps it was time to take a stand on her behalf. In 1899, philosophy and geometry finally stood in eyeball-to-eyeball confrontation. The issue was to determine what, exactly, was going on in the new geometry.

2. RUSSELL AND POINCARÉ

In 1897 Russell published a revised version of his fellowship dissertation, entitled *An Essay on the Foundations of Geometry*. The viewpoint he defended in it was an idiosyncratic combination of apriorism and empiricism. A somewhat ill-defined version of projective geometry[16] was pronounced a priori since its axioms were said to express the qualitative conditions of possibility of measurement. The metric axioms needed to determine which metric geometry is the true one were, however, not decidable a priori. It was at this point that experience was called upon to play a role.

As is well known, Russell's dissertation was written during his early, idealistic period. At this time "modern logic" was the logic of Bradley and Bosanquet (84, p. 58), and there was some glee in the discovery of the contradictions (i.e., circles and regressūs) which inevitably lie at the foundations of geometry (§§108ff.)—if you try to define and prove everything, that is.

The most interesting part of the book is, perhaps, the discussion of the philosophical relevance of Klein's results, which, in Russell's opinion, was nil. The word 'distance', Russell noted, must have precisely the same meaning in all geometries[17] "because" it is only on metrical claims that geometries disagree (84, pp. 36–37). Russell thought—quite incorrectly—that Klein was not aware of the metric-free character of projective coordinates[18] and that, because of this, he had failed to observe that no metrically significant characterization of distance could be grounded on such coordinates (84, §34). There is only one legitimate sense of 'distance', Russell argued, and by redefining that word Klein has not shown that the parallel postulate could be false, but merely that the words with which we express it could be used to mean something different. It is, no doubt, of mathematical interest to note that, under a certain redefinition of the distance relation, non-Euclidean geometries emerge, but the result has no philosophical significance. The interesting philosophical question is whether, when we use the word 'distance' in its correct sense, there is any room left for convention. And the answer is, of course, negative. In what may be the worst of his misunderstandings, Russell goes on to accuse Poincaré of not having appreciated the nature of projective coordinates and of having concluded from Klein's results that the axioms of geometry are conventional (84, §§33–34).

When his book was published in 1897, Russell was a young and promising philosopher, with a knowledge of geometry that was remarkable for a member of his profession, though far from flawless or complete. Poincaré, on the other hand, was widely regarded as one of the greatest mathematicians of his generation. In the natural course of things, he would have payed no attention to the odd criticisms of the young Mr. Russell. In 1898, however, the *Revue de Métaphysique et de Morale* published a remarkably flattering review of Russell's book by Couturat. In it, after praising Russell's text for being the first successful attempt to draw the philosophical consequences of the non-Euclidean writings, Couturat went on to say that "what was required in order to gather the fruits of these works was a mind both scientific and critical, a mind endowed with a vast mathematical erudition and an understanding of philosophical problems. That such a mind could not be found in France may be cause for regret, but not for surprise" (22, p. 354). The next volume of the *Revue de Métaphysique et de Morale* included Poincaré's review of Russell's book.

2.1. *Indefinables Defined*

In many of his writings Poincaré had opposed both empiricism and a priorism in geometry, arguing that no properly geometric axiom expresses either "an experimental fact or a logical necessity or a synthetic a priori judgment" (77, p. 127). His philosophical readers were bound to be puzzled, for what else could a judgment be if not analytic or synthetic? Poincaré's often repeated answer was that it could be a "definition in disguise."

By calling geometric axioms 'definitions', Poincaré was baptizing a difficulty in a way that invited misunderstandings. Let us see why.

Ever since modern philosophers undermined the venerable doctrine of definition by *genus* and *differentia,* a tradition had evolved according to which definitions are "best made by enumerating those simple ideas that are combined in the signification of the term defined" (72, p. 213) and are therefore, "enumerations of sufficient marks" which are "ingredients" or "constituents" of the *definiendum* (71, p. 292). During the nineteenth century this tradition became dominant among logician-philosophers. Through the influence of Bolzano, Frege, Meinong, and Russell the notions of a complex and its constituents became the central metaphysical concepts of the new semantics that inspired the doctrines of philosophical analysis and logical atomism. For them, as a melody has notes and a library, books, both propositions and concepts have constituents that analysis aims to identify. The search for the essential properties of an object was replaced by the search for the constituents of a complex. The definition of a concept, in particular, came to be viewed as structurally identical to the analysis of a chemical compound into its elements.[19]

According to this *analytic* picture of the concept (as I shall call it), there are two distinguished stages one may reach in the process of analyzing a concept. To begin with, there is a point at which we encounter a set of conceptual constituents, the so-called characteristic marks (*Merkmale*, "marks" for short) that might or might not be subject to further analysis, and whose conjunction is (not merely equivalent to but) identical with the *definiendum*. Thus, when Kant defined an analytic judgment as one in which the predicate is part of the subject, he was alluding to a process in which the predicate of the judgment is recognized (by analysis) to be a mark of the subject.

The second distinguished stage is reached when we encounter constituents that are intrinsically unanalyzable or simple. There is, in principle, no reason why the process of analysis could not go on forever, but it was widely believed among proponents of the analytic picture that, as we dig down into a complex concept or proposition, we always reach the territory of the "indefinable."

Accordingly, the definition of a concept can be seen either as the conjunction of its marks or as the reconstruction (or synthesis) of the *definiendum* from its simple constituents.[20]

Even though proponents of the analytic picture had a great deal to say concerning definitions, they had next to nothing to say concerning indefinables. This is quite puzzling, since they were also agreed on the relative ontological and semantic insignificance of definition and on the overriding importance of the indefinables. Thus, Russell writes in 1900 that "an idea which can be defined, or a proposition which can be proved, is only of subordinate philosophical interest. The emphasis should be laid on the indefinable and indemonstrable" (85, pp. 170–71).

The problem of the indefinables, of what role they play in human knowledge and how they come to play it, would soon become one of the major subjects of the philosophies of Russell and of the young Wittgenstein. But at this early stage, essential as the indefinables were, Russell did not think that there was any reason to spend too much time thinking about what and how exactly we knew about the meaning of, say, geometric primitives.

Russell's commitment to the analytic picture of the concept emerges clearly in his reply to Poincaré's review. There he notes that "the most important and difficult issue" raised by Poincaré concerns the nature of definitions. Not only had his reviewer characterized axioms as definitions but he had also challenged Russell to produce his own definitions of geometric primitives. This request, Russell thought, betrayed an appalling and deep-rooted confusion which appeared to lie at the heart of the new conventionalism.

> M. Poincaré requests "a definition of distance and of the straight line, independent of (Euclid's) postulate and free from ambiguity or vicious circle" (§20). Perhaps he will be shocked if I tell him that one is not entitled to make this request since everything that is fundamental is necessarily indefinable. And yet, I am convinced that this is the only answer which is philosophically correct. Since mathematicians almost invariably ignore the role of definitions, and since M. Poincaré appears to share their disdain, I will allow myself a few remarks on this topic. (93, pp. 699–700)

There are, Russell tells us, two kinds of definitions: mathematical and philosophical. Mathematical definitions (destined to become "knowledge by description") merely identify an object as the one and only which stands in a certain relation to already known concepts or objects (93, p. 700). For example, if we define the letter A as the one that precedes B (p. 701) or the number 1 as the one that precedes 2, what we have given is a mere mathematical definition of these objects.

> But these definitions are not definitions in the proper and philosophical sense of the word. Philosophically, a term is defined when its *meaning* is known, and its meaning cannot consist of relations to other terms. It will be readily granted that a term cannot be usefully employed if it does not mean something. Its meaning can be complex or simple. In other words, either it is composed of other meanings or it is one of those ultimate elements that are constituents of other meanings. In the first instance one defines the term philosophically by enumerating its simple elements. But when it happens to be simple, no philosophical definition is possible. . . . Definition is an operation analogous to spelling. One can spell words but not letters. M. Poincaré's request places me in the uncomfortable position of a student who has been asked to spell the letter A without allowing him to use the letter in his reply. . . . All these truths are so evident that I would be ashamed to recall them, were it not that mathematicians insist in ignoring them. (93, pp. 700–01)

And now, applying these remarks to geometry,

> These observations apply manifestly to distance and the straight line. Both belong, one might say, to the geometric alphabet; they can be used to define other terms, but they are indefinable. It follows that any proposition, whatever it may be, in which these notions occur, is either an axiom or a theorem and not a

definition of a word. When I say the straight line is determined by two points, I assume that *straight line* and *point* are terms already known and understood, and I make a judgement concerning their relations, which will be true or false, but in no case arbitrary. (93, pp. 701–02)

The extent to which these observations affect Poincaré's doctrine that geometric axioms are definitions, entirely depends, of course, on what Poincaré had meant by "definition." The obvious, and often neglected, fact is that Poincaré did not mean by that word what Russell thought he should but, unphilosophically enough, what the dictionary instructed him to mean: a process through which the meaning of an expression is identified.

The difference between Russell's and the dictionary's sense of definition is that the first sense refers only to procedures to derive new meanings from old in a way that views the latter as constituents, indeed, as marks of the former, whereas the dictionary sense refers also to procedures involving no such restrictions, including even processes in which no meaning is presupposed as being available prior to the definition (as in "ostensive definitions" and "coordinative definitions"). Thus, when Poincaré asked Russell to define geometric primitives, he was not asking for the analysis of the unanalyzable; he was simply requesting Russell to offer a sufficiently definite and geometrically acceptable characterization of what those primitive terms mean. Russell had, in fact, an answer to this question; an untenable one, as we shall soon see. But his theory of knowledge at this time was such that he couldn't even seriously entertain the idea that there could be any difficulty *in principle* (as opposed to in practice) in identifying the proper meanings of geometric primitives.

Moreover, the idea that the geometric axioms could in any way be involved in the process of assigning meanings to some of their terms appeared to him as utterly incoherent. The reasoning that led him to this conclusion involved an appeal to a principle which will figure prominently in the following pages and which I shall call *the thesis of semantic atomism.* This principle does not concern primarily propositions but the sentences that we use to express them, and it asserts that the grammatical units of a sentence *S* must have a meaning *before* they join their partners in *S*, if *S* is to be at all capable of expressing a proposition or of in any way conveying information.

The harmless appearance of this principle will soon fade away as we disclose the dominating role it came to play in these geometric debates. The basic fact to bear in mind is that *all* participants in the debates endorsed this principle; but whereas the philosophers used it to infer (by *modus ponens*) a doctrine of geometry that no geometer could accept, the geometers used it to infer (by *modus tollens*) a doctrine of geometric knowledge that no philosopher could seriously entertain. Moreover, as we shall see, the inclination to move up or down the argument chain entirely depended on one's attitude toward the character of indefinables.

At the end of our last quotation from Russell we may catch a glimpse of the train of thought which led him to regard the use of axioms in connection with definitions as obnoxious. The argument is this: since *obviously* the axioms of geometry express propositions (convey information), by the thesis of semantic atomism, the geometric primitives must *somehow* have acquired a meaning before they can contribute to the expression of a proposition via the appropriate sentence. Somehow; no matter how—that is why Poincaré's request to identify the primitives will be regarded by Russell as utterly irrelevant to the philosophical issue at hand, and, at best, as a *merely* epistemological matter of no semantic or ontological significance.

What Poincaré does, on the other hand, is to stand this argument on its head.

One of the passages that reveal most clearly the semantic dimension of Poincaré's thinking occurs in his first paper on Russell's book. Poincaré is struggling to explain why he thinks that it is an error to conceive of the axioms of geometry as bona fide propositions, in other words, as conveyors of information that attribute properties or relations to given objects. Here is what he has to say:

> If an object has two properties *A* and *B*, and if it is the only one that has the property *A*, this property can be used as a definition; and since it will suffice as a definition, the property *B* [i.e., the attribution of *B*] will no longer be a definition; it will be an axiom or a theorem.
>
> If, on the contrary, the object is not the only one that has property *A*, but it is the only one that has both properties *A* and *B*, *A* no longer suffices to define it, and the property *B* will be a complement of the definition, and not an axiom or a theorem.
>
> In a word, in order for a property to be an axiom or a theorem, it is necessary that the object that has this property has been completely defined *independently of this property*. Therefore, in order to have the right to say that the so-called distance-axioms are not a disguised definition of distance, one should be able to define distance in a way which does not involve an appeal to those axioms. But where is that definition? (79, p. 274)

The same point is made in Poincaré's discussion of free mobility. Russell had argued that the axiom of free mobility is a priori. In Russell's formulation the axiom said that "spatial magnitudes can be moved from place to place without distortion . . . shapes do not in any way depend upon absolute position in space" (84, p. 150). Here is Poincaré's comment:

> What is the meaning of "without distortion"? What is the meaning of "shape"? Is *shape* something that we know in advance, or is it, by definition, what does not alter under the envisaged class of motions? Does your axiom mean: In order for

measurement to be possible figures must be susceptible of certain motions and there must be a certain thing which will remain invariant through those motions and which we shall call shape?

Or else, does it mean:

You know full well what shape is; well, in order for measurement to be possible it is necessary that figures can undergo certain movements that do not alter this form?

I do not know what it is that Mr. Russell has meant to say; but in my opinion only the first sense is correct. (79, p. 259)[21]

With minimal sharpening, the point is this: the axioms of geometry are widely regarded as statements that convey information about spatial entities. If so (by the principle of semantic atomism), it should be possible to "define," in other words, to identify in some intersubjective manner, the meanings of their geometric primitives before they are incorporated into the axiomatic sentences. Up to this point, the geometer and the philosopher agree. But now Poincaré brings into the argument a new premise, the lesson that geometers had finally drawn from the episode surveyed in section 1: there is nothing we can say about the meanings of geometric primitives, *except what the axioms say,* that could in any way circumscribe what the primitives mean. There is, of course, nothing to prevent us from *deciding* that those primitives should be taken to mean certain physical objects, or certain mental representations, or anything else. But there is no particular meaning that geometry attaches to its primitives prior to its construction. The point is not that we have no understanding whatever of geometric primitives, but that all the meaning geometry attaches to them is given through their participation in the axioms of geometry. Under these circumstances, the thesis of semantic atomism prevents geometric axioms from conveying any sort of information. No wonder that they are neither analytic nor synthetic. No wonder, either, that they have always been regarded as extraordinary claims. They do play an irreplaceable role in geometric knowledge, but what is exceptional about them is not that they convey any privileged form of information, but that they identify (to the extent required by geometry) the meaning of geometric primitives. Geometric axioms are definitions disguised as claims; and what they define is the indefinables.[22]

The preceding quotations from Poincaré contain the first explicit acknowledgement of a distinction that was destined to play a dominating role in Carnap's and Wittgenstein's syntacticist philosophies: the distinction between sentences that genuinely convey information and sentences that, in spite of their misleading syntactic appearances are, in fact, content-free and that function as part of a system through which meaning is assigned to certain words. Poincaré's "axioms and theorems" constitute the former category, his "definitions in disguise" the latter.

It is habitual to misconstrue this debate, and the one to be reviewed in the next section, as a conflict between the proponents of implicit definition (the geometers) and the champions of explicit definition (the philosophers). Needless to say, if that were the case, explicit definition would be sure to win not merely as the better understood notion, but also as the one to which a version of the other can be reduced. But Poincaré is not offering his theory of implicit definition as a *faute-de-mieux* version of explicit definition, nor as a procedure intended to construct new meanings from already available marks. His doctrine is offered, rather, as a solution to the problem of geometric indefinables, as a procedure to assign meanings from scratch, as it were, appealing only to the meanings of purely logical expressions. Its competitor, therefore, is not the logician's theory of explicit definition but his own solution to the problem of indefinables, his theory of how we come to associate with geometric primitives the meanings that they happen to have, and of what those meanings are. When contrasted with what philosophers have to say on *this* topic, the vagaries of Poincaré's implicit definitions may appear much less offensive than they do when contrasted with its explicit counterpart.

It is now time to look at what Russell had to say on the real point at issue.

2.2. *Indefinables Intuited*

Next to the ontologico-semantic theses described earlier in this section, the analytic picture of the concept includes a corresponding set of epistemological theses aimed to answer the question, How do we ever come to be in possession of concepts? The answer was, in a nutshell, this: New concepts can be constructed (synthesized) from already available ones by treating the latter as marks of the former, in other words, by "defining" them. But definition needs concepts—like proof needs truths—in order to generate more of the same. There must therefore be a direct, noninferential source of conceptual knowledge to complement definition. Hume posed and solved the problem in a way which, in one form or another, became a standard feature of the analytic picture:

> Complex ideas may, perhaps, be well known by definition, which is nothing but an enumeration of those parts or simple ideas that compose them. But when we have pushed up definitions to the most simple ideas, and find still some ambiguity and obscurity, what resources are we then possessed of? By what invention can we throw light upon these ideas, and render them altogether precise and determinate to our intellectual view? Produce the impressions or original sentiments, from which the ideas are copied. (Hume, *Enquiries* [Oxford: Clarendon Press, 1966], p. 62)

The first explicit indication of what Russell took to be the "method" used to identify the indefinables occurs in his *Philosophy of Leibniz* (1900), where, after reminding us of the importance of indefinables and indemonstrables, Russell concludes: "The emphasis should be laid on the indefinable and indemonstrable, and here no method is available save intuition" (85, pp. 170–71).

In *Principles* (1903) philosophical definitions are demoted from their earlier, fundamental role. Now this sense of 'definition' is "inconvenient and, I think, useless" (92, p. 111). But the indefinables have only grown in significance: Part 1 of *Principles,* the philosophical centerpiece of the book, is entitled "The Indefinables of Mathematics," and its Preface tells us that "the discussion of indefinables—which forms the chief part of philosophical logic—is the endeavour to see clearly, and to make others see clearly, the entities concerned, in order that the mind may have the kind of acquaintance with them which it has with redness or with the taste of a pineapple" (p. xv). The point is more eloquently reiterated in a later chapter: "Philosophy is, in fact, mainly a question of insight and perception . . . . A certain body of indefinable entities and indemonstrable propositions must form the starting point for any mathematical reasoning; and it is this starting point that concerns the philosopher" (p. 129). Such entities, he tells us, "if we are to know anything about them, must also be in some sense perceived, and must be distinguished from one another; . . . All depends, in the end, upon immediate perception; and philosophical argument, strictly speaking, consists mainly of an endeavour to cause the reader to perceive what has been perceived by the author" (pp. 129–30).

Thus, to know or understand the meaning of a primitive term is to perceive, to be acquainted with a certain simple thing, the indefinable which it means. While we are tasting a pineapple, for example, the meaning (in Russell's sense) of 'the taste of a pineapple' is present to us, we are acquainted with it, we have an intuition of it. When we focus attention on a perceptual object and say 'this', the meaning of that occurrence of 'this' is similarly present to us. These are Russell's paradigms of what it is to understand a meaning. Upon occasion, for example, in set theory, or even in geometry, the indefinables of a theory are given to us, as it were, through a glass, darkly. Our representation of classes, of distance and straight lines, may be imprecise and vague. This may tempt one to view the vagueness of our semantic knowledge as a feature not of knowledge but of its object; a temptation to which Poincaré and Hilbert appear to have succumbed. However, the notion of an ambiguous, underdetermined object makes no more sense than the idea of a concept without sharply defined boundaries. From the fact that our understanding of some indefinables is deficient, all that follows is that our understanding should be improved, not that our understanding is appropriate to ontologically defective objects.

Poincaré had fully anticipated that this would be Russell's "solution" to the problem of indefinables. The answer which he expected was that "there is no need to define [the indefinables] because these things are directly known through intuition. I find it difficult to talk to those who claim to have a direct intuition of equality of two distances or two time-lapses; we speak very different languages. I can only admire them, since I am thoroughly deprived of this intuition" (79, p. 274; see also 80, p. 75). Russell's reply to this was the verbal equivalent of a shrug: "The meaning of fundamental terms cannot be defined but only suggested. If the suggestion does not evoque the right idea, there is nothing one can do. That is why philosophical precision is of an entirely different nature from mathematical precision. It is harder to attain, and harder to convey" (93, pp. 702–03).[23]

Much, much harder, Poincaré must have thought. At this point the debate was bound to stop.

But Russell, like Carnap—and unlike Frege and Wittgenstein—always had a hunch that scientists knew what they were doing. So doubts soon crept in.

2.3. Acquaintance Forgot

Russell's paradigms of acquaintance are not too easy to match. Meanings, he soon learned, are not always, or often, as clear as the taste of a pineapple; unfortunately, that is, it is not always clear what semantic object we are acquainted with, or even if, appearances notwithstanding, we are really acquainted with anything at all. A notorious instance of this embarrassing circumstance is provided by Russell's long and painful struggles with the notion of class. After years of unsuccessful attempts to see clearly what it was that we are acquainted with in class discourse, he decided that the answer was, Nothing. No one, certainly not Russell, had ever really been acquainted with a class. Propositions and other objects of acquaintance would soon follow the same path from self-evidence to nonbeing.

Together with this eroding loss of confidence in our relation to indefinables there was another parallel development that appeared to be unrelated to Russell's philosophy of geometry but that prepared a major shift in the direction of formalism.

In the year 1900, apparently under the impact of Peano's work, Russell conceived a grandiose anti-Kantian program in the foundations of mathematics which has come to be known as logicism. At this early stage the point of conflict with Kant's philosophy was this: According to Kant, mathematical inference always involves an appeal to constructive intuition and thereby to extralogical processes. The merely analytical character of logic, he thought, prevents it from being an adequate tool for the mathematician's reasoning. The program which Russell conceived in 1900 and to which he devoted almost a decade of research was to show Kant wrong; to show, that is, that logic, appropriately construed, suffices to draw all conclusions a mathematician may want to draw from whichever premises he may want to adopt. All proof can be carried out "formally," that is, without an appeal to intuition. That was the basic doctrine of what I have called elsewhere "conditional logicism."[24]

Russell also observed—as Frege before him—that in the case of arithmetic, broadly construed as the science of finite and transfinite numbers, the "logicist" thesis could be strengthened. In this case all primitives were definable on the basis of the indefinables of logic.

Geometry had, of course, been one of Kant's prime examples of how mathematical reasoning cannot be a mere effort to squeeze content out of axioms by means of logical inferences, but must appeal to the construction of geometric objects in intuition during the proof process. Russell's doctrine entailed the redundancy of such an appeal; for conditional logicism entailed that we need not call on *any* intuition in the course of a geometric proof. Intuition in geometry was thus relegated to play a role only at the origin of our system, in the selection of axioms, not in the proofs that we generate from them.

Even this role was soon to be curtailed. Around 1901, concerned about the inadequacy of his standpoint on the problem of the multiplicity of geometries, Russell put forth a doctrine intended to adjust his philosophy to the attitude of the geometric community. The doctrine—which we shall call "if-thenism" (Putnam's term)—was stated in the opening paragraph of *Principles:* "pure mathematics is the class of all propositions of the form '*p* implies *q*' " (92, p. 3). What this means is that the different branches of pure mathematics do not consist of a set of axioms and then a set of theorems which follow from them but, rather, of the set of implications that have the conjoined axioms as antecedent and each theorem as a consequent. As he explained decades later, what had led him to this curious doctrine was "the consideration of Geometry. It was clear that Euclidean and non-Euclidean systems alike must be included in pure mathematics, and must not be regarded as mutually inconsistent; we must therefore only assert that the axioms imply the propositions, not that the axioms are true and therefore the propositions are true" (p. 7; see also pp. 5, 8, 372–73, 429, 430, 441–42). It then occurred to him that not only geometry but topology, group theory, measure theory, and, indeed, all of mathematics could be seen as stating that a set of axioms implies appropriate consequences. Russell's conditional logicism asserts that these implications can all be stated in a purely logical language and proved on the grounds of purely logical principles. This conditional construal of pure mathematics allowed Russell to conclude that *all* of pure mathematics—and not just arithmetic—is reducible to logic.

This difference between Russell's conception of mathematics and Frege's is not significant, since it could, after all, be consigned to a merely verbal quibble on what is to be meant by 'pure'. The crucial difference, as we shall see, lies in the subject of choice in geometry. At this stage Russell is already exercising a modicum of tolerance in that he is acknowledging—however implicitly—that it is not the *pure* geometer's task to decide for this or that set of geometrical axioms. Not only should the geometer avoid an appeal to intuition in the process of proof; he should also ignore it in his choice of axioms. "Geometry no longer throws any direct light on the nature of actual space" (92, p. 373).[25]

But, even though diminished, the role of intuition in geometry was by no means eliminated. Besides pure geometry, there is, Russell thought, "geometry as the science of actual space" (90, p. 593). Here, geometric primitives are not replaced by variables but express perfectly definite indefinables. In agreement with the thesis of semantic atomism, here "we start . . . with the straightness which we have derived from analysis of perceived objects," for "it is undoubtedly by analysis of perceived objects that we obtain acquaintance with what is *meant* by a straight line in actual space" (p. 593). Once we are familiar with what straightness is, "we have to inquire what properties belong to this straightness" (p. 593), and the outcome of this inquiry is, of course, the axioms and theorems, claims which are either true or false but certainly not conventions concerning what is to be meant by "straightness."

Our acquaintance with straightness is, however, less perfect than one might have expected. It was formerly thought, Russell notes, that the inquiry concerning the properties of straightness and the other indefinables could be conducted a priori (90, p. 593). In fact, "if our representation of the straight line were absolutely exact, it alone would suffice to reveal to us the nature of space" (93, p. 695). But it doesn't, hence our representations of geometric indefinables are inexact. After all, we may be acquainted with Jones so as to know him when we meet him, yet we may know nothing of his bank account; similarly, "we may know straightness (within limits) when we see it, without knowing whether or not two straight lines can enclose a space" (90, p. 593). Experience is, in fact, needed only to turn our inexact representations into exact ones.[26]

The appeal to intuition, the associated uniqueness of the appropriate set of indefinables, and the inevitable conclusion that there is only one true geometry are now in one corner of Russell's philosophy, but they are still there throughout the first decade of the century.

The work for *Principia* led Russell away from geometry and toward logic and arithmetic. When in 1914 he returned to geometry, in the Spencer lectures on the scientific method in philosophy, acquaintance had taken yet another beating.

Now he tells us that there are three distinguishable problems of space. The *logical* problem is that of the nature of pure geometry, and its solution is if-thenism. The *physical* problem of space is "to construct from physical materials a space of one of the kinds enumerated by the logical treatment of geometry" (91, p. 87). Finally, the *epistemological* problem is to decide whether geometry is a priori or empirical. The answer to *this* problem is to notice that pure geometry is a priori (and purely logical), whereas physical geometry is empirical, and that there is no geometry other than the preceding two.

Physical geometry is, therefore, all that is left of the old "science of actual space," only now the points and straight lines of geometry are no longer the indefinables presented in acquaintance but the highly inferential and painstakingly defined constructs of Whitehead's method of extensive abstraction (91, p. 88); not the given in intuition but "complicated constructions by means of classes of physical entities" (p. 90). Intuition is no longer the imperial leader but a rather sporting follower, suspiciously willing to comply with changing fashion: "The oddity of regarding a point as a class of physical entities wears off with familiarity" (p. 88).

If the geometric indefinables are constructed by us rather than given to us in acquaintance, then, one may wonder, what is there to prevent us from constructing them so as to agree with different geometric patterns? Now that acquaintance has withdrawn from the task of identifying primitives, what could single out as uniquely justified a construction which led to a model of one geometry rather than to another? The question is not posed, let alone answered, by Russell at this stage, but it is clear that he is now ready to acknowledge that there is nothing to geometry beyond the distinction between interpreted and uninterpreted geometries, thus tacitly granting the formalist's point.[27] Acquaintance may remain as an ornament pleasant to the philosopher, but by 1914 it no longer played any effective role in Russell's philosophy of geometry.

3. FREGE AND HILBERT

Reading Frege's public and private exchanges with his contemporaries one is often tempted to imagine that he is, in fact, a time-traveler, selflessly moving from our century to the preceding one in order to set his new contemporaries straight on the central issues of logic, semantics, and epistemology. His discussions with Schröder, Boole, Cantor, Russell, Peano, and others on logical and semantic problems show a constant and staggering command of matters barely understood, if at all, by his illustrious interlocutors.

There is only one major exception to this amazing record of success, and it derives from Frege's treatment of geometry both in his correspondence with Hilbert and in the papers that he devoted to his conception of geometry.

3.1. Definitions

At the same time that Poincaré was exchanging papers with Russell, Hilbert was releasing a monograph that would soon become one of the landmarks of nineteenth-century geometry, his *Grundlagen der Geometrie.*

O. Blumenthal reports that as early as 1891, commenting on a lecture by H. Wiener, Hilbert had said that "it must be possible to replace [in geometric statements] the words 'points', 'lines', 'planes' by 'tables', 'chairs', 'mugs' " (51, pp. 402–03).[28] Several years later he decided to put the idea to work, and in the winter term of 1898–1899 he offered a course on the foundations of Euclidean geometry. This led to the writing of the *Grundlagen,* published in 1899.

Frege had had a deep interest in geometry from very early on—he once wrote that a philosopher who has nothing to do with geometry is only half a philosopher (24, p. 293)—and he had actively pursued foundational research in that field.[29] He read Hilbert's monograph as soon as it was published; shortly thereafter he secured a copy of Hilbert's lecture notes.

His immediate reaction to Hilbert's monograph was one of thorough disappointment. He wrote to a friend that the *Grundlagen* was "a failure" (44, p. 148) and decided to start a correspondence with Hilbert in order to set him straight on the revelant logical issues.

After a courteous reply to Frege's first letter, Hilbert, in effect, withdrew from the debate—even though oblique references to it occur in later correspondence. As Hilbert's interest in the issue waned, Frege's grew, eventually to the point of frenzy. In his second communication to Hilbert he proposed the publication of their letters. Whether directly or by implication, Hilbert declined, and this response did nothing to diminish Frege's involvement. In 1903 he published a two-part study of the *Grundlagen,* including the opinions he had earlier conveyed to Hilbert in correspondence. The style betrays a strained effort not to disclose his utter contempt for Hilbert's work;[30] an effort perhaps motivated by Frege's belief that Hilbert's views had changed since the time of the correspondence (38, p. 262) and that the forthcoming second edition of the *Grundlagen* might perhaps reflect those changes, and even acknowledge the corrective influence of Frege's thought.

The appearance of the second edition in 1903 without any retraction must have been a disappointment for Frege. Since polite restraint was no longer called for, Frege's next publication on the subject displays his celebrated wrath in full force. Not only is Hilbert accused of not knowing what he means by axiom or by independence; he is also charged with a deliberate obscuring of the issues, for "an obscuring of the issue may well be a condition of survival" for Hilbert's doctrine: "Mr. Hilbert, so far as I know, does not reply to my arguments at all. Perhaps in him too there is at work a secret fear, deeply shrouded in darkness, that his edifice might be endangered by closer investigation of my arguments" (39, pp. 281–82), and so on and on.

Temper tantrums aside, the Frege-Hilbert controversy is one of the most revealing documents on the conflict between the emerging conception of geometric knowledge and the classical standpoint. No philosopher before Frege had given a clearer and more articulate presentation of that standpoint, or preceived more securely its inconsistency with the developing understanding of geometry; nor has there ever been anyone more firmly committed to drawing from its premises the conclusions they entail, however unpopular or distasteful they might be.

It is hard to avoid a sense of *déjà vu* when one observes that Frege's main complaint against Hilbert is that he was thoroughly confused concerning the nature of definitions. As is well know, Hilbert had famously started the *Grundlagen* stating what he called an "Erklärung" (explanation or definition): "We conceive three different systems of things; we call the things in the first system *points* . . . the things in the second system *straight lines* . . . the things in the third system *planes*. . . . Between points, straight lines and planes we imagine certain relations that we express with words like 'lies on', 'is between' and 'it is congruent'; the exact description of these relations is given by the axioms of geometry" (52, p. 2).[31]

The five groups of axioms follow next, and right after them, the first thoroughgoing investigation of the consistency and independence properties of a variety of axiomatic systems of a broadly geometric nature. Frege was appalled. "It is high time," he wrote in his first letter, "that we began to come to an understanding about what a definition is and what it is supposed to accomplish. . . . It seems to me that at the present time complete anarchy and subjective inclination reign supreme in this area" (28, p. 62).[32] What followed was a masterful and patronizing explanation of the classical picture of knowledge: The totality of the sentences in a theory, Frege explains, is to be divided into two groups, those in which something is asserted and those in which something is stipulated. The former are the axioms and theorems of the theory, the latter are the definitions:

> It is absolutely essential for the rigor of mathematical investigations that the difference between definitions and all other sentences be maintained throughout in all its sharpness. The other sentences (axioms, principles, theorems) must contain no word (sign) whose meaning [*Bedeutung*] or, (in the case of form words, letters in formulas) whose contribution to the expression of the thought is not *already* completely settled, so that there is no doubt about the sense of the sentence—about the proposition expressed in it. Therefore it can only be a question of whether this proposition is true; and if it is, on what its truth might rest. Therefore, it can never be the purpose of axioms and theorems to establish the meaning [*Bedeutung*] of a sign or word occurring in them; rather, this meaning [*Bedeutung*] must *already* be established. (28, pp. 62–63, my italics)

The emphasis on the thesis of semantic atomism is apparent; and equally apparent is Frege's conviction that *every* sentence of a theory which is not a definition must express a proposition and thereby convey (true or false) information. Given these two assumptions, Hilbert's axioms of sections 1 and 3 should be such that "the meanings [*Bedeutungen*] of the words 'point', 'straight line', 'plane' and 'between' are not given but rather presupposed as known" (28, p. 61). It is in this conclusion that Hilbert locates "the crux of the misunderstanding."

> I do not want to presuppose anything as known. I see in my explanation [*Erklärung*] in §1 the definition of the concepts point, straight line, and plane, if one adds to these all the axioms of groups I-V as characteristics. If one is looking for other definitions of point, perhaps by means of circumscriptions such as extensionless, etc., then, of course, I would most decidedly have to oppose such an enterprise. One is then looking for something that can never be found, for there is nothing there, and everything gets lost, becomes confused and vague, and degenerates into a game of hide-and-seek. (28, pp. 65–66)

This revealing, almost impassioned account of what Hilbert thought about preaxiomatic procedures to capture the indefinables of geometry should be compared with Poincaré's remarks about those who claim to be acquainted with (to intuit) indefinables such as congruence or straightness (quoted above). For both of them, and, eventually, for all geometers, the preaxiomatic search for the indefinables is "a game of hide-and-seek" where "everything gets lost, becomes confused and vague," because, in the end, "there is nothing there."

Frege could make no sense of Hilbert's attitude, nor could he see any genuine problem at the source of Hilbert's formalism. What he saw, rather, was Hilbert's failure to understand what counts as the definition of a concept.

If we are to get clear on what it is to define a concept, Frege proceeded to explain to Hilbert in his second letter, we must first understand what a concept is. A (first-level) concept—as everybody should know—is a function that takes all objects as arguments and truth values as values. When (and only when) the concept *C* assigns Truth to the object *o*, we say that *o* falls under *C*. The concept *point,* for example, assigns to all points the value Truth, and Falsehood to everything else. To define *point,* therefore, is to specify conditions that will determine for *any* object whether it falls under that concept or not. If Hilbert's axioms defined *point,* they should determine, for example, whether Frege's pocket watch is a point or not (one hopes, in the negative). But they don't:

> I do not know how, given your definitions, I could decide the question whether my pocket watch is a point. Already the first axiom deals with two points; therefore, if I wanted to know whether it held of my pocket watch, I should first of all have to know of another object that it is a point. But even if I knew this, e.g., of my fountain pen, I should still be unable to decide whether my watch and my fountain pen together determine a straight line, because I should not know what a straight line is. (29, p. 73)

Hence, Hilbert's "axioms" cannot define *point* or, *mutatis mutandis,* any other geometric concept.

A second argument for the same conclusion derived from the analytic nature of definition. As we noted a moment ago, to define a concept is to characterize conditions when an object falls under it. But what sorts of conditions should a definition specify? Frege's answer appeals to the more traditional aspect of his theory of the concept. Besides its functional character—leading to Frege's famous demand for sharp boundaries in all concepts—there was another, more familiar element in Frege's theory of the concept. Concepts were, for him, either simple or complex; if simple they are indefinable, if complex they are capable of definition by resolution into (or by synthesis from) their constituents. Moreover, Frege lay great emphasis on the role of certain special constituents, the marks (*Merkmale*), in the process of definition. "Concepts," he writes, "are generally composed of constituent concepts, the marks. . . . I compare the individual marks of a concept to the stones which make up a house" (31, p. 150). The marks of a concept "are concepts which are logical elements of it" (38, p. 271; 25, p. 3); they are, in fact, its constituent conjuncts. Thus, every instance of a concept must be an instance of its marks (38, p. 271), and marks must be jointly sufficient conditions for their concept (see, e.g., 37, pp. 121–22). Since every (complex) concept is the conjunction of its marks, to define a concept is to enumerate the marks that constitute it.

Appealing to this second facet of his theory of the concept, Frege now argues that if Hilbert's axioms are to be regarded as definitions of concepts, they must *somehow* involve the attempt to identify marks for the concepts being defined. The most charitable proposal that occurs to Frege is this: when Hilbert says that axioms such as 'All *X* are *A*' and 'All *X* are *B*' define *X* (or give its marks), he may be taken to mean that *A* and *B* are the marks of *X* (38, p. 265). But even under this generous interpretation, Hilbert has failed to define any of the intended (first-level) concepts. For the marks of a concept must be concepts of the same level as the *definiendum*—since they must apply to the same objects; but whereas the geometric concepts Hilbert thinks he has defined (e.g., *point, straight line*) are of the first level, some of the alleged marks enumerated in his axioms (e.g., quantification) are second-level concepts. Once again, we may conclude that Hilbert's axioms define no concepts.

It is hardly necessary to emphasize either the extent to which Frege's criticisms depended on his theory of the concept or, indeed, the centrality of that doctrine within Frege's philosophy. One may, therefore, imagine his reaction when, several months after his second, unanswered letter, and after a new attempt to involve Hilbert once again in the controversy, Frege received a postcard from Hilbert with this remark: "My opinion is that a concept can only be logically determined through its relations to other concepts. These relations, as formulated in determinate assertions, are what I call axioms. I thereby come to the conclusion that axioms . . . are definitions of concepts. I have not come to this opinion for the purposes of my own amusement; rather, I have found myself forced to accept it by the requirements of rigor in logical inference and in the logical structure of a theory" (31, p. 79).

One final postcard from Hilbert would arrive three years later. It acknowledged receipt of volume 2 of *Grundgesetze;* it claimed familiarity with Russell's contradiction (attributed to Zermelo) and with many other similar ones. These contradictions, Hilbert concluded, "led me to the conviction that traditional logic is unsatisfactory, that the theory of concept construction is in need of sharpening and refinement, whereby I have come to think that the essential deficiency in the traditional construction of logic is the assumption—accepted by all logicians and mathematicians up to the present—that a concept is given when one can determine for each object if it falls under the concept or not. This is, in my opinion, insufficient. The decisive thing is the knowledge of the consistency of the axioms that define the concept" (56, p. 80).[33]

No love was lost, but at least the lines were well drawn. If the axioms of geometry express Fregean propositions, then (given semantic atomism) the primitives in the axiomatic sentences must somehow—no matter how—have been assigned a meaning before those sentences came into existence—and, therefore, independently of them. And if those meanings, that is, those concepts, are the sorts of things Frege says they are, if their proper definition is by means of conjunctions of characteristic constituents or marks, then, once again, axioms could not possibly be involved in their definitions. Whether the antecedents of these implications are true was, of course, the whole question. Hilbert's answer, like Poincaré's, was negative; but they had little more to offer as a philosophical ground for this response than an instinctive, inarticulate feeling. (Their theory of the concept, in particular, was nonexistent.) Frege's affirmative answer, on the other hand, was grounded on one of the most coherent, thorough, and appealing semantic systems ever produced.

Fortunately for us, Frege was not content with noting the inarticulateness of his opponents. He acknowledged instead that his conflict with the new geometry placed upon him two heavy burdens: to give a coherent account of what the geometers *really* meant to say and to explain why they were wrong. He proceeded to discharge those obligations as follows.

3.2. New Wine in Old Bottles

It is sometimes said that Frege recognized the possibility of transforming Hilbert's "implicit definitions" into the explicit definition of a (second level) relational concept such as *being a Euclidean space.*[34] It is doubtful, however, that Frege ever really had this interpretation in mind.[35] His preferred understanding of Hilbert's efforts went along quite different lines.

The basic idea was stated at the beginning of his second letter:

> It seems to me that you want to detach geometry completely from the intuition of space and to make it a purely logical discipline, like arithmetic. If I understand you correctly, the axioms that are no doubt usually considered the basis of the whole structure on the assumption that they are guaranteed by the intuition of space are to be carried as conditions in every theorem; not, to be sure, expressed in their full wording but as if contained in the words 'point', 'straight line', etc. (29, p. 70)

Frege says no more on this topic until 1906, when the issue is raised once again in its second installment. Hilbert's axioms and theorems—he tells us—are not propositions but improper sentences, in other words, sentences from which one or more meaningful terms have been removed and replaced by variables. Each Hilbertian "axiom" is to be seen as a conjunct of the antecedent of an implication with each Hilbertian "theorem" as a consequent. In fact, we first reach the domain of sense (i.e., of propositional knowledge) when we quantify universally the free variables in each of these implications. Thus, "what Mr. Hilbert calls a definition will in most cases be an antecedent improper sentence, a dependent clause of a general theorem" (39, p. 303).

Frege's interpretation of Hilbert's intentions is, of course, our old friend if-thenism. Part 2 of his 1906 work is almost entirely devoted to an elaboration of that doctrine which is, as might be expected, much more thorough than any Russell ever gave. Frege proceeds to examine, for example, an "alleged" proof of a Hilbertian theorem, and shows in painful detail how to reconstruct it as a proof of an implication of the appropriate sort. But whereas Russell had put forth if-thenism as a first step in his retreat from the classical standpoint, Frege does not endorse the doctrine at all, but merely states it as the best sense one can make of Hilbert's confused words. In both cases, however, the doctrine emerges as an attempt to frame the new geometric standpoint within what we might call a propositionalist framework, that is, within a framework in which only propositions are allowed as objects of the "propositional attitudes" (such as assumption, assertion, belief) and the logical operations (such as inference and proof). But whereas Russell, having endorsed if-thenism, preferred not to pursue the matter of how much sense one could make from this propostionalist perspective of the consistency and independence results of the new geometry,[36] Frege proceeded to display with fruition his inability to make any interesting sense of those results from his chosen standpoint. On this point, however, one should not fault the singer but the song.

Indeed, nowhere are the deficiencies of propositionalism more evident than in Frege's attitude towards what Hilbert regarded as the most important results of nineteenth-century geometry: the independence proofs.[37]

Hilbert's model theoretic proofs demanded a departure from Frege's propositionalist attitude, for they presuppose that logic is fully applicable to systems of only partially interpreted sentences. All talk about models and interpretations in the now standard sense would have been anathema for Frege, since the notion of an interpretation presupposes that a language can be developed, and that logic can be applied to it, without assuming that the nonlogical signs need have any meaning. Thus, when Korselt tried to explain to Frege that in order to apply a formal theory to a domain one must first determine whether its axioms are valid in the domain, Frege replied that the only sense he could make of that sloppy remark was to translate it as follows: "When in an inference from the general to the particular we derive a particular sentence from a universal one, we may detach the antecedent sentences only when they are valid [true]" (39, p. 314). In general, the best sense Frege could make of the formalist's "interpretations" was in terms of the exemplification relation. Thus, the proposition

If 2 is greater than 0, then 2 is less than 1 (1)

is, Frege thinks, an "interpretation," in the formalist sense, of the improper sentence

If *x* is greater than 0, then *x* is less than 1 (2)

(see 39, pp. 301–02).[38]

Independence proofs are to be reinterpreted accordingly. The interpretation of hyperbolic geometry that proves the independence of the parallel postulate, for example, is to be understood as follows: instead of (2), the implication to be "interpreted" has as antecedent the conjunction of Hilbert's axioms (minus the parallel postulate) and, as consequent, the denial of that postulate. Call this implication (3). Then, Klein's "interpretation" is a true proposition, call it (4), that stands to (3) precisely as (1) stands to (2).

(1) does prove the independence of something, although surely not of the proposition *2 is less than 1* from *2 is greater than 0* (since we don't know what *that* would mean) or of the improper sentence '*x* is less than 1' from '*x* is greater than 0'. What is proved is the independence of the concepts expressed by the parts of (2) that have a meaning, that is, the independence of the concept *being less than 1* from the concept *being greater than 0*. In cases like the inference from (3) to (4), Frege tells us, it is harder to describe the entities whose independence is being proved. They are "the meanings of the parts" of the antecedent and consequent of (3). In the

simpler cases we have concepts, but "we lack a short designation for the meaning of such parts" in the general case (39, p. 316).

Whatever the ultimate account of these subtleties, what should be obvious, Frege thinks, is that proofs such as Klein's or Hilbert's do nothing whatever to establish the independence of any geometric propositions. One could only be confused into thinking that they do if, like Hilbert, one failed to distinguish true axioms from sentences with only partial meaning. In fact, "all mathematicians who think that Mr. Hilbert has proved the mutual independence of the proper axioms have surely fallen into the same error. Euclid's parallel axiom simply does not occur [in his considerations] and therefore nothing is proved about it" (39, p. 317). Moreover, as far as Frege can tell, there is *no* way to prove its independence. Since the axiom is true, there is no question of finding a domain where the other axioms are true and *it* becomes false (whatever *that* might mean). So, the matter of independence is very much an open question,[39] and perhaps geometers should go back to their drawing boards to see if the parallel axiom can be proved after all.

3.3. Frege's Last Stand

Since Bernays pronounced the Frege-Hilbert debate, in effect, a draw, there has been a growing current of opinion among philosophers that Frege's case has not been given a proper hearing and that he may, in fact, have been right in many, or even all, of his criticisms.[40] Our preceding discussion was aimed to show that this is not so. Let us review the main conclusions we have reached before we turn to Frege's own vision of geometry.

We first observed that there is little reason to think that Frege understood the possibility of defining an explicit predicate by means of Hilbert's axioms, and that the passages in which he appears to hint at that idea are more naturally interpreted otherwise. His best try at an interpretation of Hilbert's words was, we argued, if-thenism. But whereas if-thenism was endorsed by Russell as a first step in his retreat from the classical standpoint, in Frege it occurred as the reconstruction of a doctrine whose untenability should become evident once so described. Having been shown that if-thenism is all that is involved in Hilbert's doctrine, Frege's reader was supposed to conclude, now by an inference too obvious to merit elaboration, that Hilbert's geometry offered a thoroughly incomplete account of geometric knowledge. Moreover, even if Frege had had a perfectly clear understanding of the way in which Hilbert's implicit definitions can be turned explicit, this would not have altered his attitude toward the new geometry. For there can be no question but that he would have regarded the identification of the appropriate "set-theoretic" predicate as an utterly insufficient characterization of geometric knowledge. In *any* case, geometry could not possibly be what Hilbert says it is. But then, what is it?

The epistemological doctrines that inspired Frege's standpoint in geometry are hardly original with him, since in one form or another (usually, in a less precise form than Frege's) they had tacitly inspired most precise philosophical thinking about science since Aristotle. The main theses of that classical standpoint which have emerged in our preceding discussion are: (i) our familiar doctrine of *semantic atomism;* (ii) *propositionalism,* that is, the demand that the axioms and theorems of a scientific discipline be construed as expressing propositions and that these should be the object of propositional attitudes and logical operations; and (iii) what we shall call the *thesis of nonvacuity,* according to which geometry, like all sciences not reducible to logic, includes propositions that are not analytic.

It is, I hope, unnecessary to insist on the role played by the last thesis in Frege's thought; the presence of the second one in his writings has been abundantly illustrated above and is, I take it, undisputed; but it may be worthwhile to provide further evidence of his endorsement of semantic atomism in view of the fact that Frege's holistic-sounding statements appear to be incompatible with that thesis.[41]

An appeal to the principle of semantic atomism, one may recall, had already occurred in the first letter to Hilbert (see above). In fact, Frege appeals to it in each one of his writings on geometry. In (38), for example, he reiterates that "all sentences which are not definitions must contain no proper name, no concept word, no relation word, no function sign whose meaning [*Bedeutung*] has not *previously* been established" (p. 263; my italics).[42]

The precedence on which Frege lays so much emphasis in these questions is not merely the precedence of meaning over the recognition of truth values, but rather, the precedence of the meaning of the parts over that of the whole. The point emerges clearly, for example, during Frege's characteristically brief discussion of the sense of the primitives of an axiomatic system or, in Russell's terms, the indefinables of the theory. The task of clarifying these senses, he writes, "does not belong to the construction of the system but must precede it. Before the construction is begun, the building stones must be carefully prepared so as to be usuable; i.e., the words, signs, and expressions that are to be used must have a clear sense" (32, p. 228). The meanings of all primitives, the "building stones," must therefore be given before we even start the construction of the system of claims that constitutes the theory.

Frege's commitment to *all* three theses of the classical standpoint is therefore beyond question. In fact, Frege's and Russell's problems with the new geometry derived entirely from its inconsistency with these three dogmas. Russell, as we saw, chose the route of compromise; his if-thenism was an attempt to solve the problem of the multiplicity of geometries within the propositionalist framework by sacrificing only the thesis of nonvacuity; for if-thenism tells us that all geometries are equally acceptable, that all of them express propositions, and that all of these propositions are, alas, logically true.

Frege's stand, on the other hand, was uncompromising; all three dogmas of classical epistemology were to be maintained, come what may. What came, of course, were questions concerning the multiplicity of geometries, and questions concerning their indefinables. Let us look first at Frege's response to the latter.

If meanings do not derive as the geometers think, from the endorsement of "axioms," where do they derive from? How does a human being ever come to be in a position to use these semantic "building blocks" of all knowledge which are given to us before we begin the construction of the system of geometry? Not surprisingly, there isn't anything remotely resembling a careful discussion of this matter in Frege's writings. All references to it could be enclosed in a few pages, and their content in no more than a paragraph; for example, this one, from a letter to Hilbert:

> We may assume that [besides definitions and claims (i.e., axioms and theorems)] there are sentences of yet a third kind: the explicatory sentences (*Erläuterungssätze*), which, however, I should not like to consider as belonging to mathematics itself but instead should like to relegate to the preamble, to a propaedeutic. They resemble definitions in that in their case, too, what is at issue is the stipulation of the meaning [*Bedeutung*] of a word (sign). Therefore, they too contain something whose meaning [*Bedeutung*] at least cannot be presupposed as completely and unquestionably familiar, simply because in the language of ordinary life it is used in a vacillating and equivocal manner. If the meaning [*Bedeutung*] to be assigned in such a case is logically simple, one cannot give a real definition, but must be satisfied with ruling out by means of hints the unwanted meanings [*Bedeutungen*] present in ordinary usage, and with indicating those that are intended. Hereby, of course, one always has to count on cooperative understanding, even guessing. (28, p. 63)[43]

And that is all.

It would be hard to ignore the contrast between the boundless disdain that Frege exhibits toward the inadequate explanations of those who dare to question the idea of what he calls "a Euclidean axiom" (i.e., geometric axioms as expressing propositions) and his total inability to explain what extra information these axioms contain beyond that transmitted in Hilbert's version of them. What is, after all, a point? Is it—as Hilbert had incredulously wondered—that which has no extension? What is a line? What does the parallel axiom, the *real* one, not Hilbert's, say? In the end, Frege tells us, these are really private matters. On the parallel postulate, for example, we are told that this question "strictly speaking, is one that each person can only answer for himself. I can only say: so long as I understand the words 'straight line', 'parallel', and 'intersect' as I do, I cannot but accept the parallel axiom" (32, p. 266). Like the author of the *Tractatus* in a similar circumstance, Frege would surely have regarded all further questions on indefinables as not his business but, perhaps, the psychologist's. All he knew is that there *had* to be a way to associate primitives with indefinables, *and* that axioms couldn't have anything to do with it. On this subject his philosophy had nothing else to offer.

Nor did it have much more to offer on the other major issue, the problem of the multiplicity of geometries. Frege's "solution," the only one compatible with the classical stand, was eloquently expressed in some of his private notes:

> One cannot serve two masters. One cannot serve the truth and the untruth. If Euclidean geometry is true, then non-Euclidean geometry is false. . . . The question now is whether Euclidean or non-Euclidean geometry should be struck off the list of the sciences and made to line up as a museum piece alongside alchemy and astrology. If one is content to have only phantoms hovering around one, there is no need to take the matter so seriously; but in science we are subject to the necessity of seeking after truth. There it is a case of in or out! Well, is it Euclidean or non-Euclidean geometry that should get the sack? That is the question. (41, p. 184; 36, p. 169)

That was definitely the question for Frege. Totally immune to the growing consensus of the geometric community, Frege relentlessly proceeded from the premises of his propositional conception of knowledge to the inevitable conclusion that all geometries but one (Euclid's) should be "counted among the pseudosciences [such as alchemy and astrology] to the study of which we still attach some slight importance, but only as historical curiosities" (41, p. 184; 36, p. 169). Who could reject the logic? Who could believe the conclusion?

It is widely acknowledged that the nineteenth-century geometric revolution had a decisive impact on epistemology. All I have tried to argue above is that an illuminating manner of describing its impact focuses on the doctrine of geometric indefinables. Frege and Russell represented the classical doctrine of indefinables at its best; Hilbert and Poincaré, the emerging new doctrine of indefinables, no doubt still shrouded in primeval darkness.

From this perspective, the effect of geometry on epistemology was threefold: it displayed the inability of the classical doctrine of indefinables to cope with the problem of the multiplicity of geometries; it forced the philosophers' attention on the problem of indefinables and on the weakness of its classical solution; and, finally, it provided—through the efforts of the new geometers—the rudiments of a new solution.

But these were, clearly, no more than rudiments. Three decades later those obscure insights had developed into full-blown philosophical theories. Revealing or ironic, the fact is that these theories were put forth by Frege's and Russell's two best students.

4. WITTGENSTEIN

In one of the least transparent sections of the *Tractatus* (the 4.12s) Wittgenstein complains about "the confusion between formal concepts and proper concepts, which pervades the whole of traditional logic," including the logics of Frege and Russell. In his view, some—perhaps most—of the entities regarded as concepts by "traditional logic" are, indeed, genuine concepts that can be represented, à la Frege, as functions. But others, in fact, the most interesting instances, can not be so represented. These "formal" or "pseudoconcepts" include the most prominent indefinables of Frege's and Russell's ontologies: *object, function,* and *complex.* They are not to be represented "by functions or classes (as Frege and Russell believed)" (4.1272).

Even though Wittgenstein is quite clear on what formal concepts are not, he is less informative on what they are. He does say that they are to be represented by variables (4.1272), but this is not much help since there is no way of telling what Wittgenstein called a variable (a notion largely unrelated to what Frege and we so call). The *Tractatus* also tells us that formal concepts have marks that are formal properties (4.126), also known as 'internal' properties. The meaning of 'internal' is, of course, another Tractarian mystery; but a decade later a further clue was dropped when Wittgenstein explained to his Cambridge audience that a better name for 'internal' might be 'grammatical' (Moore 75, p. 295). If we may put these clues together, we have that formal concepts, the concepts for which Frege's theory fails, the basic indefinables, are to be understood by means of marks that it is the purpose of grammar to identify. But, what is grammar?

4.1. Definition and Grammar

In the late 1920s and early 1930s, back from self-imposed philosophical exile, Wittgenstein slowly altered his views in ways that are still not fully understood. One of the most significant features of his new philosophy was the dominant role it assigned to the enigmatic idea of a philosophical grammar or, as he also called it—with a Tractarian expression—a philosophical syntax.

When one looks at Wittgenstein's writings of this period[44] from our geometric perspective it is hard not to detect in his remark "the axioms of, say, Euclidean geometry are rules of syntax in disguise" (101, p. 216), an echo of Poincaré's "the axioms of geometry are definitions in disguise." This suggests that the sorts of things that Wittgenstein was trying to say about certain statements (such as geometric axioms) when he called them 'rules of syntax' and the sorts of reasons that he had for saying them, might have been closely related to what Poincaré and Hilbert were trying to say about geometric axioms when they called them definitions. This suggestion is in fact strengthened as we examine more closely the nature of Wittgenstein's new theory of grammar.

Let us recall once again Poincaré's stance concerning the axioms of geometry. These statements share, he says, two features that one would not have expected to see together: the endorsement of some set of axioms or other is essential for the formulation of all our empirical knowledge, and yet neither facts nor a priori considerations can possibly establish the truth or falsehood of any such set of axioms.

The inability of a sentence to be supported by or to conflict with fact was, for Wittgenstein and later positivists, the very mark of a statement that does not convey information about the world; the very mark, that is, of what was misleadingly called a 'meaningless' statement. Wittgenstein's early discovery of the phenomenon of bipolarity already encapsulated the insight that a sentence fails to convey information unless there are conceivable circumstances (and, one may fairly presume, *observational* circumstances at that) some of which would establish the truth and others the falsehood of the sentence in question.

But already in the *Tractatus* Wittgenstein had distinguished between two radically different ways in which a statement could be meaningless; or, as we should rather say, two radically different reasons why an attempt to convey information by means of a sentence might fail; and he used the terms *sinnlos* and *unsinnig* to distinguish between the two.

The most apparent difference between these varieties of meaninglessness is that whereas *unsinnig* statements are to be banned from the domain of honorable discourse, those that are *sinnlos,* on the contrary, are regarded as somehow essential to all knowledge. The natural inclination to associate a derogatory connotation with the word 'meaningless, may have obscured the significance of this distinction; but there can be little doubt that already in the *Tractatus* the *sinnlos* included no less honorable a domain than logic and that by 1930 the meaningless-but-great encompassed most of what philosophers had traditionally regarded as worthy of their attention.

If these strange sentences fail to do what propositions are intended to do, that is, to convey information, but are nonetheless worth preserving, unlike their *unsinnig* counterparts, there must be some other worthy task they perform. Wittgenstein's philosophy circa 1930 was, to a large extent, an attempt to articulate an account of what exactly it is that these sentences do that makes them so important. Perhaps the basic idea could be explained as follows.

When a body of knowledge is conceived linguistically, there are two prominent features it displays: that it is a class of sentences apparently intended to convey information, and that they are all the object of so-called propositional attitudes such as assertion or belief. The uniformity of syntactic representation as well as the apparent uniformity of the epistemic attitudes we may display toward them, naturally suggests the idea that all the sentences that constitute a body of knowledge are alike in semantic function, their task being merely to express propositions that can become targets of the propositional attitudes. Frege and Russell were only articulating the traditional understanding of matters when they argued that the act of incorporating a sentence into a body of knowledge can only be conceived as a process in which we first recognize what information the sentence conveys and then determine or somehow come to agree that what the sentence says is the case. From this it follows (by the thesis of semantic atomism) that if some (non-variable) primitives of a sentence do not have a meaning before they join their partners in the sentence, then that sentence cannot be assumed or asserted, it cannot be the target of any "propositional" attitude, since it can express no proposition. Prior to assertion and judgment, prior to the incorporation of a sentence into our body of knowledge, it must be possible to recognize the concepts, senses or "meanings" of the constituents of the sentence..

For Wittgenstein (and, as we shall see, also for Carnap) this is true in most cases, but not in all; in fact, it is not true in any of the philosophically *interesting* cases. Most sentences do express propositions, relations of concepts and objects that we can and should understand before a decision is reached concerning what to do with the sentence. But as we move up in the hierarchy of sentences toward those that epistemology has traditionally treated as the worthiest of attention, we encounter a different situation. Here we still find, of course, sentences that are, like the others, objects of endorsement; but this endorsement can no longer be interpreted as a propositional attitude in the classical sense, since there is no way to interpret these sentences as expressing a connection involving concepts that could be recognized prior to the acceptance of the sentences in question.[45]

Wittgenstein's reasons to question the classical account of these sentences as propositions is worth comparing with Poincaré's treatment of geometric axioms. In both cases there is a verificationist element: since no facts can be adduced which could confirm or refute the sentences under consideration, they could not convey any information. It follows that they can't express propositions and, consequently, that our endorsement of these sentences cannot be construed as an attitude toward a proposition. At this point the verificationist argument fizzles out, leaving us without a way to distinguish between the good and the bad within the meaningless. What takes over is a doctrine that uncovers the unity of approach between Wittgenstein and the geometric conventionalists.

Wittgenstein, like the geometers before him, had lost faith in the

Fregean idea of a "propaedeutic" presystematic undertaking in which the meanings of indefinables are somehow identified through hints and appeals to cooperative understanding. Thus he, like the geometers, needed a different account of the way in which concepts are made available, to the extent that they are, for further use in the construction of genuine propositions. Having also lost faith in the idea that meanings come before axioms or rules ("Rules do not follow from an act of comprehension" [105, p. 50]), Wittgenstein could proceed to recognize that it is their capacity to "define" or constitute concepts that gives meaningless rules or "axioms" their attractive dimension. Thus, grammar became Wittgenstein's solution to the problem of indefinables.

Following the geometric conventionalists, Wittgenstein put forth the idea that the endorsement of the strange sentences which he called 'rules' is to be interpreted not as the recognition of the truth of what they say—for they say nothing—but as an element in the decision to incorporate certain concepts into our toolbox of expressive devices. For example, some, indeed the overwhelming majority of sentences we use involving 'not' will be accepted partly on the grounds of the meaning that we associate with that word. But the law of double negation is not accepted on those or on any other grounds. That law, Wittgenstein tells us, does not follow from the meaning of 'not'; if anything, that meaning follows from *it*. Or, rather, the acceptance of the law of double negation is part of what is involved in the recognition of the classical notion of negation, and that law itself can be said to "define" or "constitute" a meaning of negation (104, p. 4; 102, pp. 52–53, 184).

The axioms of geometry are, of course, another paradigm instance of rules of grammar; in his Cambridge lectures Wittgenstein told his students that "Euclidean geometry is a part of grammar" (75, p. 276), and he explained at some length to his Vienna Circle audience that one should resist the temptation to think that geometric axioms convey any information. In fact, he told them, "the axioms of geometry have the character of postulations concerning the language in which we choose to describe spatial objects. They are rules of snytax" (99, p. 62).[46] "Geometry," he explained elsewhere, "isn't the science . . . of geometric planes, lines and points. . . . The relation between geometry and propositions of practical life . . . about edges and corners, etc., isn't that the things geometry speaks of, though ideal edges and corners, resemble those spoken of in practical propositions; it is the relation between those propositions and their grammar" (102, p. 319).

In a striking illustration of the way he put together the diverse themes we have been exploring, Wittgenstein wrote to Schlick: "Does geometry talk about cubes? Does it say that the cube form has certain properties. . . . Geometry does not talk about cubes but, rather, it constitutes the meaning [*Bedeutung*] of the word 'cube', etc. Geometry says, e.g., that the sides of a cube are of equal length, and *nothing is easier* than to confuse the grammar of this sentence with that of the sentence 'the sides of a wooden cube are of equal length'. And yet, one is an arbitrary grammatical rule whereas the other is an empirical sentence (98, p. 5, Wittgenstein's italics).[47]

Beyond logic and geometry, the grammatical rules disguised in "the form of empirical sentences"[48] includes most of the claims that philosophers have traditionally assumed it was their job to determine, whether, why, and in what sense they are true. Among other time-honored instances, Wittgenstein mentions claims concerning the infinite divisibility and continuity of space (99, p. 230) and the order-structure of time (105, p. 14), the law of causality and the thesis of determinism (105, p. 16).

The analogy between Wittgenstein's doctrine of rules and geometric conventionalism is further emphasized by the fact that in all these instances, and in many others, Wittgenstein argues that these rules are in the nature of conventions, since it is entirely up to us to endorse one or another of apparently conflicting sets of rules. One may fully agree, regardless of facts, that time is infinitely divisible or that it is not; that the temporal structure is linear or that it is circular; that all events have a cause or that some do not; determinism and indeterminism, Wittgenstein tells us "are properties of a system which are fixed arbitrarily" (105, p. 16). In none of these cases is there a correct or an incorrect choice; correctness would only make sense if the rules were, so to speak, responsible to a preexistent meaning (105, p. 4; 102, pp. 184, 246). But they are not, and therefore grammar is arbitrary (102, pp. 184–86). It follows also that there is no genuine conflict between apparently contradictory rules, any more than there is a conflict between Euclidean and hyperbolic geometry. The person who endorses the former is not thereby contradicting someone who endorses the latter, for there is nothing that one of them says and that the other denies. The same is true, quite generally, of rules, for a change of rules is merely a change of subject or of meaning.[49]

The evidence reviewed so far strongly suggests that Wittgenstein's thesis of the arbitrariness of grammar is a generalization of geometric conventionalism that fully anticipated Carnap's doctrine of "the conventionality of language forms." There is, however, an odd element in Wittgenstein's doctrine which we have been ignoring so far and which emerges as soon as we raise the question, Why is grammar arbitrary?

4.2. Intolerance

Judging from what we have seen, Wittgenstein's answer to that question should be straightforward and obvious: grammar, like geometry, is arbitrary because its rules are not responsible to any meanings; because, that is, there are no facts that grammar could agree or conflict with before these rules are endorsed. The odd thing is that only once does Wittgenstein state anything like the preceding answer as a reason for the arbitrariness of grammar (102, pp. 184–86). In the many other places where Wittgenstein raises the issue, what he offers instead of this straightforward reason, is an oddly convoluted explanation that makes one wonder what, exactly, is going through his mind. Its opening line is harmless enough: we can't really point to anything that could be offered as a ground for a grammatical rule (99, pp. 104–05; 102, p. 185). But, to judge from the rest of the argument, this impossibility is not due to the simple *absence* of a ground but to something more difficult to explain. Here is how the argument proceeds: a statement that justifies a grammatical rule, for example, 'there are four primary colors' must take the form "since there are (in fact) exactly four primary colors, the rule must be 'there are four primary colors'." But if *this* proposition makes sense, so does its constituent part 'there are four primary colors'. From bipolarity it now follows that 'there aren't four primary colors' is meaningful, which the rule we are trying to justify denies. Hence, if we could justify a rule, it would be violated (104, p. 47; 101, pp. 53, 55; Moore 75, pp. 272–73).

As with most of Wittgenstein's arguments, it is very hard to tell what is being assumed and what (if anything) is being established in the preceding reasoning. The interesting thing about it, however, is its Tractarian aftertaste. Let us look back briefly on the Tractarian doctrines which this argument evokes.

As is well known, the *Tractatus* had put forth a correspondence theory of meaning. According to it, in order for a picture (say, a proposition) to depict a reality it is necessary for both the picture and the reality to share something, a form, a multiplicity (2.5, 2.16, 2.161, 2.17, 2.18). Hence, something that is somehow present in the fact we are trying to depict must also be present in the device we use to represent it, if the device is to serve its semantic purpose. This is, apparently, true not only locally—for atomic facts and their corresponding atomic propositions—but also globally: the world or reality has a form or an essence, and language as a whole, if it is to be able to depict it, must also have that form.[50]

The world is therefore assumed to have a form of its own. Since language is somehow unable to depict this form (2.172, 2.173), there is no condition that we could formulate in any language stating the requirement to be fulfilled in order to be a language. It follows that, according to the *Tractatus,* nothing at all can be said that can distinguish between a pictorial system capable of representing reality and one that is not. In particular, it would be impossible to justify the syntax of our language by constructing a claim to the effect that the syntax of the language has the same form as the reality it deals with, and then showing that this sentence is true.

None of this, however, has the remotest connection with conventionalism or tolerance. The inability to justify grammar is quite consistent with the Tractarian conviction that there is only one sort of proper grammar, the one that agrees in the appropriate way with the form of reality. The misleading impression of grammatical tolerance entirely vanishes once we note that in the *Tractatus* grammatical intolerance has not yet vanished but merely moved from the domain of saying to that of showing. We cannot *say* that grammar should have the form of reality; but it should. And even though language cannot say that it has the form in question, it can show it; so it must be *shown* that the forms of language and reality are the same.

How far away from this standpoint had Wittgenstein moved in the early 1930s? The many passages cited above, which were apparently inspired by the spirit of geometric tolerance, suggest one answer. But many other passages suggest another: Around 1930, for example, Wittgenstein writes, that "language derives the way it means from its meaning [*Bedeutung*], from the world" (101, p. 80) and that "the essence of the world can be captured by philosophy . . . not in the sentences of language but in the rules for this language" (p. 85). In 1932 he tells his Cambridge students that "grammar is a mirror of reality" (104, p. 9) and that it "is not entirely a matter of arbitrary choice. It must enable us to express the multiplicity of facts, give us the same degree of freedom as do the facts" (p. 8). Echoes of showing and saying are also clearly detectable: that grammar allows us to state true and false propositions tells us something about the world, but what it tells us cannot be expressed in propositions; and the reasons for the unsayability of the unsayable are still the same old awful ones (104, pp. 9–10). Moreover, rules of syntax, like logic in the *Tractatus,* are prior to the *how* but not to the *that* (99, p. 77), whatever that may mean. Worst of all, there is the frenzy of grammatical intolerance displayed in the hundreds upon hundreds of pages that Wittgenstein devoted to the extermination of all forms of nonconstructive mathematics.[51]

The sad conclusion to be drawn from all this is that there are at least three different attitudes toward grammar in the middle-Wittgenstein. To begin with, as we have seen, he often talks of rules of grammar in a conventionalist tone of voice, as if to change them were no more difficult than to stop playing chess and start playing checkers or to replace a meterstick by a yardstick (see, e.g., 102, p. 185). The impression he conveys in these passages is that we are dealing merely with different modes of representation of the same facts, and that consequently there can be no grounds for choice other than considerations of expediency, taste, and the like.

At other times, Wittgenstein talks as if it were difficult to imagine circumstances under which one might be inclined to adopt a grammatical system different from the one we actually use (e.g., 104, pp. 4, 49). The preference for a grammatical system, he appears to say on these occasions, is no less justified than the fear of fire or our belief in inductive reasoning. If so, to think of circumstances in which an alternative grammatical system will be acceptable is no easier than to think of circumstances in which people will regularly jump in a fire every time they see one or to assume that nothing will happen the way it always has. Even Scheherazade appears sometimes unequal to the task of imagining the language games and associated forms of life that would allow us to see that, sometime, somehow, people might be inclined to reject this or that grammatical rule. A change of grammar looks in those cases not like a mere change of representational system but like a paradigm change of cross-cultural proportions.

Yet, at other times, Wittgenstein appears to say that some grammatical rules are downright wrong and that those who want to endorse them should not do so. Whereas one might try to explain the difference between the first two attitudes toward grammar as the unpellucid difference between grammar regarded as a calculus and applied grammar, this distinction could surely not account for the immense amount of energy that Wittgenstein devoted to establish the absurdity of classical (nonconstructivist) mathematics. There is more than irony in the contrast between Wittgenstein's serene tolerance toward the good savage that adds two and two and gets five as a result and his refusal to grant Cantor his concept of number. One wonders how he could have failed to feel the conflict between the claims that "the axioms of mathematics are sentences of syntax. . . . The 'axioms' are postulations of forms of expression" (101, p. 189) and that "set theory is false" (p. 211).

Wittgenstein may have been the first one to turn Poincaré's insight into a point of convergence for Kantians and semanticists: a priori claims (or "rules") are constitutive; yet what they constitute is not experience or its objects but the meanings in terms of which we think about reality. And he also saw that this semantic version of the Copernican turn opened up the possibility of alternative sets of a priori principles. Yet this new-found freedom was qualified by an additional insight: Wittgenstein saw more clearly than Carnap and all the other positivists that "anything goes" could not be the guiding principle in our understanding of meaning, and that there are objective facts of meaning to which the production of sense has to conform. It is the narrow verificationist construal of these semantic facts that gives to Wittgenstein's philosophy of this period its characteristic aspect of intolerance and dogmatism. Carnap's principle of tolerance is, above all, an effort to remove the verificationist prejudice from Wittgenstein's semantic insights.

5. CARNAP

Syntax had been a topic of extensive debate at the Vienna Circle meetings. The protocols of the sessions held in 1931 indicate that before that year there had been a good deal of discussion on the connection between philosophy and syntax, but Wittgenstein's emissaries, Schlick and Waismann, had been less than successful in articulating the nature of that link. The word from the Stornborough palace was that philosophy is not a theory but an activity and that its outcome consists of certain peculiar pronouncements, so-called *Erläuterungen,* which are meaningless but enlightening. At one of the meetings Gödel had asked how one distinguishes between the meaningless *Erläuterungen* that we are encouraged to produce and the meaningless philosophy that we are instructed to avoid. No enlightening answer was available. Then, on January 15, 1931, Gödel read a paper on what appeared to be an entirely unrelated matter, for it concerned a recent discovery of his on the incompleteness of arithmetic.

In his intellectual autobiography Carnap tell us that "the whole theory of language structure and its possible applications in philosophy came to me like a vision during a sleepless night in January 1931, when I was ill. On the following night, still in bed with a fever, I wrote down my ideas on forty-four pages under the title 'Attempt at a Metalogic'. These shorthand notes were the first version of my book *Logical Syntax of Language*" (10, p. 53). Five months later he would explain to his Circle friends the intimate link between Gödel's new ideas and the nature of philosophical syntax; as a bonus, he could also offer an answer to Gödel's question.

5.1. Protosyntax

Some of the basic ideas of Carnap's syntax can be reconstructed. Judging from the earliest publications of this period, syntax was basically a corrective or prescriptive discipline, not much more than an undiscriminating stick with which to hit metaphysicians. Carnap tells us, for example, that "the task which at present occupies logicians is the construction of a logical syntax" (15, p. 228); the syntax in question, however, is not one among several alternatives but a privileged syntax, *the* logical syntax. "The possibility of the construction of pseudosentences arises from the departure of historical grammar from the logical syntax" (8, p. 184). If, as Carnap thinks, it is an "error" to use the word 'nothing' as a noun (15, p. 231), if its "correct" analysis must involve quantification and negation in a context, then Heidegger's aphorisms on nothingness have not even succeeded in making sense. Similar considerations apply to, and destroy, most of metaphysics.

One might attempt to minimize the dogmatic element in all of this by noting that, at this early stage, Carnap's attention was focused on syntax as seen, as it were, from the inside. Carnap is thinking of given languages, such as German; and these languages, he thought, already come with a given, if tacit, syntax (7, p. 8; 13, p. 43). One may hold that syntax is arbitrary, as Wittgenstein did, or even that syntax is conventional, as Carnap eventually did, without thereby denying that given a natural language, such as German, it is a question of fact and not of convention whether a certain rule is part of the language (whether, for example, the juxtaposition of two names can produce a meaningful sentence in that language).[52] Hence, under the reasonable presumption that Heidegger is attempting to convey his thoughts in German, it is proper to confront his words with the tacit syntax of that language to determine whether they conform to it or not. The evidence of intolerance in these early papers, one might argue, is not conclusive. Perhaps, but other evidence is.

As mentioned above, in June 1931 Carnap delivered three lectures to the Vienna Circle, where he explained his new conception of syntax.[53] The bulk of the technical presentation consisted of a sketch of a language form similar to the constructivist Language I of (12). Its constructivist rules, Carnap thought, were the only ones admissible in logic and mathematics. He argued, for example, that the concept *provable* is not "correct" because of its tacit unrestricted quantifier; *provable in less than n steps* (for finite *n*) is, on the other hand, perfectly meaningful.

The agreement with Wittgenstein went well beyond constructivism. For example, Carnap argued that there is, in effect, only one language, and that Gödel's techniques for the reformulation of metalanguages into their object languages show that there is no need to presuppose a hierarchy of languages.[54] He also attempted to accommodate within his system Wittgenstein's doctrine of identity. The main disagreement with Wittgenstein concerned the interpretation of syntax.

The outcome of philosophical research, Carnap argued, following Wittgenstein, consists of syntactical remarks; but these remarks are not meaningless. On the contrary, Gödel's translation techniques guarantee the meaningfulness of that portion of syntax that can be expressed in the object language. The answer to Gödel's question concerning the nature of *Erläuterungen* is this: philosophy is not an activity but a theory whose outcome, syntactic analysis, is not enlightening nonsense but coherent, syntactical remarks. *Erläuterungen* are fine as long as we regard them as descriptions of existing syntactical rules or as considerations leading to the proposal of a language form. In other words, as he would later put it, remarks in the material mode are acceptable as long as they can be translated into the formal mode.

Significant as this divergence was, the verificationism that had paralyzed Wittgenstein's grammar into a new form of dogmatism clearly held its grip over Carnap's first version of syntax. But not for long. By 1933

Carnap had banished verificationism from syntax, circumscribing its application to the scientist's object language. The outcome was a radically new conception of philosophy and of the nature of scientific knowledge.

5.2. *The Logical Syntax of Language*

The essence of the standpoint first developed in (12) can be captured in four basic theses.

The first thesis of syntax says that the class of all apparent or real knowledge claims can be divided into two large categories: the real object sentences and the pseudo object sentences (also called 'sentences in the material mode of speech' and 'quasi-syntactical sentences in the material mode'). Real object sentences are, by and large, the sentences of "science" broadly construed so as to include ordinary prescientific knowledge (7, p. 5). In other words, they include all philosophically unpolluted discourse. These sentences are basically fine the way they are. "Scientists," that is, nonphilosophers, can take good care of themselves and are not in need of advice on how to conduct their epistemic business, at any rate not from classical philosophers.

Matters are quite different with pseudo object sentences. They are, by and large, the sentences of philosophy—that is, those that people are prone to produce when they stop leading their ordinary lives and switch to a philosophical mood. Here are a few of Carnap's examples of pseudo object sentences: that time is one dimensional, that space is three dimensional, that the mathematical continuum is composed of atomic elements, and that it is not, that a thing is a complex of sensations, and that it is a complex of atoms, that the only primitive data are relations between experiences, that the sense qualities belong to the given, that the system of colors is known a priori to be three dimensional when ordered according to similarity, that its three dimensionality is an internal property of the arrangement, that every color is at a place, and that every tone has a pitch, that time is continuous, that every process is univocally determined by its causes, and, of course, that the metric structure of space is Euclidean, or that it has some other non-Euclidean structure.[55] All these are instances of the "strange sentences" of which we talked in section 1, the philosopher's turf, the habitual target of his modal hyperbole: a priori true or false, necessary or impossible, self-evident or inconceivable. In spite of their professional concern for them, Carnap tells us, philosophers have never understood what these sentences are or what role they play in knowledge.

The second thesis of syntax provides the grounds for the distinction between real and pseudo object sentences and thereby identifies the error in the traditional understanding of the latter. According to it, real object sentences deal, in fact, with objects in the extralinguistic world and convey (true or false) information about them; pseudo object sentences, on the other hand, appear to convey information but, in fact, do not. In other words, science talks about the world but philosophy doesn't.

Even though it is widely recognized that, according to Carnap's syntax, philosophy does not deal with the world, it is less widely acknowledged that, according to it, science does. In fact, Carnap's syntax does not disagree with what any sane person would want to say concerning the ability of science to deal with extralinguistic objects. Science says "everything that can be said about things and processes" (7, p. 6). In this and similar passages the emphasis is on the implicit exclusion of philosophy from the business of dealing with the world; but the implication is no less evident that in ordinary, nonphilosophical talk we normally succeed in speaking about tables, chairs, electrons, and everything there is (see, e.g., 7, p. 12).

That the sentences of philosophers, good intentions notwithstanding, fail to convey information about anything, had been a longstanding positivist dogma. The strongest inductive evidence for that claim, they thought, is contained in any good history of philosophy: just look with a sober eye at the things philosophers have been saying on the ego, the given, mind, matter, time, space, God, and the like, and contrast them with the concurrent efforts of scientists to solve their comparable problems. Properly transmuted and sublimized, this was the message that Wittgenstein passed along when he reported that philosophy is without sense. Carnap agreed—to a point.

Pseudo object sentences mislead us into believing that they deal with extralinguistic objects such as numbers, things, properties, experiences, states of affairs, space, time, and so on (11, p. 298). Positivism is right in denying that they do. Positivism is wrong, however, in concluding that these sentences are to be committed to the flames; Carnap's second thesis of syntax told us what philosophy is *not,* the third one proceeds to tell what it *is,* and why there is a point, after all, in becoming professionally involved with it.

According to *the thesis of metalogic* (apparently the "vision" of that sleepless night in January 1931), "all philosophical problems which have any meaning belong to syntax" (11, p. 280; 14, p. 435). When someone tells us, for example, that time is linear or that space is continuous, he cannot be interpreted as stating a fact of a very general or abstract character, or even as attempting to say what could only be shown by the use of language. One can make sense of these words, however, viewing them as oblique descriptions of a feature of the linguistic framework in terms of which the speaker, or his language, recognizes facts. A language may be grammatically definite to the point that its users will recognize rules such as the preceding ones spontaneously as parts of it. Usually it won't; that is, usually the facts of language use will underdetermine the tacit rules, and then a philosophical (syntactical) remark must be interpreted as part of a proposal for the constitution of a definite language form. In any case, pseudo object sentences can be seen as perfectly meaningful assertions concerning the features of a natural or artificial language.

According to the preceding theses, the totality of human cognitive speech is divided into two mutually exclusive and jointly exhaustive categories, one for normal people and the other for philosophers. It was Carnap's idea that the thorough independence of these two elements, grounded on their diversity of subject matter, should not be represented, à la Wittgenstein, by dismissing philosophical discourse from the range of the meaningful, or even by regarding it as a form of discourse detached from, but developed in, the same language as scientific discourse. It is best, he thought, to think of science and philosophy as conducted in two totally independent languages, an object language to deal with the world and a metalanguage to do philosophy—that is, to deal with the languages of science.

Up to this point one might see Carnap's syntax as the transvaluation (and, some will say, degradation) of Wittgenstein's grammatical ideas into a Hilbertian mold. Wittgenstein's well-known confusions about Hilbert's metamathematics would have prevented him from recognizing his doctrines in this form. But there is an obvious analogy between Carnap's message so far and Wittgenstein's doctrine of grammar.[56] It is with the next thesis of Carnap's logical syntax that we find a drastic departure from Wittgenstein's thought.

5.3. The Principle of Tolerance

The most important insight that Carnap gained from Wittgenstein's work, he says, is that logic does not say anything about the world and that it therefore has no content (10, p. 25).[57] Logic is not, as Russell thought, a Meinongian theory of objects (Russell 87, p. 150); there is no subsistent world of logical objects and concepts that a true logic should describe. As we saw, however, Wittgenstein's idiosyncratic deobjectivization of logic does not entail its conventionality. Since logic is, for him, the guardian of meaning, and since meaning is guaranteed through a correspondence with the form of reality, logic is also constrained to correspond, if only showingly, with circumstances beyond the range of convention. The conditions for a meaningful symbolism added yet another dimension to Wittgensteinian intolerance in mathematics.

Carnap agreed with Wittgenstein that nothing could be said to which logic and syntax should conform. But, having recognized the possibility of significant discourse in syntax, he soon came to realize that there was nothing of epistemological consequence in the doctrine of showing. After listening patiently to Schlick's and Waismann's attempt to explain that doctrine, and after having tried to make the best of it, Carnap concluded that "the concept 'unsayable' is empty. One can talk about everything. . . . At the basis of this way of speaking lies a mythology of the 'spiritual unsayable' that must be rejected" (9, p. 181; 11, §81).[58]

Moreover, even though there are, of course, limits to the sorts of sign systems that make sense (limits that Carnap did not know how to draw), any effective set of axioms and rules is as intelligible as any other, regardless of its adequacy to intuitionistic standards. Our understanding of classical quantification, for example, is no worse than that of restricted quantification. We may picture the former as an actually infinite conjunction of claims, but we don't *have* to; in particular, we need not picture the semantics of quantification at all in terms of the actual verification of each matrix instance (11, p. 163). Quantification is not defined by pictures but by the rules to which it must conform. After all, "the truth of a universal sentence is established by a proof of that sentence itself" (p. 163) and not by anything we do in connection with its instances. *Mutatis mutandis,* the same is true of all mathematical primitives. Decades later Carnap would extend this criterion, noting that all meaningful (internal) questions of existence are to be decided in the affirmative precisely when the rules of the appropriate language authorize the inference to an existence claim (under possibly contingent circumstances). This criterion stands in clear opposition to the "realistic" (not to be confused with "semantic") answer to this question.[59]

Free from the bonds of the standard correspondence theory of logic and from Wittgenstein's oblique version of it, free also from metalinguistic verificationism, Carnap's philosophy was ready to incorporate the lessons of geometric conventionalism. For the first time the deobjectivization of logic and mathematics had become complete.

> Up to now, in constructing a language, the procedure has usually been, first to assign a meaning to the fundamental mathematico-logical symbols, and then to consider what sentences and inferences are seen to be logically correct in accordance with this meaning. . . . The connection can only become clear when approached from the opposite direction—let any postulates and any rules of inference be chosen arbitrarily; then this choice, whatever it may be, will determine what meaning is to be assigned to the fundamental logical symbols. (11, xv, see also pp. 18, 131)

But if the basic statements of logic, mathematics, and, more generally, grammar, are not responsible to ontological or semantical elements identifiable prior to their endorsement, and if these semantical elements are constituted, to the extent that they are, by the endorsement of those principles, then we have the same freedom to endorse alternative mathematical or metaphysical systems as we have in our choice of geometries.

> The first attempts to cast the ship of logic off from the *terra firma* of the classical forms were certainly bold ones, considered from the historical point of view. But

they were hampered by the striving after correctness. Now, however, that impediment has been overcome and before us lies the boundless ocean of unlimited possibilities. (11, p. xv)

The range of language forms and, consequently, of the various possible logical systems is incomparably greater than the very narrow circle to which earlier investigations in modern logic have been limited. Up to the present, there has been only a very slight deviation, in a few points here and there, from the form of language developed by Russell. . . . The fact that no attempts have been made to venture still further from the classical forms is perhaps due to the widely held opinion that any such deviations must be justified—that is, that the new language form must be proved to be 'correct' and to constitute a faithful rendering of 'the true logic'. (11, pp. xiv–xv)

We have finally reached the principle of tolerance. Here our story ends.

6. LEGACY

In one of his celebrated animadversions on Carnap's conventionalism, Quine once raised the issue of the link between that doctrine and geometry. His verdict was that conventionalism had, indeed, been encouraged in the philosophy of mathematics "by the non-Euclidean geometries . . . with little good reason." His survey of the relevant facts is worth reproducing *in extenso.*

In the beginning there was Euclidean geometry, a compendium of truths about form and void, and its truths were not based on convention. . . . The truths were there, and what was conventional was merely the separation of them into those to be taken as starting point (for the purposes of the exposition at hand) and those to be deduced from them. The non-Euclidean geometries came from artifical deviations from Euclid's postulates, without thought (to begin with) of true interpretation. . . . Non-Euclidean geometries have, in the fullness of time, received serious interpretations. This means that ways have been found of so construing the hitherto unconstrued terms as to identify the at first conventionally chosen set of non-sentences with some genuine truths, and truths presumably not by convention. . . . Uninterpreted systems became quite the fashion after the advent of non-Euclidean geometries. This fashion helped to cause, and was in turn encouraged by, an increasingly formal approach to mathematics. Methods had to become more formal to make up for the unavailability, in uninterpreted systems, of intuition. . . . The tendency to look upon non-Euclidean geometries as true by convention applied to uninterpreted systems generally, and then carried over from these to mathematical systems generally. A tendency indeed developed to look upon all mathematical systems as, qua mathematical, uninterpreted. This tendency can be accounted for by the increase in formality, together with the use of disinterpretation as a heuristic aid to formalization. Finally, in an effort to make some sense of mathematics thus drained of all interpretation, recourse was had to the shocking quibble of identifying mathematics merely with the elementary logic which leads from uninterpreted postulates to uninterpreted theorems.[60] What is shocking about this is that it puts . . . geometry qua interpreted theory of space, outside mathematics altogether. (82, pp. 108–10)

The reader who has followed me this far may have more than a quibble to raise about a few of Quine's points. Our business here, however, is with only one of these. By Quine's reckoning, all that is needed in order to avoid the conventionalist's confusions is to keep the interpreted-versus-uninterpreted distinction firmly in mind as we survey the evolution of geometry. Uninterpreted geometry is surely not what inspired conventionalism, since everyone agrees that there are no propositions, no truths or falsehoods, in it; hence, no truths by convention. All that is left as inspirational matter is interpreted geometry which, of course, expresses propositions—or truths or falsehoods—which, of course, have their truth value quite independently of (relevant) linguistic convention. And that is all.

The burden of this paper has been to show that this is not all. The heart of the geometric debates, we have argued, was not the odd question whether propositions (or truths or falsehoods) can be made true by convention—they can't—but whether the classical account of geometric indefinables was tenable. Geometers were the first to recognize its shortcomings. Faithful to semantic atomism they sought a different solution to the problem of indefinables and found it in conventionalism, the doctrine that the endorsement of some sentences does not reflect the recognition of their truth but only a partial commitment to an assignment of meanings to some of their terms. Since such endorsements could not be described as the recognition of facts, they were often labeled 'conventions'. Since the sentences endorsed could not be false, they were sometimes labeled 'true by convention'. The phrase may be inexcusable but the underlying idea is not.

In the late 1920s and 1930s philosophers came to recognize the wide-ranging implications of the geometers' point; not, of course, that the axioms of geometry could be interpreted only as definitions in disguise, even if that was the best account of the information that they conveyed circa 1900. One philosophical point that philosophers borrowed from geometers was that a possible source of indefinable meanings is the endorsement of certain sentences that collectively embody not the acknowledgment of certain facts but the endorsement of a language form. Another point was that some—indeed, most—of the sentences that have traditionally concerned philosophers *can be* and *are often* best interpreted as vehicles of meaning rather than of information and that their endorsement is better construed as a definition ("constitution") than as an assertion.

Broadly speaking, linguistic conventionalism is grounded on a discriminating attitude toward the semantic status of certain terms. This attitude emerges from a sensitivity to semantic nuances that both pragmatism and holism instruct us to blunt. A conventionalist, for example, will not fail to notice, and regret, Quine's failure to explain what he means by "geometry qua interpreted theory of space" or by "the science of form and void" in the preceding quotation. Frege and the early Russell (and, apparently, Quine) would have agreed that these expressions call for no further explanation either because their meanings are clear enough, or because meanings come in much, much larger units. The conventionalist knows better.

Of course, there is no dearth of explanations that could be offered as to what those phrases mean—even if none is likely to produce anything that is both mathematical and a theory of space, as Quine expects. We could say, for example, that our straight lines are light rays or linear functions, or that congruence is to be defined by the behavior of a rod, and then take it up (and down) from there in standard ways. Or we could follow Hilbert and forgo all explanations as irrevelant to geometry. What we could *not* do is forgo all explanations and then insist that we do not mean to deal with geometry as uninterpreted but, rather, as a theory of space—as if that meant anything definite. To put the same point differently, for Frege and the early Russell a rule such as " 'straight line' designates straight lines" would not be worth stating because it is too obvious; for the conventionalist it is not worth stating because it is synonymous with " 'Glub' designates glubs." What the conventionalist has learned from the history of geometry is that expressions such as 'straight line' and its geometric cognates suffer from semantic ailments that do not affect other expressions such as 'light ray', 'meter stick', and the like. No one (no reader of this paper, anyhow) could seriously say that he has no idea of what a light ray is and that he needs that word defined before he can use it responsibly. On the other hand (coherence theories of reference notwithstanding), since around 1900 no one could seriously insist that it is sufficiently clear to any educated person what a straight line is.

There are people—more and more every day, apparently—who find such semantic elitism obnoxious. Some of them infer from the (true) premise that the conventionalist distinctions are a matter of degree to the conclusion that they cannot be legitimate or useful. Others avoid this egregious blunder and appeal, instead, to a doctrine that we might call semantic holism (or, in our nastier moods, Münchhausen semantics), which involves the denial of semantic atomism. In spite of his frequent incursions into the Frege-Russell speech style, the holist's semantics is no closer to that of Frege or Russell than to that of conventionalists. The semantic holist joins the conventionalists in his distrust of Frege's and Russell's propaedeutic endeavors; for him, however, nothing need be said, propaedeutically or otherwise, concerning the primitives of geometric sentences before they join their linguistic partners in meaningful sentences. One may, perhaps, have to conjoin them with other sentences, such as those of a physical theory, in order for all of them to acquire full semantic propriety; but no particular sentence in that lot is to be singled out from the others as involved only in the meaning-giving process.

The basic conflict between holists and conventionalists might be schematized as follows: Whenever two sentences *A* and *B* are such that the endorsement of *A* signifies in the conventionalist's eyes the agreement to use certain words in *B* with a certain meaning, the conventionalist draws a clear-cut line between *A* and *B*, calling the former a convention and the latter an assertion. Whether he labels *A* a "coordinative definition" or a "definition in disguise" or a "rule of grammar," or something else, the point is always the same: some sentences must sacrifice their hopes to say anything at all in order that others could stand a chance of fulfilling their semantic dreams.

Holism, on the other hand, promises semantic success for everyone without semantic sacrifices. If *A* and *B* suffer from some semantic deficiency when isolated—if, for example, they have no empirical content or some of their terms have no meaning or no referent—somehow these difficulties are solved through the labors of conjunction.

In the most popular version of holism, we are urged to believe that A & B may acquire the requisite semantic virtues (it may convey information, refer, be true or false) without its conjuncts being in any way similarly redeemed by conjunction. Even if no sense may be made of the idea that A and B say something, or of what they say, there is no comparable problem with A & B, or with a sufficiently extended conjunction.

The conventionalist can make no sense of this. His main problem is not with the obscurity of the holist's doctrine but with its apparent consequences. The failure to detach meaning from fact had traditionally led to two complementary errors: the classification of every epistemological issue as ultimately factual, and the classification of every epistemological issue as ultimately conventional. The episode reviewed in the preceding essay suggests a way to pursue a middle course, based on the distinction between language and theory, between processes in which meaning is contributed and processes in which claims are made. Carnap consistently reiterated his conviction that one should distinguish between those readjustments of knowledge elicited by conflicts with experience that lead to a change of language form and those that lead to a mere reassignment of truth value to a sentence: "A change of the first kind constitutes a radical alteration, sometimes a revolution, and it occurs only at certain historically decisive points in the development of science. On the other hand, changes of the second kind occur every minute" (10, p. 921). The first type of change, by altering the structure of what is regarded as analytic—as true in virtue of meanings—alters the very fabric of concepts in terms of which we think about the world. This "paradigm shift" or "categorical change" (as others would call it) may be motivated by pragmatic reasons, and thus, by reasons qualitatively similar to those leading to a mere refutation. But a common origin does not determine a common outcome. Carnap and Wittgenstein thought it essential to distinguish between pragmatically motivated changes of grammatical framework, and mere conflicts between theory and experiment. Talk of agreement or conflict with fact is reasonable in the latter case but not in the former; talk of conversion is reasonable in the former case but not in the latter. Pragmatism finds this distinction fictitious and is thereby tempted, once again, to see every case of theory-change uniformly, either in the model of "refutation" or in the crypto-idealist model of "conventions." Carnap offered a middle route, with a place for the constitution of meaning, and another for reference to fact. It still looks like an idea worth pursuing.

NOTES

I should like to express my gratitude to Linda Wessels for her comments on a draft of this paper and to Gregorio Klimovsky and Thomas Oberdan for a number of very enlightening discussions, especially on the topics of sections 4 and 5. Their expert advice on matters concerning the evolution of logical positivism has been invaluable. I am also indebted to Kenneth Blackwell (Bertrand Russell Archives, McMaster University), Professor Henk Mulder (Schlick Archives, University of Amsterdam), and the Carnap Committee (Carnap Archives, University of Pittsburgh) for permission to consult and/or quote some of the materials from their archives. Thanks are also due to the National Science Foundation for its support.

Citations in the text, as well as in the notes, give the number of the item in the reference list and the page number. When the context identifies the author, his or her name is not included.

1. See, for example, Beth's *Foundations of Mathematics* (Amsterdam: North-Holland, 1959), chap. 2.

2. Neither the mathematicians I am alluding to nor the author of this essay is very good at distinguishing between the meaning and the reference of *mathematical* names such as 'two' or 'the square root of minus one'. So, I will follow them—and some logicians, such as Russell—in using the word 'meaning' for whichever semantic object one chooses to associate with mathematical names. Whenever the sense-reference distinction is essential for the formulation of an intended point, I will, of course, observe it.

3. The best works known to me on this episode are Freudenthal (45), Nagel (76), and Torretti (97). In one of its drafts, this paper included a much more detailed account of the evolution of geometric ideas in the nineteenth century. The publication of Torretti's superb book made that material redundant.

4. The analogy had been detected, apparently, by Lambert (see Stäckel and Engel [94, pp. 145–47]), and Schläfli, also, had been close to recognizing the full force of this analogy: see H. M. S. Coxeter, "The Space-Time Continuum," *Historia Mathematica,* 1975, esp. pp. 292–93.

5. Hilbert showed in app. 5 of *Grundlagen* that the attempt to represent the *whole* hyperbolic plane in a Euclidean surface of constant negative curvature by means of Beltrami's methods is doomed to failure.

6. Little or no attention was paid to hyperbolic geometry until Gauss's favorable opinion of it became known through the publication of his correspondence with Schumacher in 1860–63.

7. Cayley did not appeal to projective coordinates, which were introduced in von Staudt's *Geometrie der Lage* (1847) as part of his project to provide a purely projective foundation to projective geometry. (Freudenthal argues in [45] that this goal was achieved only much later, through the work of the Italian school.) For Cayley's odd attitude concerning the nature of coordinates and von Staudt's definition see Cayley (16, vol. 2, pp. 605–06) and Klein (70, pp. 153–54).

8. See Stäckel and Engel (94, p. 26). Wallis's assumption refers to all geometric figures and not only to triangles, even though his proof applies it only to triangles. Saccheri noted that one need only assume that there are two similar triangles in order to prove the parallel postulate, but *he* did not regard even this assumption as evident. For Cataldi's claim, see Bonola's *Non Euclidean Geometry* (New York: Dover, 1955), p. 13.

9. Proclus had argued that since there are lines (such as asymptotes) that approach a straight line without ever intersecting it, we cannot assume without proof that the same cannot happen with straight lines (81, p. 151).

10. See, e.g., Contro (21, p. 290), Steiner (96, p. 36), Freudenthal (46, pp. 114–15).

11. If the reader thinks he can define 'red' or 'colored' so as to turn the implication from *a is red* to *a is colored* into a logical truth, then let him try to do the same with *a is a meter long* and *a is not two meters long*, or *a is red* and *a is not blue*. These are, by the way, the sorts of examples that led Wittgenstein away from his Tractarian standpoint and into the new grammatical philosophy of the early 1930s. If logical truths were truths in virtue of the meanings of logical words (as I think they are—see Coffa [18]) then the notion of analyticity could be easily vindicated; for no one could seriously argue that *only* logical words are sufficiently clear to ground claims by virtue of their meaning. For example, it would be absurd to pretend that knowledge of the meaning of disjunction and implication suffices to establish that 'if *p*, then *pvq*' is true, but that knowledge of the meaning of color expressions does not suffice to ground the truth of 'if something is red, then it is colored'. Thus, Quine's rejection of analyticity naturally led him to argue against Carnap's linguistic doctrine of logical truth. One problem with the holistic picture of knowledge that Quine attempted to substitute for Carnap's is that it blinds us to the element of surprise in Klein's independence argument, and it prevents us from recognizing the dramatic change in attitude of the geometric community toward geometric descriptive words.

12. The propriety of this translation, as well as its relevance to independence proofs, is discussed in greater detail in Coffa (18). See especially the last section: "What did Klein prove?"

13. This was, as we shall see, Frege's opinion; it has been recently revived in Hunter (59).

14. But even he was willing to admit that the "Latin and Hebrew races," with their more developed critical-logical sense, were less attached to the "strong naïve space intuition" that is predominant in the Teutonic race (64, p. 228).

15. The best account of this development is in Torretti (97).

16. The link between Russell's projective geometry and what was so called at the time was less than clear. Some of the "errors" in his presentation are listed in Poincaré (79, § 3); a longer list appears in Brouwer's doctoral dissertation (6). The main problem appears to be that Russell missed (understandably at the time) the purely conventional character of coordinatization and concluded that the projective points to which imaginary coordinates are assigned have a merely fictitious being. (He says, for example, that "the circular points at infinity are not to be found in space," p. 45.) This led him to think that the projective plane, for example, differs from the affine plane, if at all, only in the fact that in the former it is not possible to draw metrical correlations between parallels (i.e., lines intersecting "at infinity"). Klein's planimetric constructions of hyperbolic and elliptic geometry are also

misinterpreted. Russell thinks that they "are all obtained from the Euclidean plane" and that, in general, "Euclidean space is left in indisputed possession, and that the only problem remaining is one of convention and mathematical convenience concerning the definition of distance" (p. 30).

17. Incongruously, Russell appears to have thought that the remaining geometric primitives have different meanings in different geometries. Euclidean points, straight lines, and planes differ so much from geometric objects in other spaces that the only space that we can, in fact, imagine is the actual one, whatever it may happen to be (84, p. 74; 93, pp. 694–95). If our space were hyperbolic, for example, we could have absolutely no idea of what a Euclidean straight line looked like. See Poincaré's penetrating objection to this (79, p. 276).

18. On the contrary, this was one of the points Klein had emphasized from the very beginning, since it was essential to the noncircular character of his metric construction. As Klein often noted, his ideas relied on von Staudt's metric-free coordinates. One might also note, in passing, that von Staudt's coordinatization procedure breaks down if one assumes, with Russell, that the straight lines to which his process applies are "Euclidean" (i.e., affine), for then the process fails to associate any coordinate to the points lying on one side of "the point at infinity." The "straight lines" to which von Staudt's procedure applies are projective and therefore closed.

19. Logico-philosophical commentators, who tend to be generous toward the logician-philosophers in these debates, often take it for granted that the notion of definition to which Frege and Russell were appealing was, in effect, not significantly different from what we now call 'explicit definition'. As we shall see, this is not so. One could hardly deny that Frege's and (to a lesser extent) Russell's work made possible our current understanding of explicit definition—even though the decisive impetus came from the Italian school. But, as will emerge in the following pages, Frege's and Russell's conceptions of definition were thoroughly committed to the much narrower chemical picture of definition as the analysis (or synthesis) of a complex.

20. One should not confuse the marks of a concept with its simple constituents, even though, upon occasion, they might coincide. If *Fxy, Gx,* and *Hx* are simple, then in the complex concept *(y) (Fxy ⊃ Gx) & Hx, (y) (Fxy ⊃ Gx)* is a mark but not a simple constituent, whereas *Gx* is (presumably) a simple constituent but is certainly not a mark.

21. Russell entirely missed the point. His reply to the preceding remark was this: "M. Poincaré seems to think that one can determine through experiments whether bodies preserve or change their shape without knowing anything about this shape, without even being able to assign any meaning to the word *shape.* If measurement is not an unambiguous operation designed to uncover something, rather than to create it, it is difficult to see how the measurement of bodies could reveal that they move in approximate agreement with the Euclidean group" (93, p. 688).

22. Notice that Poincaré was not arguing that intuition can *never* be used to "define" the primitives. That is what it does, according to him, in arithmetic, for example.

I might as well add at this point that my concern here is with only an aspect of Poincaré's philosophy of geometry. The ontological dimension of Poincaré's conventionalism has been thoroughly explored in Grübaum's writings (see, for example, 47, chap. 1). I do not know whether this aspect of Poincaré's philosophy—as well as others that are also present in it—can be coherently synthesized with the semantic dimension we are exploring.

23. A few years later, while dealing with the same subject but in the field of ethics, Moore would argue: "If I am asked 'What is good?' my answer is that good is good, and that is the end of the matter. Or if I am asked 'How is good to be defined?' my answer is that it cannot be defined and that is all I have to say about it" (74, p. 6).

24. See Coffa (20), where I also discuss the standard (and erroneous) view that Russell rejected the Kantian doctrine that mathematics consists of synthetic a priori judgments.

25. See also (88, pp. 60, 71), (86, p. 673).

26. Poincaré duly registered his surprise at this extraordinary feature of Russellian semantics (80, pp. 75–76).

27. This is, perhaps, the time to challenge a widespread misunderstanding concerning the role of the interpreted-uninterpreted distinction in geometry. The debates surveyed in this paper were often thought to be largely verbal, because their participants were talking about different sorts of geometries. According to this line of thought, Hilbert and Poincaré were talking about uninterpreted geometry and were, of course, right in denying that there are any propositions in it; whereas Frege and (initially) Russell were talking about interpreted geometry, and they were right in viewing it as a set of true or false propositions. The problem with this resolution of the conflict is that it entirely grants the formalist point, for to endorse the interpreted-uninterpreted distinction as a *sufficient* account of the nature of geometry is, in effect, to grant Hilbert's whole case. If all there is to geometry besides its uninterpreted form is the populous democracy of geometric models, then the noble class of propositions that constitute true geometry according to Frege and Russell, is lost, forever unnoticed, among the uncountable multitude of swindlers that pass for interpretations. The idea that propositions about his watch could be part of geometry was, for Frege, unspeakably silly. Thus, an account of geometry that did not include devices to distinguish between propositions about points, real points, that is, and propositions about watches, was hopelessly inadequate.

28. Wiener's paper was later published as "Über Grundlagen und Aufbau der Geometrie," in *Jahresbericht der Deutschen Mathematiker Vereinigung* 1 (1892): 45–48.

29. Frege's dissertation (1873) dealt with the representation of imaginary magnitudes in plane geometry. In his letter to Hilbert of 29 Dec. 1899 Frege reports that he has done some work on the foundations of geometry which leads him to believe that it is possible to use a smaller number of "primitive forms" than those employed by Hilbert (28, p. 60). Frege may have been referring here to a system such as the one that he develops in (34).

30. It may be worth noting that Russell's attitude toward Hilbert's logico-geometric researches was not much more enthusiastic than Frege's. In a letter to Couturat of 4 June 1904, he writes, "As to Hilbert, I haven't read him with much attention. He has offended my aesthetic sensibility, for example, through his habit of discussing projective theorems assuming Euclidean space. He does not lack merit, but the problems with which he deals do not seem to me to be the most interesting ones" ("Pour Hilbert, je ne l'ai pas lu avec beaucoup d'attention. Il a outragé mon goût ésthétique, *p. ex.*, par son habitude de discuter les théorèmes projectifs en supposant l'espace euclidean. Il n'est pas sans mérit, mais les problèmes dont il traite ne me paraissent pas les plus interessants"). On 9 Feb. 1905, referring, perhaps, to Hilbert (57), Russell writes again to Couturat "I have read the memoir by Hilbert that you were kind enough to send me, but I have found in it nothing of value. It shows the same absence of system and order as his geometry, the same penchant for what is complicated and twisted and against fundamental principles" ("J'ai lu le mémoire de Hilbert que vous avez eu la bonté de m'envoyer, mais je n'y trouve rien qui ait aucune valeur. J'y trouve la même absence de système et d'ordre que dans sa géométrie, le même goût pour tout ce qui est compliqué et détourné au préférence aux principles fondamentaux"). These letters are now at the Russell Archives in McMaster University, Hamilton, Ontario.

31. Later editions substitute 'Definition' for 'Erklärung'. The change is not significant, since the word 'Erklärung' was commonly used to mean 'definition'. For example, in Stäckel and Engel's *Theorie der Parallellinien,* Euclid's 'definitions' *(joroi)* is translated as 'Erklärungen' (94, p. 6).

32. The translations of the correspondence and of (39) follow Kluge's (35) except for minor departures, mostly on technical terms such as *Gedanke* (proposition), *Satz* (sentence), *Bedeutung* (meaning), *uneigentlicher Satz* (improper sentence), etc.

33. See also Hilbert's remark on Frege (57, p. 175), translated in van Heijenoort's *From Frege to Gödel* (Cambridge, Mass.: Harvard University Press, 1967), pp. 130–38. When Russell claimed that mathematical definitions (as opposed to philosophical definitions) gave no real knowledge of the things whose relational structure was being characterized, Poincaré replied: "I do not know whether outside mathematics one can conceive a term independently of relations to other terms; but I know it to be impossible for the objects of mathematics. If one wants to isolate a term and abstract from its relations to other terms, what remains is *nothing*" (80, p. 78). See also (78, pp. 27, 28).

34. See, for example, (61, p. xix), (5, pp. 92–93), and (23, p. 235).

35. Second-level concepts (which are functions from first-level concepts to truth-values) are introduced by Frege in the geometric debate in order to argue that, insofar as Hilbert's axioms are to be taken as lists (however incompetently stated) of marks, the concept that they define could not possibly be a first-level concept, like *point, line,* etc., but a second-level one. "If they [i.e., the axioms] define a concept at all, it can only be a second-level concept" and then he adds: "It must, of course, be doubted whether any concept is defined at all" (38, p. 272). These are not words that would be written by someone who thinks that something *is,* after all, defined. The recognition of the role played by the appeal to marks in Frege's argument may have been obscured by the fact that the only version of Frege's second letter to Hilbert (1900) that was published before 1976 included a sentence that says that we can (the original letter says "can not") conjoin first- and second-level concepts as marks of a concept. See (33, p. 416), (35, p. 19) and compare with (44, p. 73).

36. Russell pointed out on several occasions that arguments such as Klein's established the independence of the parallel postulate (see, e.g., 84, pp. 12–13; 90, p. 593). However, this is an opinion that he first endorsed when "logic" was, for him, the contents of Bradley's text. The opinion was not therefore the outcome of Russell's considered logical opinion but an echo of the by then near-unanimous opinion of the geometric community on whose judgment Russell placed much more trust than Frege. The matter calls for a much more detailed analysis than I could offer here. Suffice it to say at this point that if Russell had understood the independence results in the standard formalist manner, there would have been no reason to question their connection with consistency results, the way he does in the case of elliptic geometry (84, §40). Russell's ability to make sense of independence proofs for what he called "geometry as the science of actual space" must have been seriously impaired by his conviction that it makes no sense to say of a true proposition that it might have been false (92, p. 454).

37. Hilbert had written to Frege that his primary purpose in the *Grundlagen* had been to give a satisfactory account of "the most important results of geometric research" (53, p. 75), "the most beautiful and most important propositions of geometry (indemonstrability of the parallel axiom, of the Archimedean axiom, provability of the Killing-Stolz axiom, etc.)" (54, p. 68).

38. In the most puzzling passage of his correspondence with Hilbert, Frege says: "When a universal sentence contains a contradiction, the same is true of all particular sentences included under it. One can therefore conclude from the consistency of the latter to that of the universal one, although not conversely" (29, p. 75; see also 43, p. 398). Frege appears to be saying that if *(x)Fx* is inconsistent, so is *Fa,* for any *a.* Since this is obviously not so (let *'Fx'* be *'Gx & - (y)Gy'),* one may conjecture that Frege was led to this remark by his feeling that there was something right about proving consistency by means of models, plus his erroneous view that interpretation is inference from the general to the particular.

39. In the last section of (39), Frege explores a possible path to establish the independence of the parallel axiom. His propositionalism leads him to introduce meanings where Hilbert had excluded them—his procedure applies to propositions, not to partially interpreted sentences—but his method is designed to neglect this additional element of meaning, by applying rules that are indifferent to its presence. Independently of this excess

baggage, Frege's method suffers from the problem that since he only allowed true propositions as premises of logical inferences, he could not prove the mutual independence of false propositions. In any case, Frege was obviously dissatisfied with this proposal, and in 1910 he wrote to Jourdain reiterating his old opinion that "the indemonstrability of the parallel axiom cannot be proved" (30, p. 119; see also 43, p. 398).

40. See (5) and references in note 34.

41. This is particularly important in view of the fact that semantic atomism is inconsistent with the view that the unit of meaning is the sentence, or that it makes no sense to talk of meanings outside propositional contexts. Frege may have been tempted to think something like this. His famous remark that "words have meaning only in the context of a sentence" (265, §60) has often been interpreted along such lines. Dummett pointed out that this version of holism vanishes from Frege's writings after 1884. Actually, it reappears once again (38, p. 270). Hacker (49) has argued that it reappears in the debate with Hilbert and offers one of my preceding quotations as evidence of this. But the quotation clearly states a version of semantic atomism that is inconsistent with the holism that Hacker reads into it. The other three quotations that Hacker cites as evidence of Frege's holism (49, p. 229) bear no relation to that doctrine, since they refer to variables, not to primitive constants, and assert that even though they have no meaning in isolation, they contribute to the meaning of the sentence as a whole. If holding this thesis were enough to make someone a holist, then (1) anybody with any knowledge of logic would be a holist and (2) Russell's theory of descriptions would suffice to make *him* a holist. Under Hacker's interpretation, holism has become a trivial doctrine.

The fact is that even though Frege occasionally made holistic pronouncements of an ontological nature, this doctrine played no significant role in his philosophy. See, for example, Angelelli's remarks (1, §2.7). The doctrine is, in fact, inconsistent with the semantic atomism displayed in the geometric debates; moreover, the ontological version expressed, e.g., in Frege (38, p. 270) implies that there must be wholes with concepts as constituents, and there are no such wholes in Frege's ontology.

42. See also (39, pp. 265, 267).

43. See also (39, p. 288), (32, p. 224), and (24, p. 290).

44. The basic sources for the period that interests us here (roughly from 1929 to 1932) are these: *Philosophische Bemerkungen* (1964), written between February 1929 and April 1930, *Philosophische Grammatik* (1969), written in the early 1930s; manuscript and typescript volumes (from which the preceding two texts were selected) in the Wittgenstein *Nachlass* Microfilms; Waismann's notes of conversations with members of the Vienna Circle from December 1929 to July 1932, in (99); notes by Moore of lecture courses at Cambridge from 1930 to 1933, in Moore (75); Lee's notes of the Cambridge Lectures for 1930–32, in Wittgenstein (104); Ambrose and MacDonald's notes of the lectures for 1932–35, in Wittgenstein (105).

45. Wittgenstein talked about these two different sorts of sentences, e.g., in his Cambridge lectures. See Moore (75), esp. pp. 256–57, sec. 2, p. 300. At one point he said of grammatical "propositions" that "they are not propositions at all" (p. 262). See also Wittgenstein (99, p. 114) and (105, pp. 50–51).

46. See also (102, p. 320) and (101, pp. 216–17).

47. See another statement of the same point in (105, p. 51), (102, p. 52). See also (103, pp. 113–14).

48. See (100, sec. 308, 309, 401, 402).

49. (102, pp. 111, 184–85; 104, pp. 57, 58; 99, p. 71; 101, p. 178, with 'syntaktischen' instead of 'synthetischen' in line 20).

50. In spite of what commentators will tell you, nobody really knows what 'form' and 'multiplicity' mean in these passages; but here is a guess, no worse than most others. An on-off switch has the same multiplicity as the (standard) states of a coin, but not the same

multiplicity as the states of a die; a map has the same multiplicity as the facts of geographic location on a surface, but English has a greater multiplicity since we can say in English—though not in map-language—'*A* is North of *B* and *B* is North of *A*', which corresponds to no possible state of affairs in geography. (One of the jobs of syntax is to cut down the multiplicity of language to size [101, p. 74; 99, pp. 79–80].). Form is said to be the possibility of structure (2.033). I take this to be an effort to convey the following claim: A structure is each of the states a system can display: the on-off switch has two structures, "on" and "off"; the die has six structures; the arrangements of objects expressed by "*A* is North of *B*" and "*B* is North of *A*" are two more structures. The totality of structures or states that a system can display determines (or perhaps is determined by, or is) its form. Two systems can represent each other only when they can display the same number of corresponding structures, and when all pairs of corresponding structures are isomorphic; when they have, that is, the same form or multiplicity.

51. In their review of the *Philosophical Grammar* (50) Hacker and Baker attempt to show that the thesis of the autonomy of grammar (Hacker's name for grammatical conventionalism) is consistent with Wittgenstein's constructivism, since the latter doesn't really circumscribe the class of allowable rules of grammar but merely helps us recognize that the only rules that make any sense *at all* in mathematics are constructivist rules. The authors do not explain in detail how this conclusion is reached, but apparently it derives from the observation that "any rule of grammar must establish the correctness or incorrectness of some use of symbols, and hence we must be able to describe what would constitute such uses" and whether or not they are correct (p. 289). Even granting all this, it is not clear how the authors conclude that classical mathematics "specifies a logically impossible use of symbols" in view of the well-known, logically possible use of symbols in standard accounts of classical mathematics. The only piece of evidence brought forth for that conclusion is a quotation from Wittgenstein (102, p. 452), where he assumes that the classical understanding of quantification over an infinite class stands or falls with the idea of a verification procedure actually applied to each instance of the universal claim. The argument could carry any weight only for someone who thought, like Wittgenstein did, that to understand the meaning of a sentence is to know how to decide whether it is true or false (Wittgenstein 101, p. 77). But, as an argument to explain why nonconstructivist symbolic systems are meaningless, this is, of course, entirely begging the question.

52. Moreover, when seen from the inside or in use or, as Wittgenstein might put it, when we consider rules as applied, they do not present themselves to us as conventions, but as statements too obvious for words. It was, no doubt, this internal aspect of rules that Wittgenstein was referring to when he replied to Schlick's question on how to decide whether a rule is valid (99, p. 77). His answer was not that there is no such thing as validity for rules, but that the discovery of a rule is the discovery of what, in a way, we always knew. Like Carnap, at this early stage Wittgenstein also thought that natural languages come with a tacit syntax. It is not easy to tell how he managed to make this view consistent with his equally strong conviction that in matters of meaning "there is nothing hidden." Upon occasion he attempted to water down his idea of a tacit syntax by claiming that the precisely constructed syntactical rules were idealizations of the syntactic facts of language (e.g., 102, p. 76); but one is then left to wonder how this version of syntactic tolerance agrees with Wittgenstein's attitude in the foundations of mathematics.

53. A record of these lectures is preserved among the protocols of the Circle's meetings at the Carnap Archives in the University of Pittsburgh.

54. By the time of (12), Carnap had realized that his own syntactic definition of mathematical truth (analytic) could not be formulated in the object language for which it was defined. See (12; 11, §§60c; 18). Carnap had reached this result independently of Tarski's work on the notion of truth. See Coffa (17) and my forthcoming "Tarski, Carnap and the Search for Truth."

55. These examples come from (13, pp. 81–82) and (11, pp. 305–07).

56. In a letter to Schlick of June 5, 1932 (available at the Schlick Archives in Amsterdam), Wittgenstein accused Carnap of plagiarism. The charge refers to Carnap (14), the first paper in which his basic "metalogical" ideas are presented. The accusation, ludicrous as it is, shows that Wittgenstein recognized what his Cambridge and Oxford followers have been unable to see, that there is a strong similarity between Wittgenstein's views at that time and Carnap's syntacticism. The charge is, of course, ungrounded. First of all, as Carnap patiently explained to Schlick in response to the accusation, he had access to Wittgenstein's new ideas only through the reports of people (Schlick and Waismann) who were totally unaware of the extent to which this new doctrine departed from Tractarian dogmas. In the Vienna Circle meeting of Feb. 12, 1931, for example, when Waismann's *Thesen* were discussed, Waismann reported on the unjustifiability of grammar, saying that syntax is not fortuitous *[zufällig]* and that we cannot think what it would be like if the rules of syntax did not apply. (From Carnap's protocols of the Vienna Circle meetings, at the Carnap Archives, Pittsburgh.) Waismann's *Thesen* (printed in Wittgenstein 99) are a clear indication of the way in which Wittgenstein's doctrines were being transmitted to the Circle. Quite apart from this, there is the even more significant fact that, as we are arguing here, even though Wittgenstein's doctrines represented a major first step in the direction of tolerance, that is *all* they were.

57. Although Carnap was also aware that the point had been anticipated by Weyl in *Das Kontinuum* (see 11, p. 186).

58. If we are to judge from the results of several decades of Wittgenstein scholarship, Carnap was right. In his *Wittgenstein* (New York: Viking Press, 1970), for example, Pears tells us that the doctrine of showing recognizes that we cannot formulate the "truths" of ethics, religion, and the like in empirical propositions, but that we can nonetheless state them in nonfactual discourse (p. 96) which is not nonsense but merely non-sense (p. 72). These "truths" (p. 57) are, he tells us, neither true nor false (p. 37) since they say nothing, and only what can be said can be true. They are, however, "valid" (p. 53) and they convey what "lies beneath" and "is trying to get out" in such deep tautologies as 'there is what there is' and 'what is reflected in the mirror of language is reflected in the mirror of language' (p. 86). (Wittgenstein's penchant for deep tautologies was amply illustrated in his conversations with Waismann; see, e.g., 99, pp. 33, 102, 104, 106, 112, 114, 119, 120, 139, 145.) In his article "Wittgenstein" for Edwards's *Encyclopedia of Philosophy,* Malcom explains the status of the metaphysical insights that can be shown only by noticing that they "cannot be stated in language, but *if* they could be, they would be true" (vol. 8, p. 334). Anscombe's spurious defense (2, pp. 83–86) of the idea that some things cannot be said is based entirely on the equivocation between "saying informatively" (a notion that never occurs in Wittgenstein's writings) and "saying." The doctrine that she succeeds in vindicating is the triviality that there must be a way to explain a language without presupposing that the person to whom it is being explained already understands a language. Incredibly, she thinks that Carnap (12) rejects that opinion (but see, e.g., 11, p. 227). Kenny's *Wittgenstein* (62), an otherwise sober and enlightening account, assumes, like Anscombe, that when Wittgenstein said "*A* can't be said" he must have meant "You can't say *A* unless you already know it is true" (see, e.g., pp. 46, 66; see also the equivocation between saying and explaining in p. 44). According to them, apparently, the doctrine of solipsism can, in fact, be said, i.e., stated in perfectly meaningful language, only not informatively, i.e., the sentences stating the doctrine can be understood only by people who—like Wittgenstein—have already recognized its meaning in some other way. Neither Kenny nor Anscombe provide a single piece of evidence in support of this singularly implausible interpretation. Hacker's excellent (48) avoids this pitfall. His failure to make a case for the epistemological relevance of Wittgenstein's distinction must be counted as strong evidence that there isn't any.

59. A major subject that we shall not deal with here is the extent to which Carnap

further departed from the doctrine of explicit definition at this stage of his thought. Once meanings are removed from their role as primary determinants of rules and occur only as the outcomes of the endorsement of rules, there is but a small step to the conclusion that the meaning in question is an unnecessary fiction and that all that really counts for the purpose of knowledge and communication is what is given at the level of our behavior toward linguistic rules. If languages are games and if, in Sellars's happy phrase, their "*esse est ludi,*" then those meanings that were thought to be implicitly defined by the endorsement of rules become fictions whose only underlying reality are the facts of linguistic behavior. Wittgenstein had told his Vienna Circle audience that the "ludic" approach to knowledge was an alternative to Frege's referentialist standpoint (see, e.g., Wittgenstein 99, p. 105). Carnap's syntax incorporates a similar approach that was developed even further in Carnap's later philosophy. (Recall, for example, his contention in "Empiricism, Semantics and Ontology" that the only sense one can make of existence claims is in those circumstances in which the rules of the appropriate language authorize an assertion of existence—rather than, as a realist would put it, when the things in question exist. In particular, the existence of a domain is no longer thought to be decided by the consistency of the corresponding syntactic [or semantic] rules. Rather, questions of existence are significant only when they are "internal.")

60. A footnote at this point indicates that Quine is referring to Russell's if-thenism.

REFERENCES

1. Angelelli, I. *Studies on Gottlob Frege and Traditional Philosophy.* Dordrecht: Reidel, 1967.
2. Anscombe, G. E. M. *An Introduction to Wittgenstein's Tractatus.* London: Hutchinson, 1959.
3. Beltrami, E. "Saggio di Interpretazione della Geometria Non-Eculidea" (1868). In *Opere Matematiche,* vol. 1, pp. 374–405. Milan: V. Hoepli, 1902.
4. ———. "Teoria Fondamentale degli Spazii di Curvatura Costante" (1868–69). In *Opere Matematiche,* vol. 1, pp. 406–29. Milan, 1902.
5. Bernays, P. Review of Max Steck, "Ein unbekannter Brief von Gottlob Frege." *Journal of Symbolic Logic,* 7 (1942): 92–93.
6. Brouwer, L. E. J. *On the Foundations of Mathematics* (1907). In *Collected Works,* vol. 1, pp. 13–101. Amsterdam: North Holland, 1975.
7. Carnap, R. *Die Aufgabe der Wissenschaftslogik, Einheitswissenschaft.* Vienna, 1934.
8. ———. "Ergebnisse der logischen Analyse der Sprache" (abstract). *Forschungen und Fortschritte,* pp. 183–84. Berlin: Kerkhof, 1931.
9. ———. "Erwiderung auf die vorstehenden Aufsätze." *Erkenntnis* 2 (1932): 177–88.
10. ———. "Intellectual Autobiography" and "Replies." In *The Philosophy of Rudolf Carnap,* ed. P. A. Schilpp, pp. 3–84, 858–1013. LaSalle, Ill.: Open Court, 1963.
11. ———. *The Logical Syntax of Language.* London: Routledge & Kegan Paul, 1937.
12. ———. *Logische Syntax der Sprache.* Vienna, 1934.
13. ———. *Philosophy and Logical Syntax.* London: Psyche Miniatures, Kegan Paul, Trench, Trubner & Co., 1935.
14. ———. "Die Physikalische Sprache als Universal Sprache der Wissenschaft." *Erkenntnis* 2 (1932): 432–65.
15. ———. "Überwindung der Metaphysik durch logische Analyse der Sprache." *Erkenntnis* 2 (1932): 219–41.
16. Cayley, A. *The Collected Mathematical Papers.* Vol. 1. Cambridge: Cambridge University Press, 1889.

17. Coffa, A. "Carnap's Sprachanschauung circa 1932." In *PSA,* 1976, ed. F. Suppe and P. Asquith, pp. 205–41. East Lansing: Philosophy of Science Association, 1977.
18. ______. "Machian Logic." *Communication and Cognition* 8 (1975): 103–29.
19. ______. "Notas para un Esquema de la Filosofía de la Ciencia Contemporánea." *Crítica* 6 (1972): 15–51.
20. ______. "Russell and Kant." *Synthese* 46 (1981): 247–63.
21. Contro, W. S. "Von Pasch zu Hilbert." *Archives for the History of the Exact Sciences* 15 (1976): 285–95.
22. Couturat, L. "Essai sur les Fondements de la Géometrie, par Bertrand Russell." *Revue de Métaphysique et de Morale* 6 (1898): 354–80.
23. Dummett, M. "Frege on the Consistency of Mathematical Theories." In *Studien zu Frege,* ed. M. Schirn, vol. 1, pp. 229–42. Stuttgart: Frommann-Holzboog, 1976.
24. Frege, G. "Erkenntnisquellen der Mathematik und Naturwissenschaften" (1924–25). In *Nachgelassene Schriften,* pp. 286–94. *See* 33.
25. ______. *Grundgesetze der Arithmetik.* Hildescheim: Olms, 1893.
26. ______. *Die Grundlagen der Arithmetik.* Breslau: M. & H. Marcus, 1884.
27. ______. *Kleine Schriften.* Ed. I. Angelelli. Hildescheim: Olms, 1967.
28. ______. Letter to Hilbert of 27 Dec. 1899. In *Wissenschaftlicher Briefwechsel,* pp. 60–64. *See* 44.
29. ______. Letter to Hilbert of 6 Jan. 1900. In *Wissenschaftlicher Briefwechsel,* pp. 70–76. *See* 44.
30. ______. Letter to Jourdain, item xxi/9 (1910). In *Wissenschaftlicher Briefwechsel,* pp. 114–24. *See* 44.
31. ______. Letter to Liebmann of 29 July 1900. In *Wissenschaftlicher Briefwechsel,* pp. 147–49. *See* 44.
32. ______. "Logik in der Mathematik" (1914). In *Nachgelassene Schriften,* pp. 219–70. *See* 33.
33. ______. *Nachgelassene Schriften.* Ed. H. Hermes et al. Hamburg: Felix Meiner Verlag, 1969.
34. ______. "Neuer Versuch der Grundlegung der Arithmetik" (1924–25). In *Nachgelassene Schriften,* pp. 298–302. *See* 33.
35. ______. *On the Foundations of Geometry and Formal Theories of Arithmetic.* Ed. E. H. W. Kluge. New Haven: Yale University Press, 1971.
36. ______. *Posthumous Writings.* Trans. P. Long and R. White. Oxford: Blackwell, 1979.
37. ______. "Über den Begriff der Zahl" (1891–92). In *Nachgelassene Schriften,* pp. 81–127. *See* 33.
38. ______. "Über die Grundlagen der Geometrie, I, II" (1903). In *Kleine Schriften,* pp. 262–72. *See* 27.
39. ______. "Über die Grundlagen der Geometrie, I–III" (1906). In *Kleine Schriften,* pp. 281–323. *See* 27.
40. ______. "Über einige geometrische Darstellung der imaginären Gebilde in der Ebene" (1873). In *Kleine Schriften,* pp. 1–49. *See* 27.
41. ______. "Über Euklidische Geometrie" (1899–1906). In *Nachgelassene Schriften,* pp. 182–84. *See* 33.
42. ______. "Unbekannte Briefe Freges über die Grundlagen der Geometrie" (1899). In *Kleine Schriften,* pp. 400–22. *See* 27.
43. ______. "Ein unbekannter Brief von Gottlob Frege" (1900). In *Kleine Schriften,* pp. 395–99. *See* 27.
44. ______. *Wissenschaftlicher Briefwechsel.* Ed. G. Gabriel et al. Hamburg: Felix Meiner Verlag, 1976.
45. Freudenthal, H. "The Impact of Von Staudt's Foundations of Geometry." In *For Dirk Struick,* ed. R. S. Cohen et al., pp. 189–200. Dordrecht: Reidel, 1974.

46. ———. "Zur Geschichte der Grundlagen der Geometrie." *Nieuw Archief voor Wiskunde* 5 (1957): 105–42.
47. Grünbaum, A. *Philosophical Problems of Space and Time.* New York: Knopf, 1963.
48. Hacker, P. M. S. *Insight and Illusion.* Oxford: Clarendon Press, 1972.
49. ———. "Semantic Holism: Frege and Wittgenstein." In *Wittgenstein: Sources and Perspectives,* ed. C. G. Luckhardt, pp. 213–42. Brighton: Harvester Press, 1979.
50. ———; and Backer, G. P. Review of Wittgenstein, *Philosophische Grammatik. Mind* (1976): 269–94.
51. Hilbert, D. *Gesammelte Abhandlungen.* Vol. 3. Berlin: Springer, 1935.
52. ———. *Grundlagen der Geometrie.* Leipzig: Teubner, 1922.
53. ———. Letter to Frege of 29 Dec. 1899. In *Wissenschaftlicher Briefwechsel,* pp. 65–8. *See* 44.
54. ———. Letter to Frege of 29 Dec. 1899 (draft). In *Wissenschaftlicher Briefwechsel,* pp. 68–9. *See* 44.
55. ———. Letter to Frege of 22 Sept. 1900. In *Wissenschaftlicher Briefwechsel,* p. 79. *See* 44.
56. ———. Letter to Frege of 7 Nov. 1903. In *Wissenschaftlicher Briefwechsel,* pp. 79–80. *See* 44.
57. ———. "Über die Grundlagen der Logik und Arithmetik." In *Verhandlungen des 3. Internationalen Mathematiker-Kongresses in Heidelberg,* pp. 174–85. Leipzig, 1905.
58. Hinst, P. "Freges Analyse der Hilbertschen Axiomatik." *Grazer Philosophische Studien* 1 (1975): 47–57.
59. Hunter, G. "What Do the Consistency Proofs for Non-Euclidean Geometries Prove?" *Analysis* 40 (1980): 79–83.
60. Kambartel, F. "Frege und die Axiomatische Methode." In *Studien zu Frege,* ed. M. Schirn, vol. 1, pp. 215–28. Stuttgart: Frommann-Holzboog, 1976.
61. ———. "Zur Formalismuskritik und zur Logikbegriff Freges." In *Nachgelassene Schriften,* pp. xvii–xxiv. *See* 33.
62. Kenny, A. *Wittgenstein.* Cambridge, Mass.: Harvard University Press, 1973.
63. Klein, F. *Gesammelte Mathematische Abhandlungen.* 3 vols. Berlin: J. Springer, 1921–23.
64. ———. "On the Mathematical Character of Space Intuition" (1893). In *Abhandlungen,* vol. 2, pp. 225–31. *See* 63.
65. ———. "Über die geometrieschen Grundlagen der Lorentz-Gruppe" (1910). In *Abhandlungen,* vol. 1, pp. 533–52. *See* 63.
66. ———. "Über die sogenannte Nicht-Euklidische Geometrie, I" (1871). In *Abhandlungen,* vol. 1, pp. 254–310. *See* 63.
67. ———. "Über die sogenannte Nicht-Euklidische Geometrie, II" (1873). In *Abhandlungen,* vol. 1, pp. 311–43. *See* 63.
68. ———. "Vergleichende Betrachtungen über neuere geometrische Forschungen" (1872). In *Abhandlungen,* vol. 1, pp. 460–97. *See* 63.
69. ———. "Vorbemerkungen zu der Arbeiten über die Grundlagen der Geometrie" (1921). In *Abhandlungen,* vol. 1, pp. 241–43. *See* 63.
70. ———. *Vorlesungen über die Entwicklung der Mathematik im 19. Jahrhundert.* New York: Chelsea, 1967.
71. Leibniz, G. W. "Meditations on Knowledge, Truth and Ideas." In *Philosophical Papers and Letters,* ed. L. E. Loemker, pp. 291–95. Dordrecht: Reidel, 1976.
72. Locke, J. *An Essay Concerning Human Understanding.* New York: Dover, 1959.
73. Moore, G. E. *Philosophical Papers.* New York: Collier, 1959.
74. ———. *Principia Ethica.* Cambridge: Cambridge University Press, 1903.
75. ———. "Wittgenstein's Lectures in 1930–33" (1954–55). In *Philosophical Papers,* pp. 274–318. *See* 73.
76. Nagel, E. "The Formation of Modern Conceptions of Formal Logic in the Development of Geometry." *Osiris* 7 (1939): 142–224.

77. Poincaré, H. "Analyse de ses travaux scientifiques." *Acta Mathematica* (Stockholm) 38 (1921): 3–135.
78. ———. "Le Continue mathématique." *Revue de Métaphysique et de Morale* 1 (1893): 26–34.
79. ———. "Des Fondements de la géométrie." *Revue de Métaphysique et de Morale* 7 (1899): 251–79.
80. ———. "Sur les Principes de la géométrie." *Revue de Métaphysique et de Morale* 8 (1900): 73–86.
81. Proclus. *A Commentary on the First Book of Euclid's Elements.* Trans. Glenn R. Morrow. Princeton: Princeton University Press, 1970.
82. Quine, W. O. "Carnap and Logical Truth." In *The Ways of Paradox,* pp. 100–25. New York: Random House, 1966.
83. Resnik, M. "The Frege-Hilbert Controversy." *Philosophy and Phenomenological Research* 34 (1974): 386–403.
84. Russell, B. *An Essay on the Foundations of Geometry* (1897). New York: Dover, 1956.
85. ———. *A Critical Exposition of the Philosophy of Leibniz* (1900). London: Allen & Unwin, 1937.
86. ———. "Geometry, Non-Euclidean" (1902). In *The New Volumes of the Encyclopedia Britannica,* vol. 4, pp. 664–74. London, 1902.
87. ———. Letter to Meinong of 15 Dec. 1904. In *Philosophenbriefe, Aus der Wissenschaftlichen Korrespondenz von Alexius Meinong,* ed. R. Kindinger, pp. 150–51. Graz: Akademischen Druck, 1965.
88. ———. "Mathematics and the Metaphysicians" (1901). In *Mysticism and Logic,* pp. 59–74. *See* 89.
89. ———. *Mysticism and Logic.* London: Allen and Unwin, 1963.
90. ———. "Non-Euclidean Geometry." *The Athenaeum,* 29 Oct. 1904, pp. 592–93.
91. ———. "On Scientific Method in Philosophy" (1914). In *Mysticism and Logic,* pp. 75–93. *See* 89.
92. ———. *The Principles of Mathematics* (1903). New York: Norton, 1950.
93. ———. "Sur les Axiomes de la géométrie." *Revue de Métaphysique et de Morale* 7 (1899): 684–707.
94. Stäckel, P., and Engel, F. *Die Theorie der Parallellinien von Euklid bis auf Gauss.* Leipzig: Teubner, 1895.
95. Steiner, H. G. "Frege und die Grundlagen der Geometrie I." *Mathematisch-Physikalische Semesterberichte,* n.s. 10 (1963): 175–86.
96. ———. "Frege und die Grundlagen der Geometrie II." *Mathematisch-Physikalische Semesterbertichte,* n.s. 11 (1964): 35–47.
97. Torretti, R. *Philosophy of Geometry from Riemann to Poincaré.* Dordrecht: Reidel, 1978.
98. Wittgenstein, L. "Diktat für Schlick" (1931–33). Unpublished typescript, item 302 in von Wright's catalogue of Wittgenstein's *Nachlass.*
99. ———. *L. Wittgenstein und der Wiener Kreis.* Ed. B. F. McGuinness. Oxford: Blackwell, 1967.
100. ———. *On Certainty.* New York: Harper, 1969.
101. ———. *Philosophische Bemerkungen.* Frankfurt: Suhrkamp Verlag, 1964.
102. ———. *Philosophische Grammatik.* Oxford: Blackwell, 1969.
103. ———. *Remarks on the Foundations of Mathematics.* Oxford: Blackwell, 1967.
104. ———. *Wittgenstein's Lectures, Cambridge 1930–1932.* Ed. D. Lee. Totowa, N.J.: Rowman and Littlefield, 1980.
105. ———. *Wittgenstein's Lectures, Cambridge 1932–1935.* Ed. A. Ambrose. Totowa, N.J.: Rowman and Littlefield, 1979.

JOHN A. WINNIE
Indiana University

Invariants and Objectivity: A Theory with Applications to Relativity and Geometry

Whenever you have to do with a structure-endowed entity Σ *try to determine its group of automorphisms. . . .* You can expect to gain a deep insight into the constitution of Σ in this way.

—Weyl
Symmetry

1. THE PROBLEM OF STRUCTURE

1.1. Structure and Determinate Expansions

The theories of contemporary mathematics all have this in common: they describe sets, or "universes," structured by other distinguished sets—relations, functions, or operations. A differentiable manifold, for example, is a set paired with a set of "charts," a Minkowski space-time is a set together with the metric ω^2 giving the space-time interval between any pair of events, and a topology is a set together with a set of its "open" subsets. But whatever the theory, its formulation is deliberately indifferent to the actual entities that realize its account. *Any* set: elementary particles, tones—even numbers—will do, provided the cardinality is right and the appropriate relations or functions are singled out for distinction. This is at least part of what is meant by saying that theories describe or define *structure* and that structure is *abstract.*[1]

Only rarely, of course, will a theory determine a "single" structure. For one thing, models isomorphic to any given model of a theory will also be models of that theory—but all of these may be considered structurally identical, and essentially "the same" model. The more significant and common case is when a theory has nonisomorphic models, models differing in structure. For example, first-order theories having an infinite model must always, by the Lowenheim-Skolem theorem, have models of arbitrary infinite cardinality (although some theories may be strong enough to ensure that any two of its models of a given infinite cardinality are isomorphic).

Precision, however, does not require a first-order formulation; instead, we may use the richer languages of contemporary logic and mathematics to generate a class of only isomorphic structures.[2] Such are the theories of (globally flat) Galilean and Minkowski space-times or, for a fixed number *n*, the theory of *n*-dimensional vector spaces or Euclidean spaces.

In many cases, however, restriction to merely isomorphic models is not the theoretician's intent. Newtonian or relativistic systems of particle mechanics are clear examples. The structure of such systems will depend upon the number or configurations of their particles. All Newtonian systems, say, will contain a common *space-time* structure, but the structures of the total systems—particles and all—will vary greatly. The models of general relativistic particle mechanics differ even more widely, some having only their local manifold structure in common. What such systems share is their conformity to certain dynamical laws, not their overall structure.

Yet, while the same theory may describe very different structures, the objectively determinate features of particular structures may be of considerable mathematical or physical interest. The theory of finite-dimensional real vector spaces, for example, has nonisomorphic models, corresponding to each dimension *n*. Nevertheless, each vector space has its unique dimension as an objectively determinate feature.[3] Some models of general relativity permit the definition or determination of a total mass function, whereas for other models this is not possible. What is or is not objectively determinate in each (type) of relativistic model can be a matter of considerable physical and philosophical importance (see Earman 10).

For these reasons, the following investigations deal with the problem of objectivity from the standpoint of the individual structure. Applications are then immediately forthcoming to the sorts of cases mentioned above. Furthermore, as will soon be clear, linguistic approaches to objectivity seem to contain irremediably arbitrary elements that are due to the choice of language used to formulate the theory under consideration. Better, then, to proceed algebraically in terms of the models themselves instead of the linguistic apparatus used to generate these models.

These and the ideas to follow are best explained by having some more or less concrete illustrations of structures before us. Some of the simplest structures are those described by first-order (elementary) languages; they are called *relational systems*.

A relational system consists of a set A, called the *universe* or *domain* of the system, together with a sequence of relations $R_1, \ldots, R_k$ defined on the set A. Relations are here considered as sets of ordered n-tuples of elements of A, and subsets of A are identified with one-placed relations. (Individual elements of A may also be exhibited, but, for simplicity's sake, this possibility will be ignored.) The resulting system A is usually written as:

$$A = (A, R_1, \ldots, R_k), \tag{1.1}$$

and if the predicates of a specific first-order language are listed in a given fixed order $\underline{R}_1, \ldots, \underline{R}_k$, and the relation terms are of the appropriate syntactical type, then a structure like *A* above is said to be a *possible model* of the sentences of that language.

Equivalence of possible models is just set-theoretical isomorphism: the existence of a one-one mapping (a bijection) from one universe to the other that preserves all the relations involved. The fundamental semantical fact about isomorphic structures is that they are truth-preserving: A sentence true in a relational system is also true in any isomorphic system. It is not difficult to show that isomorphism of structures is indeed an equivalence relation, and the smoothness of the formal semantics of first-order languages demonstrates that it is a useful equivalence relation. As *the* structural equivalence relation, however, it has a number of serious shortcomings. Let me begin with a more or less trivial flaw and lead up to more serious matters.

Consider a structure $A = (A, R^2, S)$, where R^2 is a two-placed relation on A and S is simply a subset of A. Next, let $A' = (A, S, R^2)$, that is, A' is just like *A* except that the relations are listed in the reverse order. Surely, the set A is in both cases given the "same" structure; yet *A* and A', being systems of different types, cannot be isomorphic.

In this case, there would seem to be a simple remedy. Since the order in which the relations are listed is arbitrary, why not regard a relational system as a universe A paired with a *set* of relations, instead of listing the relations in sequence? Thus, instead of $A = (A, R^2, S)$, we would have $A = (A, \{R^2, S\})$, and $A' = (A, S, R^2)$ would now become $A' = (A, \{S, R^2\})$. Not only would *A* and A' be isomorphic structures, they would also be *identical* structures—which in *this* case, at least, accords well with our intuitions about the matter.

But now consider the system

$$I = (\{\ldots, -2, -1, 0, 1, 2, \ldots\}, <, >). \tag{1.2}$$

Since any isomorphism of *I* onto itself must preserve the 'less than relation' '$<$,' the reflection of the integers about zero ($\gamma(x) = -x$) is not such an isomorphism. But if *I* is reconstrued according to the above proposal, *I* becomes:

$$I' = (\{\ldots, -2, -1, 0, 1, 2, \ldots\}, \{<, >\}). \tag{1.3}$$

But now the reflection mentioned above, and others like it, will be isomorphisms of I' onto I' since they will take $<$ to $>$, $>$ to $<$, and so take the *set* of these relations to itself. Sequences of relations, however arbitrary the chosen order, at least served to preserve the individuality of their elements. Sets of relations, however, may confuse relations we intend to distinguish. There is, fortunately, a way out; but first, let us consider another, more serious, obstacle to taking isomorphism as necessary for structural sameness.

Consider the pair of systems

$$N = (\{0,1,2, \ldots\},<) \tag{1.4}$$

and

$$N' = (\{0,1,2, \ldots\},\leq), \tag{1.5}$$

where '$<$' and '$\leq$' are the standard 'less than' and 'less than or equal to' relations, respectively. Clearly, the two systems are not isomorphic, and, equally clearly, this is not owing to the way the components of the system have been listed. It is rather that the two relations displayed are logically very different; for example, 'less than' is irreflexive, 'less than or equal to' is reflexive.

Yet, despite such differences between the component relations of these two structures, most logicians and mathematicians would probably agree that the two systems are "essentially" the same structure. And they could and would provide a seemingly good reason for their view: *the two relations are interdefinable,* and while interdefinable relations need not be the same, they nevertheless in some sense "codetermine" one another. As a result, the set A is structured by each of these relations in exactly the same way: The set $\{0,1,2, \ldots\}$ ordered by the 'less than' relation is *thereby* ordered by the 'less than or equal to' relation, and conversely.

This idea, which expresses one of the basic intuitions of this essay, can also be put in a slightly different form. Consider the structure

$$N^{+} = (\{0,1,2, \ldots\},<,\leq) \tag{1.6}$$

an *expansion* of the structure N. Again, since the relation '$\leq$' is definable in terms of '$<$,' it is natural to regard N^{+} as having *added* no more structure to the set A, and so, in a sense that remains to be defined precisely, it is natural to regard N^{+} as "equivalent" to N. Similarly, the structure N'^{+} given by:

$$N'^{+} = (\{0,1,2, \ldots\},\leq,<) \tag{1.7}$$

is also a "superficial" expansion of N'. Finally, since the order in which a system's relations are listed is structurally irrelevant, we may as well take N'^{+} as equivalent to

$${}^{+}N' = (\{0,1,2, \ldots\},<,\leq),$$

which brings us back to N^+ once again. From this standpoint, the "equivalence" of our original pair of structures N and N' is the result of our being able to "superficially" or *determinately* expand each structure to obtain a pair of identical structures.

A general strategy emerges: Superficially expand, permute, and then compare. In this example, we were able to extend N and N' to *identical* structures. When this is possible, the two structures will be called *codeterminate* structures. More generally, when two structures may be extended to *isomorphic* structures, they will be said to be (invariantly) *equivalent* structures. The general situation is illustrated by figure 1, where the squiggly arrow depicts a superficial expansion.

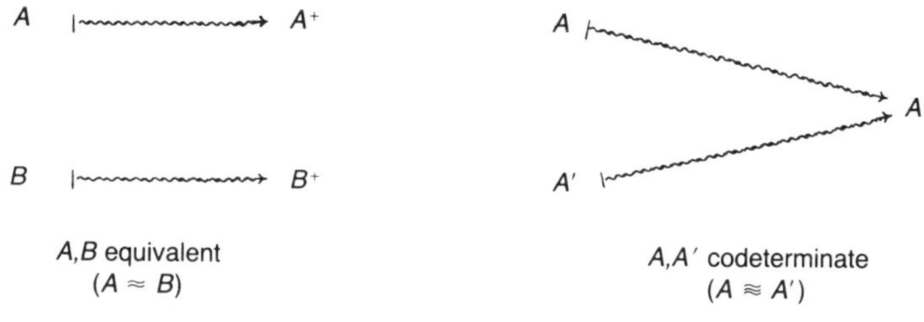

Figure 1

Equivalent structures, on this account, are those that *have the same structure;* this idea is merely a generalization of isomorphism as *the* structural equivalence relation. Codeterminate structures are, of course, equivalent as well, since A^+ is isomorphic to itself. But codetermination is a much stronger relation on structures than equivalence. The intuitive idea here is that codetermine structures not only *have* the same structure, they *are* the same structure. Codeterminate structures, like alternative axiomatizations of the same theory, differ only in their choice of exhibited relations. Thus, $N = (\{0,1,2, \ldots\},<)$ displays the "less than" relation, while $N' = (\{0,1,2, \ldots\},\leq)$ displays the "less than or equal to" relation. Since N and N' are codeterminate, the suggestion now is that both be regarded as generators of one and the same "superstructure" N. (For now, this superstructure can be regarded, if you like, as the class of all structures codeterminate with N.) What have been called structures or relational systems up to now might, from this stance, better be called "prestructures"; in what follows, the context should make the intended meaning clear.

So far, I have outlined how a new theory of structure and structural sameness can be developed, once we know how to superficially expand relational systems. The same general strategy may also be used for structures other than relational systems, say, for topologies, where a set A is structured, not by a sequence of relations on A, but by a set of subsets of A, and generalization to all of the structures encountered in contemporary mathematics, as we shall see in the next chapter, presents no serious difficulties.

But what is to count as a superficial extension of a structure? Up to this point, only an example or two has been given; a general account of superficial or determinate extensions must now be provided and justified.

In the examples, the definability of a relation was assumed to guarantee its objective determination. Since 'less than or equal to' is definable in terms of 'less than', the addition of '$\leq$' to N was regarded as an inessential or superficial extension of N. But why this faith in definitions as sources of convenient, but harmless, additions?

In the case of linguistic theories, the answer is more or less obvious: the explicitly defined term can be eliminated from any sentence in which it occurs, and when this is done in a theory, no new theorems couched in the original vocabulary appear. Explicit definability is dispensability, and what better reason for seeing definitions as matters of convenience?

But in the case of N and N^+ above, definition played a different role. The added relation '$\leq$' was not a linguistic entity, a predicate, but a *set.* Add *any* relation R to N, and it is eliminable (drop it, and you are right back where you started). So if we are content to supplement $N = (\mathrm{N},<)$ by '$\leq$', it is not "eliminability" that guarantees the innocence of the addition; it is rather, I suggest, that we regard the definability of '$\leq$' as a *sign* of its "implicitly being in" the structure N. Hence, nothing is added to the ordering of the numbers by distinguishing '$\leq$' as well as '$<$'. Definition, in this case anyway, is used in the sense of "singling out" an entity implicitly present for explicit consideration.

This is not to say that definability is a satisfactory basis on which to build a theory of superficial or determinate expansions. For one thing, the definability of a relation will depend upon the language used. Languages with infinite conjunction or quantification, for example, will permit relations to be defined that are not definable in standard first-order languages. Hence, in some cases at least, failure of definability in a standard first-order language is more a matter of linguistic limitations than symptomatic of a lack of structural determination.[4]

On the other hand, some possible languages are so powerful, but so artificial, that they seem to allow *too* many entities to be "defined." Infinite conjunctions are bad enough, but are conjunctions or quantifications of arbitrary ordinality to be taken "seriously"?

What we need to know amounts to this: When a definition *does* work, what does it *do* to deserve such a respect? Thus, suppose we say that a pair (x,y) is in $\leq$ just in case the formula "$(x<y) \vee x = y$" is satisfied by (x,y) in system $N = (\mathrm{N},<)$. Something seems to be preserved or transferred by this definition, something, indeed, that would be preserved *regardless* of the relation with which we started. This leads naturally to looking at "definitions" as creating operators: Feed in a relation, out comes another. In this case, we have, where R is a two-placed relation:

$$D_1(\mathrm{R}) = \{(\mathrm{x,y}) : (\mathrm{x,y}) \in \mathrm{R} \text{ or } \mathrm{x} = \mathrm{y}\}.$$

Our question now becomes: What is so "nice" about the operator D_1?

An answer to this question was given some time ago by the mathematician, physicist, and philosopher Hermann Weyl (see 39, §13; 40). To understand Weyl's basic idea, we need a few preliminaries.

Consider the system

$$K = (\{\ldots, -2, -1, 0, 1, 2, \ldots\}, <),$$

the integers ordered by the usual 'less than' relation. An isomorphism of a structure onto itself is called an *automorphism* or *symmetry* of the structure. For example, let

$$\sigma = \begin{pmatrix} \ldots, -2, -1, 0, 1, 2, \ldots, \\ \ldots,\ 0\ ,\ 1\ , 2, 3, 4, \ldots, \end{pmatrix}.$$[5]

Then σ preserves the 'less than' relation and so is an automorphism of *K*. On the other hand, the permutation

$$\tau = \begin{pmatrix} \ldots, -2, -1, 0,\ 1\ ,\ 2\ , \ldots, \\ \ldots,\ 2\ ,\ 1\ , 0, -1, -2, \ldots, \end{pmatrix}$$

changes '<' into '>', and hence is *not* an automorphism of *K*. Automorphisms, then, are "changes that make no difference."

Now consider *K* again, and all the structure "implicitly" generated by the 'less than' relation. Weyl's basic idea is this: *A change that makes no difference in K can make no difference in the structure implicit in K*. This gives us the following criterion of implicit structure: S is implicit in *K* just in case S is preserved by every change in *K* that makes no difference, that is, S is preserved by every automorphism of *K*. In this case, we shall say that S is an *invariant* of the system *K*.[6]

For example, the automorphisms of *K*, it is easy to see, are just those permutations that shift the integers by some fixed amount. In other words, those of the form $\sigma_k(x) = x + k$. So among the invariants of *K* we have the less than or equal to relation '≤' and the greater than relation '>'; each is preserved by all such automorphisms, and also, it turns out, each is definable in terms of the original 'less than' relation.

Now it is easy to show that in first and higher order languages, definability preserves invariance. In other words, if a relation S is definable in terms of a set of relations R^β, then every permutation that preserves all the relations R^{β} must also preserve S. The converse, however, does not hold: an invariant of a system need not be (first-order) definable in the system. There is a result, however, that comes close to being a converse.[7]

Let $T(\underline{R}^{\beta},\underline{S})$ be a first-order theory containing predicates $\underline{R}^{\beta}$ and the predicate $\underline{S}$. Call the predicate $\underline{S}$ an $\underline{R}^{\beta}$-invariant of the theory just in case, for every model $A = (A,R^{\beta},S)$ of the theory, S is an invariant of the structure $A^{-} = (A,R^{\beta})$. It can now be shown, using a theorem from Svenonius, that if $\underline{S}$ is an $\underline{R}^{\beta}$-invariant of $T(\underline{R}^{\beta},\underline{S})$, then not only is S an invariant of any reduced model $A^{-} = (A,R^{\beta})$, but S is also definable in A^{-}.

From the invariant standpoint, a superficial or determinate expansion is just an *invariant expansion,* and the standard structured set $A = (A,R^{\beta})$ is analogous to an axiom system of a deductive theory. Just as the same theory may be axiomatized in various ways for various purposes, the same structure may be generated by different choices of distinguished relations on a set. Choices are then equivalent when they yield the same or isomorphic invariants, or—what amounts to the same thing—when they have the same or isomorphic automorphism groups.[8]

Furthermore, just as the deductive consequences of a set of axioms are in some sense implicitly contained within the axioms themselves, superficial or determinate extensions of structures are intended merely to make explicit previously implicit structures. At this time it is not possible, I think, to say with any confidence whether or not this can be carried out in a way that is clearly the "best" way, like the situation with standard logical implication. Invariance, as a *sufficient* condition for objective determination, has its drawbacks.

The major problem with invariance arises when structures having only a single automorphism (the identity) are considered. These *rigid structures,* as they are called, will thus have every relation or operation on their universe as an invariant, since every relation or operation will be (trivially) preserved by the identity permutation. As a result, any two rigid structures of the same cardinality will be invariantly equivalent (Proposition 2.21)—a result which, while not outright absurd, is somewhat disquieting.[9]

For this reason, an alternative to invariance, *free invariance,* is presented in the last section. To be a free invariant, a relation has to be preserved by all of a structure's *local* as well as global automorphisms. This criterion is very severe, and indeed perhaps too severe for use in a general theory of structure. However, in section 5 it is shown that for some structures, invariance and free invariance will coincide. Euclidean spaces (and some non-Euclidean ones as well) turn out to be such structures.

In the next chapter, these ideas are made precise and developed in some detail. First of all, the notion of a (pre) structure—including all those structures studied in contemporary mathematics and logic—is defined. Roughly, these will be ordered pairs consisting of a set A together with a sequence $R_1, \ldots, R_\mu, \ldots$ of sets (called "structors") built up from elements of A. The notion of being "built up" from A is defined so that the σ-image of any structor of A is well defined for any permutation σ of the set A. Thus, talk of automorphisms and invariants will always make sense, permitting the constructions outlined above to proceed.

Invariantly equivalent structures are defined in section 2.2, and this relation is shown to be an equivalence relation. In section 2.3, invariants and how they are represented by coordinate systems are studied.

Codeterminate structures are introduced in section 2.4, and a new theory of natural isomorphisms is introduced. Section 2.5 provides a critique of the standard, category-theoretic approach to natural isomorphisms. The objections raised there are, I believe, decisive.

Section 2.6 introduces a still more general relation of structural equivalence motivated by the idea that the "same" structure can be specified starting with different universes. Euclidean and other geometries, for example, can be formulated taking points as fundamental, treating lines as subsets of points; or, instead, lines may be taken as basic, and points defined as subsets of "intersecting" lines. The constructions of this section provide, I believe, the basis for a constructive answer to Nelson Goodman's antirealistic arguments in *Ways of World Making* (15).

Sections 3 and 4 apply the theory of section 2 to some outstanding issues in the philosophy of geometry and relativity physics. Both treat the matter of convention, and both attempt to separate the conventional from the nonconventional ingredients in the two areas. They contain some new technical results in both areas, and also, I hope, some conclusions of philosophical importance.

Section 5, as already stated, outlines an alternative, stronger, theory of objective determination based on *free invariants.* I show there that when a structure satisfies a certain free mobility condition, then, and only then, will invariance and free invariance coincide. Most of the spaces of Klein's Erlanger Program, it turns out, do satisfy this condition.

Section 1.2 examines the relation between invariance and definability in an area well known to many philosophers: first-order languages and their model theory. In a sense, it provides some justification for the algebraic approach to objectivity taken throughout this essay, by showing the limitations of definability and, more generally, a linguistic approach to the problem of objectivity. While the details, especially the proofs, may be skipped upon first reading without a serious loss of continuity, the distinctions, examples, and general approach provide a basis for the developments of the following chapter. The scheme for numbering definitions and propositions should be obvious. The appendix provides proof-sketches of the theorems of section 2 when these have not been provided in the main text.

I am greatly indebted to the editor of this volume, Robert Colodny, for his almost divine patience, and to Adolf Grünbaum for his encouragement when it was needed most. I would like to thank Burke Townsend for many helpful conversations, and Gordon Brittan for his reading and comments on section 2. Comments by George McRae (Department of Mathematics, University of Montana) were of great help with the work on category theory.

The late Alberto Coffa was of inestimable help in developing the ideas of this essay. He will long be missed by philosophy and by me.

Finally, I would like to dedicate this essay to the memory of Grover Maxwell, a true philosopher and a beloved teacher and friend.

1.2. Definability and Invariance in First-Order Languages[10]

The relation between definability and invariance in first-order theories is fairly straightforward once the appropriate distinctions are introduced; then, a number of elegant results are forthcoming.[11] The language used here will be a standard first-order language *L* with identity, and, for simplicity's sake, it will be taken to contain only two nonlogical signs, the relation predicates $\underline{R}$ and $\underline{S}$ of degree m and n, respectively.[12] A possible model *A* of *L* will then be triple $A = (A,R,S)$, where A is a set and R and S are relations on A of degrees m and n, respectively. Throughout, A^- shall be the reduct of *A* obtained by omitting the (second) relation S, that is, $A^- = (A,R)$.

Definability, in the sense used by Beth (4), is a relation between predicates in *L* with respect to a theory $T(\underline{R},\underline{S})$ involving those predicates. Thus the predicate $\underline{S}$ is said to be $\underline{R}$-definable in the theory $T(\underline{R},\underline{S})$ containing $\underline{R}$ and $\underline{S}$ as nonlogical signs just in case $T(\underline{R},\underline{S})$ logically implies a sentence

$$\delta = (x_1) \ldots (x_n) \, [\underline{S}_{x_1}, \ldots, {}_{x_n} \equiv (\ldots, \underline{R}, \ldots)]$$

having the form of an explicit definition of $\underline{S}$ in terms of $\underline{R}$. Call a pair of possible models, *A* and *A′*, a *Padoa pair* when $A = (A,R,S)$, $A' = (A,R,S')$ and $S \neq S'$. Thus the relational systems of a Padoa pair differ only in the relation assigned to the predicate $\underline{S}$ in the language *L*. The fundamental result on explicit definability, which will here be called "Beth's theorem."[13] amounts to the following.

PROPOSITION 1.1 (Beth 4). $\underline{S}$ *is* $\underline{R}$*-definable in* $T(\underline{R},\underline{S})$ *just in case the theory* $T(\underline{R},\underline{S})$ *fails to have a countable Padoa pair as two of its models.*

An *automorphism* of a relational system $(A, R_1, \ldots, R_k)$ is a bijection σ: $A \to A$ of the universe A onto itself which preserves each relation R_i in A; that is, for each R_i (having degree k) in *A* and $a_1, a_2, \ldots, a_k$ in A, $(a_1, a_2, \ldots, a_k)$ is in R_i if and only if $(\sigma(a_1), \sigma(a_2), \ldots, \sigma(a_k))$ is in R_i. A relation S on A is said to be an *invariant* of $A = (A, R_1, \ldots, R_k)$, in symbols, $A \mapsto S$, just when S is preserved by every automorphism of *A*. In particular, where the system $A = (A,R,S)$ is a possible model of the language *L*, S is an R-*invariant of A* just in case S is an invariant of (A,R), that is, just when $A^- \mapsto S$. And S will be said to be *(L) definable* in $A = (A,R,S)$ if and only if there is a sentence $\delta = (x_1) \ldots (x_n)$ $[S_{x_1}, \ldots, _{x_n} \equiv (\ldots, R, \ldots)]$ of *L* having the form of an explicit definition of *S* in terms of $\underline{R}$, and $A = (A,R,S,)$ is a model of δ. It is easy to show that the definability of S in $A = (A,R,S)$ implies that S is an R-invariant of *A*. Only in some cases, however, will an R-invariant relation also be definable in *A*.

Definability in a *possible model* must be distinguished from definability in a *theory* of *L*. The former, as the previous definition shows, involves the relations in a structure. The latter, as we have seen, is a relation between predicates and a theory formulated using these predicates. Nevertheless, the two notions are closely related: the relation S will be R-definable in $A = (A,R,S)$ just in case the predicate $\underline{S}$ is $\underline{R}$-definable in the complete theory of *A*, Th(*A*,) where Th(*A*) is defined as the set of all sentences of *L* that are true in *A*. Since the proof is both brief and instructive, it will be presented here.

PROPOSITION 1.2 (De Bouvére 8). *The relation* S *is* R-*definable in* $A = (A,R,S)$ *iff* $\underline{S}$ *is* $\underline{R}$-*definable in* Th(*A*).

Proof. ($\Rightarrow$) Suppose S is R-definable in *A*. Then a suitable δ_0 exists which is true in *A*. Hence, δ_0 is in Th(*A*)—the complete theory of *A*. Thus, Th(*A*) logically implies δ_0, and so *S* is *R*-definable in Th(*A*). ($\Leftarrow$). Suppose Th(*A*) logically implies δ_0, where δ_0 is an explict definition of $\underline{S}$ in terms of $\underline{R}$. Then δ_0 is true in *A*, and hence S is R-definable in *A*.

There is also, for definability in a model, a Padoa pair result analogous to Beth's theorem. First, recall that two possible models $A = (A,R,S)$ and $B = (B,R',S')$ of *L* are said to be *elementary equivalent* ($A \equiv B$) just in case exactly the same sentences are true in each of them. It will now be shown that S is R-definable in $A = (A,R,S)$ if and only if there is no elementarily equivalent Padoa Pair *B,B'*, one of which (and, hence, each of which) is elementarily equivalent to *A*.

PROPOSITION 1.3. S *is* R-*definable in* $A = (A,R,S,)$ *iff there is no Padoa pair* B,B' *such that* $B \equiv B' \equiv A$.

Proof. ($\Rightarrow$) Suppose such a pair B,B' existed. Then δ_0—the sentence that explicitly defines S in A, would be true in both B and B'. So, since B and B' agree in what they assign to the predicate $\underline{R}$, they must also assign the same interpretation to $\underline{S}$ (since δ_0 holds in both), which is impossible, since they are a Padoa pair.
($\Leftarrow$). Suppose S is not R-definable in A. Then, by Proposition 1.2, $\underline{S}$ is not $\underline{R}$-definable in Th(A). Hence, by Beth's theorem (Proposition 1.1), there is a Padoa pair B,B' which is a model of Th(A). Since A,B, and B' are all models of the complete theory Th(A), they are elementarily equivalent.

Invariance can be given a Padoa pair formulation also, as the following proposition shows:

PROPOSITION 1.4. S *is an* R-*invariant of* A = (A,R,S) *iff there is no Padoa pair* B,B' *such that* $B \simeq B' \simeq A$.

Proof. A simple exercise using the fact that invariants are preserved by isomorphisms.

A comparison of Propositions 1.3 and 1.4 shows why definability in a model is not implied by invariance: elementarily equivalent models, even of the same cardinality or having the same universe, need not be isomorphic. Under conditions where the implication does obtain, say for the finite models of first-order languages, or the countably infinite models of a language $L_{\omega_1,\omega}$ having countably infinite conjunctions, invariance does imply definability within a possible model.[14]

Actually, even more is needed if invariance is to imply definability. As the proof of Proposition 1.3 shows, Beth's theorem is used to guarantee the existence of the required Padoa pair. Now Beth's theorem holds in $L_{\omega_1,\omega}$, but there are no known cases of its holding in larger infinitary languages, and there are proofs of its failure in an extensive class of these languages (Dickman 9, p. 132 and Theorem 2.4.9).

The failure of analogues to Beth's theorem amounts to the inability of T($\underline{R}$,$\underline{S}$) to entail a definition of $\underline{S}$ even though no Padoa pair for T($\underline{R}$,$\underline{S}$) exists, that is, even though the fact that A^- = (A,R) is a model of the $\underline{R}$-consequences of T($\underline{R}$,$\underline{S}$) suffices to fix a unique extension A = (A,R,S) of A^-, which is a model of T($\underline{R}$,$\underline{S}$). Intuitively, the interpretations of $\underline{S}$ in T($\underline{R}$,$\underline{S}$) is fixed by the interpretation of $\underline{R}$, yet T($\underline{R}$,$\underline{S}$) contains no sentence that expresses this determination.

Similarly, when two structures A and B are elementarily equivalent but not isomorphic (A $\equiv B$ but $A \not\simeq B$), this is obviously due to the nonexistence of any sentence in the language that holds in the one structure but fails in the other. In particular, if a language contained a "Scott sentence" ϕ_A which was such that for every possible model A, ϕ_A, "said" in effect "This structure is isomorphic to *A*,"[15] then elementarily equivalent structures would always be isomorphic as well. Once again, the ascent to infinitary languages is no help here. The general situation is still unclear; however, the results obtained by workers in this area so far indicate that equivalent but nonisomorphic models are almost everywhere (see Dickman 9, chap. 5, §3).

The fact that an invariant of a possible model is not always definable in that model contrasts nicely with the fact that an invariant of a *theory* is *always* definable in each of its models. More precisely, let us say that $\underline{S}$ is an $\underline{R}$-invariant of T($\underline{R}$,$\underline{S}$), just in case S is an R-invariant of *A* = (A,R,S), for every model *A* of T($\underline{R}$,$\underline{S}$). Svenonius has proved the following remarkable theorem:

PROPOSITION 1.5 (Svenonius 36). $\underline{S}$ *is an* $\underline{R}$*-invariant of* T($\underline{R}$,$\underline{S}$) *iff* T($\underline{R}$,$\underline{S}$) *logically implies* $\delta_1 \vee \delta_2 \vee \ldots \vee \delta_k$ (*some* $k \in \omega$), *where each* δ_i *is an explicit definition of* $\underline{S}$ *in terms of* $\underline{R}$.

It follows at once that if $\underline{S}$ is an $\underline{R}$-invariant of T($\underline{R}$, $\underline{S}$) and *A* is a model of T($\underline{R}$,$\underline{S}$), at least one of the S_i sentences must hold in *A;* hence, S must then be definable in *A*. (The definition used, however, may vary with the chosen model of T($\underline{R}$,$\underline{S}$).) Hence we have the immediate result:

COROLLARY 1.6. $\underline{S}$ *is an* $\underline{R}$*-invariant of* T($\underline{R}$,$\underline{S}$) *iff in every model* *A* = (A,R,S) *of* T($\underline{R}$,$\underline{S}$), S *is* R*-definable in A*.

There is even a "Padoa" theorem for theoretical invariance.[16]

PROPOSITION 1.7. $\underline{S}$ *is an* $\underline{R}$*-invariant of* T($\underline{R}$,$\underline{S}$) *iff* T($\underline{R}$,$\underline{S}$) *fails to have an elementarily equivalent* ($\equiv$) *Padoa pair as models.*

Proof. ($\Rightarrow$) Assume $\underline{S}$ an $\underline{R}$-invariant of T($\underline{R}$,$\underline{S}$). Suppose that *A* = (A,R,S) and *A*′ = (A,R,S′) are an elementarily equivalent Padoa pair, each a model of T($\underline{R}$,$\underline{S}$). Since A is a model of T(*R*,*S*), by Corollary 1.6, there is an explicit definition of δ_0 of S in *A*. Since $A \equiv A'$, δ_0 is true in *A*′ also. But then S′ = S, contradicting the fact that *A* and *A*′ are a Padoa pair.
($\Leftarrow$) Suppose $\underline{S}$ is not an $\underline{R}$-invariant of T($\underline{R}$,$\underline{S}$). Then there is some model *A* = (A,R,S) of T($\underline{R}$,$\underline{S}$) in which S is not an R-invariant of *A*. But then there must exist an automorphism ϕ of A^- = (A,R) such that $\phi[S] \neq S$.

Now let *A*′ = (A,R,ϕ[S]). Clearly, ϕ is an isomorphism of *A* and *A*′, since $\phi[S] \neq S$, *A* and *A*′ are a Padoa pair. And since $A \simeq A$, they are also an elementarily equivalent Padoa pair.

Notice that $\underline{S}$ may be an $\underline{R}$-invariant of T($\underline{R}$,$\underline{S}$) but need not be definable in T($\underline{R}$,$\underline{S}$)—although definable in each model of T($\underline{R}$,$\underline{S}$). But this can only happen, the next theorem reveals, when the theory $\mathrm{T}(\underline{\mathrm{R}},\underline{\mathrm{S}})$ is incomplete.

PROPOSITION 1.8. *If* $\mathrm{T}(\underline{\mathrm{R}},\underline{\mathrm{S}})$ *is semantically complete, then* $\underline{\mathrm{S}}$ *is an invariant of* $\mathrm{T}(\underline{\mathrm{R}},\underline{\mathrm{S}})$ *iff* $\underline{\mathrm{S}}$ is definable in $\mathrm{T}(\underline{\mathrm{R}},\underline{\mathrm{S}})$.

Proof. (One way is trivial; hence it need only be shown that invariance implies definability.)
Suppose $\underline{\mathrm{S}}$ is an invariant of $\mathrm{T}(\underline{\mathrm{R}},\underline{\mathrm{S}})$, with $\mathrm{T}(\underline{\mathrm{R}},\underline{\mathrm{S}})$ semantically complete.

Assume $\underline{\mathrm{S}}$ is not, however, $\underline{\mathrm{R}}$-definable in $\mathrm{T}(\underline{\mathrm{R}},\underline{\mathrm{S}})$. Then, by Beth's theorem (Proposition 1.1), there exists a Padoa pair A,A', each a model of $\mathrm{Th}(\underline{\mathrm{R}},\underline{\mathrm{S}})$. But then A and A' are models of a complete theory and so are elementarily equivalent. Hence, by Proposition 1.7, $\underline{\mathrm{S}}$ is not an $\underline{\mathrm{R}}$-invariant of $\mathrm{T}(\underline{\mathrm{R}},\underline{\mathrm{S}})$, a contradiction.

We can now improve on De Bouvère's result (Proposition 1.2) considerably. Earlier, definability in a model was shown to hold just in case there was explicit definability in the complete theory of that model; now, it turns out, invariance in that theory will do.

COROLLARY 1.9. S *is* R-*definable in* A = (A,R,S) *iff* $\underline{\mathrm{S}}$ *is an* $\underline{\mathrm{R}}$-*invariant of* Th(A)—*the complete theory of* A.

Proof. At once from Proposition 1.2 and 1.8.

Finally, we are in a position to return to the relation between definability in a model and invariance in that model. It turns out that while definability in a model is not—as we have seen—equivalent to invariance in that model, it *is* equivalent (in a sense) to invariance in all elementarily equivalent models. Precisely:

PROPOSITION 1.10. S *is* R-*definable in* A = (A,R,S) *iff for all* B = (B,R′,S′), $B \equiv A$, S′ *is an* R′-*invariant of* B.

Proof. (⇒) Suppose $B \equiv$ (B,R′,S′) and $B \equiv A$. Let δ_0 be the explicit R-definition of S in A. Then, since $B \equiv A$, δ_0 is true in B as well. Hence, S′ is R-definable in B, and is thus an R-invariant of B.

(⇐) First we show that $\underline{\mathrm{S}}$ is an $\underline{\mathrm{R}}$-invariant of Th(A).

Suppose B is a model of Th(A). Then, since Th(A) is complete, $B \equiv A$. Hence, S′ is an R′-invariant of B, and so (by definition) $\underline{\mathrm{S}}$ is an $\underline{\mathrm{R}}$-invariant of Th(A). Thus by Corollary 1.9, S is R-definable in A.

Suppose that A = (A,R,S) and S is an R-invariant of A. By the preceding theorem, the existence of a possible model B = (B,R′,S′), elementarily equivalent to A, but where S′ is *not* an invariant of B, implies that S is not R-definable in A. Notice that if we had to find a possible model B, *isomorphic* to A, with S′ not an R′-invariant, we could not—since invariance is preserved by isomorphism. Once again, it is the existence of elementarily equivalent but nonisomorphic structures that permits some invariants to be indefinable.

But what if a structure A is such that only structures isomorphic to A are elementarily equivalent to A? Clearly, in this case, invariance will then entail definability. The trouble is that only finite structures have this property. For all other structures there will always exist another, elementarily equivalent to the first, but not isomorphic to it (say, a special model of higher cardinality; see Chang and Keisler 7, p. 218). Fortunately, not all structures A of infinite cardinality have elementarily equivalent but nonisomorphic structures *of the same cardinality as* A. Hence, we may define, without fear of triviality, a structure A to be *lucid* just in case every structure B *of the same cardinality as* A that is elementarily equivalent to A is also isomorphic to A. We can now show, as the following proof demonstrates, that when A is lucid, invariance implies definability.

PROPOSITION 1.11. *If* $A = (\mathrm{A},\mathrm{R},\mathrm{S})$ *is lucid and* S *is an* R-*invariant of* A, *then* S is R-*definable in* A.

Proof. When $|\mathrm{A}| \in \omega$, every possible model B that is elementarily equivalent to A is also isomorphic to A. Since $(\mathrm{B},\mathrm{R}',\mathrm{S}') \simeq (\mathrm{A},\mathrm{R},\mathrm{S})$, then, since $\mathrm{A}^- \mapsto \mathrm{S}$, S′ is an R′-invariant of B. Hence, by Proposition 1.10, S is R-definable in A.

Next, suppose that $|\mathrm{A}| \geqslant \omega$. But now assume that there is a $B = (\mathrm{B},\mathrm{R}',\mathrm{S}') \equiv A$, but S′ is *not* an R′-invariant of B. Then, clearly, S′ is not R′-definable in B, and so, by Propositions 1.1 and 1.2, there exists a countable Padoa pair $C = (\mathrm{C},\mathrm{R}_1,\mathrm{S}_1)$, $C' = (\mathrm{C},\mathrm{R}_2,\mathrm{S}_2)$ of $\mathrm{Th}(B) = \mathrm{Th}(A)$. Hence, $C \equiv C' \equiv B \equiv A$, and since $|\mathrm{A}| \geqslant \omega$, $|\mathrm{C}| = \omega$. (That is, C is not finite.)

Now suppose $|\mathrm{A}| = \omega$. Then $|\mathrm{C}| = |\mathrm{A}|$, and since A is lucid, we must have $A \simeq C \simeq C'$. But then S_1 must be an invariant of C^-, and since any isomorphism $\phi: C \simeq C'$ is also an automorphism of C^-, $\mathrm{S}_2 = \mathrm{S}_1$—contradicting the fact that C,C' are a Padoa pair.

Next, suppose $|\mathrm{A}| > \omega$. Let $C^+ = (\mathrm{C},\mathrm{R}_1,\mathrm{S}_1)$. Then C^+ is a model of $\mathrm{K} = \mathrm{Th}(A) \cup \mathrm{Th}(A) \cup \{\sim(x_1)\ldots(x_n)\ (\underline{S}_{x_1,\ldots,x_n} \equiv \underline{S}'_{x_1,\ldots,x_n})\}$, where $\mathrm{Th}(A)'$ is the result of replacing all occurrences of $\underline{S}$ in $\mathrm{Th}(A)$ by the new predicate $\underline{S}'$. (Obviously, we have first extended the language to L to L^+.)

Now let $D^+ = (\mathrm{D},\mathrm{R}_2,\mathrm{S}_2,\mathrm{S}_2')$ be a model, $D^+ \equiv C^+$, with $|\mathrm{D}| = |\mathrm{A}|$ (using, for example, the upward Lowenheim-Skolem theorem on K.) Since $D^+ \equiv C^+$, D^+ is also a model of K, and so $\mathrm{S} \neq \mathrm{S}'$. Hence, we may define the Padoa pair $D = (\mathrm{D},\mathrm{R}_2,\mathrm{S}_2)$ and $D' = (\mathrm{D},\mathrm{R}_2,\mathrm{S}_2')$, and, since D and D' are both models of $\mathrm{Th}(A)$, $D \equiv D' \equiv A$. Since $|\mathrm{D}| = |\mathrm{A}|$ and

A is lucid, $D \simeq D' \simeq A$. Hence, as in the case where $|\mathrm{A}| = \omega$, $\mathrm{S}_2 = \mathrm{S}_2'$, contradicting the fact that *D,D'* are a Padoa pair.

Hence, when $B = (\mathrm{B},\mathrm{R}',\mathrm{S}) \equiv A$, S' is an R'-invariant of *B*. So, by Proposition 1.10, S is R-definable in *A*.

Finite models are lucid, and so are all models of theories categorical in the power of that model. What is needed, however, is not more results analogous to Proposition 1.11, but results that give us interesting algebraic conditions under which a structure of type $A^- = (\mathrm{A},\mathrm{R})$ will have its invariants definable.

To sum up: A *predicate* $\underline{\mathrm{S}}$ that is an invariant term in a theory $\mathrm{T}(\underline{\mathrm{R}},\underline{\mathrm{S}})$ will always be definable in each of the models of that theory. And when the theory in question is *complete,* an invariant predicate will also be Beth-definable in that theory as well. Nevertheless, a *relation* S that is an invariant of a possible model (A,R) need not be definable in that model. It *will* be R-definable in $A = (\mathrm{A},\mathrm{R},\mathrm{S})$, however, just in case S' is an invariant of $B = (\mathrm{B},\mathrm{R}',\mathrm{S}')$, whenever *B* is elementarily equivalent to *A*.

First-order languages, although they have the expressive power to guarantee explicit definability when a theory $\mathrm{T}(\underline{\mathrm{R}},\underline{\mathrm{S}})$ fixes a unique interpretation for $\underline{\mathrm{S}}$ when $\underline{\mathrm{R}}$ is interpreted in a model (Beth's theorem), nevertheless fail to distinguish nonisomorphic models of infinte cardinality. In fact, for any model *A* of infinite cardinality, there are models of all higher and lower infinite cardinalities whose complete theories are identical with the complete theory of *A*. Were there, for each structure *A* of *L,* a sentence ϕ_A that held in an arbitrary model *B* just in case *B* and *A* were isomorphic, then Proposition 1.10 would yield the result that every invariant of a model is definable.[17] Thus the failure of invariance to entail definability in first-order languages and their possible models need not be seen as a failure of the structure $A = (\mathrm{A},\mathrm{R})$ to objectively determine all its invariants, but instead, as due the expressive incompleteness of elementary languages. When it is seen also that the expansion of *L* to $L_{\omega_1,\omega}$, which allows infinite conjunctions, results in every invariant of a countably infinite structure being definable, it would seem to be prudent not to allow a *failure* of definability in some first-order language to count against objectivity.

On the other hand, it is by no means obvious that definability in any language, especially in the infinitary languages, should be routinely seen as a guarantee of a relation's objectivity. For if a structure is rigid (has the identify permutation as its only automorphism), then every relation on its universe will be an invariant. But, as we have seen, in the infinitary language $L_{\omega_1,\omega}$, any invariant of a countably infinite structure, is definable. Hence for any rigid, countably infinite structure, all relations on its universe will be definable in $L_{\omega_1,\omega}$. The result, if we identify definability and objectivity, is that any two rigid, countable structures having the same invariants become structurally identical.[18]

2. STRUCTURES: A GENERAL THEORY

2.1. Structures, Automorphisms, and Invariants

The possible models of elementary languages are just a few of the structures studied in contemporary mathematics. A topology, for example, consists of a set together with a set of subsets of that set, while vector spaces, Lorentz space-tmes, and many other structures involve distinguished functions from sets to real numbers (or R^n). Distinguished relations are but one of many ways to structure a set.

Intuitively, any such structure is a set—its universe—together with other sets "built up" from that universe and singled out for distinction. Hence, a precise general account of structure must specify just what entities are available for distinction, given an arbitrary universe A. If we call all such candidates *structors* of the set A, a structure will then consist of A, together with a selected sequence of its structors. Intuitively, we want these structors to be sets built up from A, but how—exactly?

The basic idea of the following account is simple enough. We begin by taking as structors of a set A the elements of A, and, for convenience, add the empty set here as well. Then we add to these structors the set of all subsets of A (the power set $P(A)$), next add the power set of all these, and so on. We thus obtain ever larger sets of structors:

$$A,\ A \cup P(A),\ A \cup P(A) \cup P(A \cup P(A)),\ \ldots.$$

A structor of A is then any set that eventually appears in the sets so obtained. The following definition makes this precise.[19]

DEFINITION 2.1. *Where* A *is any set,* μ *an ordinal number, and* $P(x)$ *is the power set of set x, we inductively define the set of tier* μ *structors of* A *as follows:*

(i) $\text{tier}_A(0) = A \cup \{\phi\}$

(ii) $\text{tier}_A(\alpha + 1) = \text{tier}_A(\alpha) \cup P(\text{tier}_A(\alpha))$,

and where ν *is a limit ordinal,*

(iii) $\text{tier}_A(\nu) = \bigcup_{\mu<\upsilon} \text{tier}_A(\mu)$.

A *structor* of A will then be any set that appears in some tier of A. Notice that the tiers are cumulative, and the empty set first appears in $\text{tier}_A(0)$, $\{\phi\}$ in $\text{tier}_A(1)$, $\{\phi, \{\phi\}\}$ in $\text{tier}_A(2)$, and so forth. Thus, the ordinal number μ will first appear in $\text{tier}_A(\mu)$, and if we define the *rank of set* x ($\text{rnk}_A(x)$) to be the first ordinal α so that x is in $\text{tier}_A(\alpha)$, every ordinal is equal to its rank (see Montague and Vaught 24). Since in Zermelo-Fraenkel set theory without individuals (but with the axiom of regularity) every set appears in some tier of the empty set (see Bernays and Fraenkel 3, chap. 8, §2), and $\text{tier}_\phi(\mu) \subseteq \text{tier}_A(\mu)$, the *pure sets* available in that set theory (e.g., the ordinal, cardinal, real numbers) will be among the structors of *any* universe A. Hence, the same will be true for functions from A to R^n and similar "mixed" functions found in current mathematical practice.

The following lemma states some properties of tiers and ranks that will be needed later on.[20]

LEMMA 2.2. *Where x and y are structors of set* A, *and* μ,ν *ordinal numbers:*
(i) $rnk_A(x) = 0$ *iff* $x \in A$ or $x = \phi$,
(ii) $\mu \leqslant \nu$ *iff* $tier_A(\mu) \subseteq tier_A(\nu)$,
(iii) *if* $x \in y$ *and* $y \in tier_A(\mu)$, *then* $x \in tier_A(\mu)$, *and*
(iv) *if* $x \in y$, *then* $rnk_A(x) < rnk_A(y)$.

Having defined those entities—structors—which are available for distinction, the motivation for the following definition of a structure should be obvious.

DEFINITION 2.3. *A set A* $= (A, \{R_\alpha : \alpha < \beta\})$ *is a structure iff:*
(i) A *is a nonempty set containing no pure set, and*
(ii) $\{R_\alpha : \alpha < \beta\}$ *is a sequence of structors of* A, *where* β *is an ordinal. (The set* A *is called the universe or domain of A, and the sequence* $\{R_\alpha : \alpha < \beta\}$ *will often be abbreviated as* "R^β." *Whenever possible,* A *will be taken as the universe of A,* B *as the universe of B, etc.)*

The relational systems of elementary theories are special cases of structures, with each R_α an n-placed ($n \in \omega$) relation on A.[21] A topology on A, being a set of subsets of A, first appears in $tier_A(2)$. It seems that all the spaces studied in contemporary mathematics are structures in the above sense, models (standard or nonstandard) of the various set theories included.

Notice that *A* is also a structor of A, that is, every structure is a structor of its own universe. Nevertheless, *A* cannot be in (the range of) R^β, otherwise there would be a descending chain of the form $A \in \ldots \in A$, which is forbidden by the axiom of regularity. Definition 2.3 circumvents this difficulty by requiring that *A* be a set.

In order to define isomorphisms of structures, we need the notion of the *image* of a structor of A under a function f from A to some other set.

DEFINITION 2.4. *Where* $f\colon A \to B$ *is a function and* R *is a structor of* A, *the* f-*image of* R, $f[R]$, *is defined as follows:*
(i) $f[R] = f(R)$, *when* $R \in A$; *otherwise,*
(ii) $f[R] = \{f[x] : x \in R\}$.

The definition proceeds by a kind of "downward induction." Any structor R of rank μ contains only members of lesser ranks (Lemma 2.2, iv). Hence the process of taking images will come to ground after finitely many steps, regardless of the rank or R. Such is a consequence of requiring that structors be built up from A.

With images on hand, the definition of isomorphic structures is straightforward.

> DEFINITION 2.5. *Let* $A = (\mathrm{A}, \{\mathrm{R}_\alpha: \alpha < \beta\})$ *and* $B = (\mathrm{B}, \{\mathrm{S}\mu: \mu < \nu\})$ *be structures. Then the function* f: A → B *is an isomorphism of A and B iff:*
> (*i*) f *is a bijection,*
> (*ii*) $\beta = \nu$, *and*
> (*iii*) *for all* $\mathrm{R}_\gamma \in \mathrm{R}^\beta$, $\mathrm{f}[\mathrm{R}_\gamma] = \mathrm{S}_\gamma$.
> *A and B are said to be isomorphic just in case such a function exists.*

When f is an isomorphism, this will often be written as: f: $A \simeq B$. Notice that when f is a bijection, f: $A \simeq B$ iff $\mathrm{f}[\mathrm{A}] = B$ iff $\mathrm{f}^{-1}[B] = A$.

As usual, an *automorphism* of a structure is an isomorphism taking that structure to itself. In that case, $\mathrm{f}[A] = A$, so when f is not merely the identity on A, f might be viewed as a change that makes no difference. The set of all automorphisms of a structure *A* will be written "Auto(A)." Intuitively, the automorphisms of a structure reveal its symmetries, the structural equivalences of its constitutents. A structor S of A is said to be an *invariant* of $A = (\mathrm{A},\mathrm{R}^\beta)$ just in case every automorphism f of *A* leaves S intact, in other words, f[S] = S. When S is an invariant of *A*, we write A ↦ S. A structure is said to be *homogeneous* (or *transitive*) when, for any two elements of its universe, there is an automorphism taking one to the other. Euclidean space is an example. When the identity mapping is a structure's only automorphism, no element may be automorphically taken to another, so the structure will be said to be *rigid*. A rigid structure is also said to *totally individuate* its universe. The various number systems (naturals, reals, etc.) are examples of rigid structures. Two structors of A are said to be *congruent* (or *indiscernible*) when there is an automorphism of *A* that takes one to the other as its image.[22] Thus, any two elements of a homogeneous structure's universe will be congruent, while a rigid structure's universe contains no pair of distinct congruent elements. Congruent figures (subsets of points) in Euclidean geometry are examples of congruent structors.

It is easy to see that the automorphisms of any structure *A* form a group when functional composition is taken as the group operation. This group will be called the *symmetry group* of *A*, written Sym(*A*) = (Auto(*A*), ∘), where Auto(*A*) is the set of all automorphisms of *A* and ∘ is functional composition. The structure of Sym(*A*) often provides crucial information about the structure *A*, and can even—as will be seen later—determine *A* completely.

The class of all permutations of a set A—written S_A—is called the *symmetric group* of A. When the automorphisms of a structure *A* are just the symmetric group S_A, the structure *A* will be called *trivial*. Intuitively, in a trivial structure the distinguished structors function vacuously, leaving A structured merely by its cardinality. For example, we can show that any finite relation that is a distinguished structor of a trivial relational system is (elementarily) definable in terms of the identity relation.

Let us call a structor of A a *pure structor* if it is built up from the empty set alone, that is, if it is in $\text{tier}_{\phi}(\mu)$ for some ordinal μ. If f is a function from A to A, then the f-image of the empty set is the empty set itself, and the same will be true for any pure structor. Hence, if $A = (A, R^{\beta})$ is a structure in which (the range of) R^{β} contains only pure sets, any permutation of A is an automorphism of the structure *A*, and so *A* is trivial. This result might be seen as a partial explication of the view that pure sets—such as real numbers—belong to the "logic" of theory construction: distinguished pure structors fail to add structure.

Since taking images of structors of a set A under functions from A to a set B is crucial to the developments to follow, this process must now be examined in some detail. The upshot is that the process is well behaved; the details, unfortunately, are technically tedious, predictable, but important.

First of all, image-taking preserves ranks; more precisely,

LEMMA 2.6. *Where* $\psi : A \rightarrow B$ *is a function, and* S *is a structor of* A, *then* $\text{rnk}_A(S) = \text{rnk}_B(\psi[S])$.

Furthermore, the process "grounds" properly; that is, when B is a structor of A, a permutation of A will induce a function defined on B. This induced function may now be used to take images of structors of B. The result is the same, the following lemma asserts, as using the original permutation of A.

LEMMA 2.7. *Let* σ *be in* S_A *and let* B *be a structor of* A. *For all* $b \in B$, *define* $\sigma^*(b) = \sigma[b]$. *Then for any structor* S *of* B, $\sigma^*[S] = \sigma[S]$.

Using this result, we may now show that invariance is a transitive relation between structures. More precisely:

PROPOSITION 2.8. *Let structures B, C be structors of the universe of A. Then, if* $A \mapsto B$ *and* $B \mapsto C$, $A \mapsto C$. *(Clearly, we also have* $A \mapsto A$.*)*

Image-taking is also well behaved with respect to functional composition and has the expected cancellation properties; these are summarized by the following:

LEMMA 2.9. *Let* S *be a structor of set* A, *and* $\phi: A \mapsto B$, $\psi: B \rightarrow C$ *be arbitrary functions. Then:*

(i) $(\psi \circ \phi)[S] = \psi[\phi[S]]$,

and when ϕ, ψ *are bijections,*

(ii) $\phi^{-1}[\phi[S]] = S$; $\phi[S] = T$ iff $S = \phi^{-1}[T]$,

and when B *and* C *are structors of* A,

(iii) $\sigma[\psi \circ \phi] = \sigma[\psi] \circ \sigma[\phi]$, *for all* $\sigma \in S_A$.

It is also useful to have available the special results of taking the σ-image of a function ϕ: A $\rightarrow$ B when σ is a permutation of A (i.e., $\sigma \in S_A$) and B is a structor of A. In this case we have:

LEMMA 2.10. *Where* $\sigma \in S_A$, B *is a structor of* A, *and* σ: A $\rightarrow$ B *is a function:*
(i) *for any structor* S *of* A, $\sigma[\phi]S = \sigma[\phi\sigma^{-1}[S]]$,
(ii) *when* $\phi \in S_A$, $\sigma[\phi] = \sigma\phi\sigma^{-1}$, *and*
(iii) *when* ϕ *a bijection,* $\phi[\sigma] = \phi\sigma\phi^{-1}$.

Proof-sketch.
(i) $\phi = \{(a,\phi(a)): a \in A\}$. Hence, $\sigma[\phi] = \{(\sigma(a),\sigma[\phi(a)]): a \in A\}$.
Let $a' = \sigma(a)$; hence, $a = \sigma^{-1}(a')$. Thus, $\sigma[\phi] = \{(a', \sigma[\phi\sigma^{-1}(a')]): a' \in A\}$, since $a \in A$ iff $a' \in A$. So $\sigma[\phi](a') = \sigma[\phi\sigma^{-1}(a')]$, for all $a' \in A$.
Next, the result by induction of the rank of S in A.
(ii) Follows at once from (i).
(iii) Compute, as in (i).

It is now easy to show that an isomorphism of two structures preserves automorphisms and invariants, and also yields an induced isomorphism of their symmetric groups.

PROPOSITION 2.11. *Let* ϕ: $A \simeq B$ *and* σ,τ *be permutations of* A *and* B, *respectively. Then:*
(i) σ *is an automorphism of A iff* $\phi[\sigma]$ *is an automorphism of B,*
(ii) τ *is an auto of B iff* $\phi^{-1}[\tau]$ *is an automorphism of A,*
(iii) $A \mapsto$ R *iff* $B \mapsto \phi[R]$, *and*
(iv) *where* $\phi^*(\sigma) \equiv \phi[\sigma]$, ϕ^*: (Auto *(A)*, $\circ$) $\simeq$ (Auto*(B)*, $\circ$).

Proof of (i). By Lemmas 2.9 and 2.10(iii), $\sigma \in$ Auto (A) iff $\sigma[A] = A$ iff $\sigma[\phi^{-1}[B]] = \phi^{-1}[B]$ iff $\phi\sigma\phi^{-1}[B] = B$ iff $\phi[\sigma][B] = B$ iff $\phi[\sigma] \in$ Auto*(B)*.

Proof of (ii). As in (i).

Proof of (iii). ($\Rightarrow$). Suppose $A \mapsto$ R and $\tau \in$ Auto(B). By (ii), $\phi^{-1}[\tau] \in$ Auto(A). Thus, $\phi^{-1}[\tau][R] = R$, in other words, $\phi^{-1}\tau\phi[R] = R$, that is, $\tau[\phi[R]] = \phi[R]$.
($\Leftarrow$) As in ($\Rightarrow$).

Proof of (iv). ϕ^* is one-one: Suppose $\phi^*(\sigma_1) = \phi^*(\sigma_2)$; then $\phi[\sigma_1] = \phi[\sigma_2]$, and so by Lemma 2.9 (ii), $\sigma_1 = \sigma_2$. From (ii), ϕ^* is onto; hence, ϕ^* is a bijection. From Lemma 2.9 (iii), ϕ^* preserves functional composition, and from Lemma 2.10 (iii), $\phi^*(1_A) = 1_B$. Hence, ϕ^*: Sym $(A) \simeq$ Sym (B).

With these results out of the way, let us go on to more interesting matters. Recall that the general strategy is to regard what we have been calling structures as mere devices for generating invariants. What is important about a structure $A = (A, R^\beta)$ is not the particular structors R^β we have chosen for this purpose, but the invariants of A that result.[23] These invariants, in turn, are fixed by the automorphisms of *A;* hence, the following two issues naturally arise. First, if G is a subgroup of S_A, is there always structure $A = (A, R^\beta)$ whose automorphisms are just those permutations in G? And second, how extravagantly must we exploit the structors of A to obtain such a structure? More precisely, how high up in the tiers of A need we go to find such an R^β? The following proposition answers both of these questions.

PROPOSITION 2.12. *Let* G *be a subgroup of* S_A. *Then there is a structor* S *of* A *of rank three, and, where* $A = (A, \{<0,S>\})$, *the automorphisms of A are just those permutations in* G (*i.e.*, Auto(*A*) = G).

The details of the proof may be found in the appendix. The basic idea is that we well-order A and use this well-ordering to define a set $A^\nu = \{\{a_0\},\{a_0,a_1\}, \ldots, \{a_0,a_1, \ldots\}, \{a_0,a_1, \ldots, a_\omega\}, \ldots\}$. We now close under the permutations in G, that is, we let $S = \{\tau[A^\nu]: \tau \in G\}$. It turns out that when $\tau \in S_A$, $\sigma[S] = S$ if and only if $\sigma \in G$. Furthermore, since the members of A^ν are subsets of A, they are all of rank 1 in A. Hence, A^ν itself is of rank 2, making S of rank 3. Whereas initially we countenanced structors of the rank of any (finite or transfinite) ordinal, the above result shows there was no need for so liberal an ontology.

This result justifies some new notation. When G is a subgroup of S_A, we now know that there is a structure *A* with G as its automorphisms; we use *A* = (A;G) to denote any such structure. At once, Auto((A;G)) = G, that is, *A* = (A;Auto(*A*)).

While we are at the business of constructing structures having some specified automorphism group, it might be well to dispel a possible confusion. Let *A* be a structure and Auto(*A*) its automorphisms. Now, Auto(*A*)—being just a set of bijections from A to A—is also a structor of *A;* hence, we may define another structure

$$A_S = (A, \mathrm{Auto}(A))$$

having Auto(*A*) as its sole distinguished structor.[24] Let us call A_S the *permutation structure* of *A*. Like any structure, A_S has its automorphisms, Auto(A_S). Now we ask: How are these two automorphism groups, Auto(*A*) and Auto(A_S), related?

In order to answer this question, we need some preliminary notions from group theory. Where G is a subgroup of S_A, any permutation $\tau \in S_A$ is said to *commute* with G just in case $\tau\sigma\tau^{-1}$ is in G whenever σ is. (Notice that by Lemma 2.9 (ii), $\tau\sigma\tau^{-1}$ is just $\tau[\sigma]$, the τ-image of σ.) A permutation τ is said to be in the *normalizer of* G (in S_A) just in case both τ and τ^{-1} commute with G. It is then easy to show that this amounts to saying that $\tau G = G\tau$, that is, $\tau G \tau^{-1} = G$, which, from Lemma 2.9 (ii) again, amounts to $\tau[G] = G$: In other words, τ takes the set G to itself.

Since $A_S = (A, \text{Auto}(A))$, the automorphisms of A_S are just those permutations of A that leave Auto(A) intact. Hence, from the above discussion, these are just those permutations of A that are in the normalizer of Auto(A). Hence, we have the result

$$\text{Auto}(A_S) = \text{Norm}\ (\text{Auto}(A)). \tag{2.1}$$

It now follows at once that A and A_S will have the same automorphisms just in case Norm (Auto(A)) = Auto (A)—a not unusual situation. In such cases, $A = (A, R^\beta)$ may be considered as $A = (A, \text{Auto}(A))$, for both structures will have the same invariants.

In general, we see that in order to replace $A = (A, R^\beta)$ by an equivalent $A' = (A, G)$, G must be a subgroup of Auto(A) whose normalizer is just Auto(A). Hence, the following proposition:

PROPOSITION 2.13. *Where* G *is a subgroup of* S_A, *let* $A' = (A, G)$ *and* $A = (A, R^\beta)$. *Then:* Auto(A') = Auto(A) *iff the normalizer of* G (*in* S_A) *is equal to* Auto(A).

Let us say that a structure $A = (A, R^\beta)$ is *group-theoretically reducible* (or just g-*reducible*) when there exists a sub-group G of permutations of A in which Auto(A) = Auto ((A,G)). The above proposition implies at once that A is g-reducible just in case Auto(A) contains a subgroup whose normalizer is just Auto(A) itself. Clearly, when the normalizer of Auto(A) *is* just Auto(A), this condition is met. The classical geometries of constant curvature construed as having all similarities as automorphisms (congruence geometries) all meet this condition and so are g-reducible. Not all structures are g-reducible, however. Euclidean metric spaces $E = (E, d)$,[25] where Auto(E) is just the isometry group, are familiar counterexamples. Such an isometry group fails to contain a subgroup whose normalizer is just the set of isometries.[26]

So while the automorphism group of a structure A determines its invariants, the structure A cannot always be regarded as a set A *structured* by Auto(A). Indeed, for some structures, there is no group of permutations of its universe that can be used to structure that universe and obtain just the original automorphisms of these structures. In the next section, where we explicitly characterize structural equivalence in terms of invariance, we shall return to these matters in an expanded setting.

2.2. *Invariantly Equivalent Structures*

Aside from its generality, there is little new in the developments thus far. The preceding definition of isomorphic structures is merely a generalization of the standard equivalence relation between structures.[27] From the invariant standpoint, however, what we have called "structures" might more aptly be termed *prestructures,* useful only insofar as they serve to determine a set of automorphisms—and, hence, invariants—of a universe. The next step, then, is to define a broader equivalence relation between structures (prestructures), which has the effect of identifying prestructures whose invariants are structurally alike.

As a first try, it would seem natural to replace a structure $A = (A, R^{\beta})$ by a counterpart $A' = (A, \mathrm{Inv}(A))$, where Inv(A) is a sequence containing exactly those structors that are invariants of *A*. The structures *A* and *B* would then be equivalent just when (A, Inv(*A*)) and (B, Inv(*B*)) were isomorphic when the sequence Inv(*B*) is suitably ordered.

The problem with this approach is that the sequence Inv(*A*) of invariants of any structure *A* does not exist—at least in Zermelo-Fraenkel set theory. For recall that each ordinal appears in $\mathrm{tier}_A(\mu + 1)$ and thereafter. Furthermore, the ordinals, being pure sets, are invariants of every structure *A*. Were sequence Inv(*A*) then to exist, it would contain all of the ordinals, and the set containing exactly the ordinals would then exist as well. But in Zermelo-Fraenkel it is a theorem that the set of ordinals does not exist. So, too, then for Inv(*A*).[28]

But there is another way. Although we cannot pair A with *all* the invariants of A, we may pair A with as many invariants as needed to show equivalence with another structure *B*, to which we have done the same. Since *any* invariant of *A* may be chosen to supplement *A*, the effect is the same as if we had a set Inv(*A*) available.

DEFINITION 2.14. *Let* $A = (A, R^{\beta})$ *be a structure. Then the structure* $A^{+} = (A, S^{\mu})$ *is an invariant expansion of A iff:* R^{β} *is a subsequence of* S^{μ}, *and every structor* S *in (the range of)* S^{μ} *is an invariant of A.*

DEFINITION 2.15. *Structures A and B are invariantly equivalent* ($A \simeq i\ B$) *iff: A and B have isomorphic invariant expansions.*

It follows at once that a structure is invariantly equivalent to each of its invariant expansions and that the relation of invariant equivalence is reflexive and symmetrical. The transitivity of $\simeq_i$ is shown below (Corollary 2.18), and is based upon the fact that an isomorphism preserves automorphisms, and so invariants (see Proposition 2.11). Clearly, isomorphic structures are invariantly equivalent.

If structures *A* and *B* are invariantly equivalent, then there must exist a bijection ϕ: A → B which is such that $\phi: A^{+} \simeq B^{+}$, where A^{+} and B^{+} are invariant expansions of *A* and *B*, respectively. In general, ϕ will not be an isomorphism of *A* and *B* as well; nevertheless, we write $\phi: A \simeq i\ B$ and take this to mean that ϕ is an isomorphism of suitable invariant expansions of *A* and *B*. The question now arises: Under what conditions will a bijection ϕ: A → B be such that $\phi: A \simeq i\ B$? In other words, when will suitable invariant expansions of A and B exist so that ϕ is an isomorphism of these expansions?

It turns out that there is a simple test that settles this matter.

PROPOSITION 2.16. *Let* $A = (\mathrm{A},\mathrm{R}^\beta)$ *and* $B = (\mathrm{B},\mathrm{S}^\nu)$. *Then* $\phi\colon A \simeq\mathrm{i}\, B$ *iff* $B \mapsto \phi[\mathrm{R}^\beta]$ *and* $A \mapsto \phi^{-1}[\mathrm{S}^\nu]$. *In other words,* $\phi\colon A \simeq\mathrm{i}\, B$ *just in case* ϕ *takes the distinguished structors of A to invariants of B and* ϕ^{-1} *takes the distinguished structors of B to invariants of A. As a corollary we have:*

COROLLARY 2.17. *Where* $A = (\mathrm{A},\mathrm{R}^\beta)$, $B = (\mathrm{B},\mathrm{S}^\nu)$, *and* $\phi\colon \mathrm{A} \to \mathrm{B}$ *is a bijection, the following conditions are equivalent:*

(i) $\phi\colon A \simeq \mathrm{i}\, B$,
(ii) ϕ *takes every invariant of A to an invariant of B and* ϕ^{-1} *takes every invariant of B to an invariant of A,*
(iii) $\phi[\mathrm{Auto}(A)] = \mathrm{Auto}(B)$,
(iv) $B \mapsto \phi[\mathrm{R}^\beta]$ and $A \mapsto \phi^{-1}[\mathrm{S}^\nu]$.

Thus, invariant equivalences are just those bijections that, along with their inverses, take invariants to invariants.

An example—useful later on—might help at this point. Let $\mathrm{A} = \{a,b,c,d\}$, and let the structures A, A' be given by:

$$A = (\mathrm{A},\mathrm{R}) = (\mathrm{A}, \{\langle a,b,c,d\rangle,\langle b,a,d,c\rangle\}),$$
$$A' = (\mathrm{A},\mathrm{S}) = (\mathrm{A}, \{\langle a,b,c,d\rangle,\langle a,c,b,d\rangle\}).$$

We now can show that A and A' *fail* to be invariantly equivalent as follows. Let $\phi\colon \mathrm{A} \to \mathrm{A}$ be a bijection. Then

$$\phi[\mathrm{R}] = \{\langle\phi(a),\phi(b),\phi(c),\phi(d)\rangle,\langle\phi(b),\phi(a),\phi(d),\phi(c)\rangle\}.$$

Hence, $\phi[\mathrm{R}]$ is a set of type

$$\{\langle s_1,s_2,s_3,s_4\rangle,\langle s_2,s_1,s_4,s_3\rangle\},\quad s_i \neq s_j$$

when

$$i \neq j.$$

Now a check of the permutation

$$\tau_1 = \begin{pmatrix} a & b & c & d \\ a & c & b & d \end{pmatrix}$$

shows it to be an automorphism of A'. But τ_1 interchanges just a single pair of elements of A (namely, b and c), and an examination of the type of $\phi[\mathrm{R}]$ shows that the τ_1- image of $\phi[\mathrm{R}]$ cannot be $\phi[\mathrm{R}]$ itself; hence, $\phi[\mathrm{R}]$ is not an invariant of A'. Since ϕ was arbitrary, no bijection ϕ can take R to an invariant of A'. Hence, by Proposition 2.16 (or Corollary 2.17), A and A' are not invariantly equivalent.

Since the transitivity of $\simeq$i is an easy consequence of Corollary 2.17, we state, for the record:

PROPOSITION 2.18. *The relation* $\simeq$i *is an equivalence relation.*

Invariantly equivalent structures can at first sight seem very different, indeed. The following simple results provide some idea of what structures are so equated.

Recall that a structure A is so to be trivial when Auto(A) is the set of *all* permutations of the universe A. We then have:

PROPOSITION 2.19. *Any two trivial structures with universes of the same cardinality are invariantly equivalent.*

Proof. Let A,B be trivial, with $|A| = |B|$. Let ϕ:A→B be any bijection. Since Auto(A) = S_A and Auto(B) = S_B, and $\phi[S_A] = S_B$ (exercise), we have the result by Corollary 2.17 (i) and (iii).

This result should not be surprising, since, as was mentioned earlier, trivial structures are distinguished only by their cardinality. The triviality of trivial structures is perhaps most clearly revealed by the following proposition.

PROPOSITION 2.20. *If* S *is an invariant of a trivial structure,* A = (A,R^β), *then* S *is an invariant of every structure—trivial or no—with universe* A.

Proof. Suppose $A \mapsto$ S, A trivial. Then for all $\sigma \in S_A$, $\sigma[S] = S$. Now, if A is the universe of A', Auto(A') $\subseteq S_A$. Hence, S is preserved by all σ e Auto(A').

The next result is no less obvious (see the proof), but perhaps more unsettling.

PROPOSITION 2.21. *Any two rigid structures with universes of the same cardinality are invariantly equivalent.*

Proof. Assume A,B rigid with $|A| = |B|$. Let ϕ:A → B by any bijection. Since Auto(A) = 1_A and Auto(B) = 1_B, and $\phi[1_A] = 1_B$ (exercise), ϕ[Auto(A)] = Auto(B). Hence, by Corollary 2.17 (i) and (iii), ϕ:$A \simeq$i B.

For example, consider the structures:

$$N = (\{0,1,2, \ldots\}, <_N)$$
$$N^* = (\{\ldots, 2,1,0\}, \breve{<}_{N^*}),$$

where $<_N$ and $\breve{<}_{N^*}$ are the (standard) orderings shown. Both structures are rigid and have the same universe; hence, by Proposition 2.21, they are invariantly equivalent. However, the two structures clearly are not isomorphic and exhibit what would customarily be regarded as different types of order (type ω and ω^*, respectively): N contains a minimal but no maximal element, while for N^* the reverse is true. Even worse, N is well ordered while N^* is not; the undoubtedly useful distinction between well-orderings and their (generally) non-well-ordered converses would seem to be obliterated.

Some reassurance that we have not really committed set-theoretical *sepuku* may be provided by noting that N and N^* are invariantly equivalent because of the existence of

$$N^+ = (\{0,1,2, \ldots\}, <_N, \breve{<}_{N^*}),$$

which is an invariant extension of both. Intuitively, $<_N$ objectively determines $\breve{<}_{N^*}$, and conversely. From the invariant standpoint, just which of the two relations is signaled out for display is a matter of indifference. The relations $<_N$ and $\breve{<}_{N^*}$ are merely different *aspects* of the same structure, and—like different axiomatizations of the same deductive theory—are rightly regarded as equivalent since they generate the same invariant structure.[29]

What it comes to is this: From the invariant view, simple isomorphism relates *correlated aspects* of a structure. And while either N or N^* above may be taken as a representative of the same invariant structure, they do not represent the same *aspects* of that structure—as failure of isomorphism shows. So there is no need to abandon distinctions (such as order type) based on simple isomorphism. Instead, we may view such distinctions as applying to aspects (prestructures) of a structure, rather than to the way the universe is structured in its entirety.[30]

The above results on trivial and rigid structures show that in some instances a structure $A = (\mathrm{A}, \mathrm{R}^\beta)$ is completely determined by the automorphism group of A. Suppose we now regard a structure to be completely determined by its invariants. Since these are in turn determined by the structure's automorphisms, we should expect there to be a close connection between two structures A and B being invariantly equivalent and the automorphism groups, Auto(A) and Auto(B), of these structures. Of course, we cannot in general expect these to be the *same* group of permutations, since A and B will generally have different universes. If A and B are invariantly equivalent, however, then $\phi: A^+ \simeq B^+$ will take the automorphisms of A to those of B while preserving functional composition; hence, the two symmetry groups $<\mathrm{Auto}(A), \circ>$, $<\mathrm{Auto}(B), \circ>$ will be *isomorphic groups*, and we have:

PROPOSITION 2.22. *If* $A \simeq\mathrm{i}\, B_1$, *then* $(\mathrm{Auto}(A), \circ) \simeq (\mathrm{Auto}(B), \circ)$.

Proof. Assume $\phi: A \simeq\mathrm{i}\, B$. Then by Corollary 2.17, $\phi[\mathrm{Auto}(A)] = \mathrm{Auto}(B)$. Using Lemma 2.10 (iii), ϕ preserves functional composition and inverses. Hence, $\phi: (\mathrm{Auto}(A), \circ) \simeq (\mathrm{Auto}(B), \circ)$.

The converse, however, fails. Two structures—indeed two structures A and A' with identical universes—may have isomorphic symmetry groups yet fail to be invariantly equivalent. The difficulty is that the same abstract group may act on a set A in essentially different ways. The structures A and A' of (2.18) above provide a simple example of this. There, it is easy to see that

$$\text{Auto}(A) = \left\{\sigma_0 = 1_A, \quad \sigma_1 = \begin{pmatrix} a & b & c & d \\ b & a & d & c \end{pmatrix}\right\},$$

and

$$\text{Auto}(A') = \left\{\tau_0 = 1_A, \quad \tau_1 = \begin{pmatrix} a & b & c & d \\ a & c & b & d \end{pmatrix}\right\},$$

Both groups are clearly isomorphic *as abstract groups*. But σ_1 in Auto(A) permutes the elements of each *two* pairs of elements of A_1, while τ_1 in Auto(A') permutes only a single such pair. As we have previously shown, A and A' fail to be invariantly equivalent.

We obtain the general result when we require that Auto(A) and Auto(A') be isomorphic as *concrete* groups of permutations, that is, when we consider (A, Auto(A)) and (B, Auto(B)), that is, A_S and B_S, respectively. We then have:

PROPOSITION 2.23. *Where A and B are structures,*

(*i*) ϕ: $A \simeq\text{i } B$ *iff* ϕ: (A, Auto(A)) $\simeq$ (B, Auto(B)), *in other words,* ϕ: $A_S \simeq B_S$, *and*

(*ii*) A $\simeq$i B *iff* $A_S \simeq B_S$.

Proof.

(*i*) Since $\phi[A] = B$ and by Corollary 2.17 (ii), (iii), $\phi[\text{Auto}(A)] = \text{Auto}(B)$, the result follows immediately.

(ii) At once from (i).

Thus, a bijection is an invariant equivalence of structures just in case it is an isomorphism of their corresponding permutation structures.

We have encountered this replacement of A by A_S = (A, Auto(A)) in the previous section. There we saw (equation 2.1 and Proposition 2.13) that a structure A and its automorphism structure A_S = (A, Auto(A)) need not have the same automorphisms. In general, Auto(A_S) is just the normalizer of Auto(A). Similarly, we may now ask: Just when are A and A_S invariantly equivalent?

We obtain the answer at once by substituting A_S for B in Proposition 2.23 (ii) above. At once, $A \simeq$i A_S just in case $A_S \simeq$ (A, Auto(A_S)). Earlier, we showed that Auto(A_S) is just the normalizer of Auto(A). Hence, we have:

COROLLARY 2.24. *Where A and B are structures, $A \simeq i A_S$ iff $A_S \simeq (A_S)_S$ iff* (A, Auto(A)) $\simeq$ (A, Norm(Auto(A)).

Clearly, (A, Auto(A)) and (A, Norm(Auto(A)) will not always be isomorphic, and so a structure A will not generally be invariantly equivalent to its automorphism structure A_S. And for much the same reasons discussed at the conclusion of the previous section, some structures will not be invariantly equivalent to their universe structured by some subgroup of its permutations.

Structures—more accurately, prestructures—are defined in terms of distinguished structors on a universe. Coordinate systems on a set are also structors of that set, as we shall soon see, and while a structure A cannot always be considered as a set A structured by a group, it can be reduced to set A structured by special sets of coordinate systems. Such matters are the subject of the following section.

2.3. Invariants and Distinguished Coordinate Systems

When a coordinate system for a structure is characterized as a set of "mere labels" for the items in its universe, at least two things are misleading about such a description. First of all, to be a set of labels for the items in a universe is no small matter, especially when that universe is infinite. Since one purpose of labeling a universe is to keep tabs on each of its members, in order to be useful coordinates must individuate elements that are often intrinsically indistinguishable. For example, in homogeneous structures (say, Euclidean spaces), the elements (points) of the universe are intrinsically alike. But the coordinates of the items in the universe, in order to be "labels," must be individually distinguishable. (Hence, the futility of using the points of a Euclidean space as the coordinates of some other universe.) For this reason, a coordinate system of a universe is not merely a one-one function from that universe into an arbitrary set of suitable cardinality, but the set of coordinates must be the universe of some *rigid* structure. This well accords with actual practice, since the most commonly used coordinates—the natural numbers, the real numbers, or R^n—are all universes of rigid structures.

This individuality of coordinates, however, while necessary if they are to serve their purpose as labels, makes for a situation fraught with logical perils. Since the (rigid) structure of the coordinates is generally richer than that of its coordinatized universe, it is easy to confuse objective relations between *coordinates* with objective relations on the coordinatized, structured *universe.* This is especially easy—and this is the second reason why calling coordinates "mere labels" is misleading—when coordinate systems are used to define the very structure they coordinatize. For example, consider the following definition.

DEFINITION. A = (A,R) *is an open-ended, discrete, strict ordering* = df.
(i) A *is a nonempty set,*
(ii) R *is a two-placed relation on* A, *and*
(iii) there is a bijection ϕ: A $\rightarrow$ *Integers where by for any* a,b $\in$ A, <a,b> $\in$ R *iff* $\phi(a) < \phi(b)$.

The coordinate function ϕ is here used to "pull back" the 'less than' relation on the (positive and negative) integers to the universe A, thereby giving A the structure of a discrete, open-ended string of elements. Notice that the defined structure A is not rigid, but homogeneous: any element of A may be "shifted" to any other by an automorphism. However, the coordinates—the integers—are the universe of a rigid structure. Also, the choice of that rigid structure was not arbitrary, but was based on: (i) its cardinality, and (ii) the easy availability of standard structors suitable for pulling back to A (in this case, the 'less than' relation on the integers). Finally, notice that the existence of the bijection ϕ permits us to view the structure $\phi[A]$ = (Integers,<) as *representing* A, so that we may now investigate the structure of A by turning attention to $\phi[A]$ and using a variety of developed numerical, algebraic, or analytical techniques.

It is at this point that complete clarity is required, since some relations *on the coordinates* will correspond to determinate relations on A, whereas others will not. Instead, these others will belong to the surplus structure inevitably had by the coordinate structure itself. Hence, what we require is a means of determining what structors of the *coordinate* universe "pull back" to determinate features of A—in our terminology, which structors *represent* invariants of A. (For example, with $\phi[A]$ as above, we might be interested in determining whether or not the relation $C = \{<i_1,i_2,i_3,i_4>: i_1 - i_2 = i_3 - i_4\}$ represents an invariant of A.)

While it is true that current mathematical practice at its best deals with these issues clearly and correctly, still, there is a tendency to handle such questions by cases, depending on the kind of structure—group, topology, space-time, and so on—involved. The result is a proliferation of special notation and terminology, and only a dim glimpse of the underlying logic of the situation. The following account seeks to clarify the relations of coordinates and their structures in a general way; it also paves the way for applications, in a later section, to the problem of simultaneity in Minkowski space-time.[31] The development is inspired by Weyl (39, §13); some of the results, however, appear to be new.

Let A = (A,R^β) be a structure. Then a *B-coordinate system of* A is a pair (ϕ,B), where B is a rigid structure and ϕ is a bijection from A to the universe (B) of B. Often (ϕ,B) will be written ϕ_B, or, when the context is clear, simply ϕ. If S is a structor of A, then the *ϕ-representative of* S is simply the ϕ-image of S, $\phi[S]$. Similarly, we may represent the entire structure A by means of ϕ, obtaining

$$\phi[A] = (B, \{\phi[R_\alpha]: \alpha < \beta\}),$$

a structure on B which is isomorphic to A (since ϕ is an isomorphism).

When σ is a permutation of A—such as an automorphism—then, as noted earlier, $\phi[\sigma] = \phi\sigma\phi^{-1}$. Hence, structure $\phi[A]$ will have as its corresponding automorphism structure

$$(\phi[A])_S = (B, \text{Auto}\ (\phi[A])) = (B, \{\phi\sigma\phi^{-1}: \sigma \in \text{Auto}(A)\}),$$

and, clearly, ϕ is an isomorphism of $A_S = (A, \text{Auto}(A))$ and $\phi[A]_S$.

Suppose that (ϕ,B) and (γ,B) are both coördinate systems of A. Then the coordinate transformation from ϕ to ψ is the function $\Sigma_{\psi,\phi}$: B→B defined by

$$\Sigma_{\psi,\phi} = \text{df. } \psi\circ\phi^{-1},$$

and where K is any set of coordinate systems from A to B, the corresponding set of coordinate transformations $\Sigma(K)$ is given by

$$\Sigma(K) = \text{df. } \{\Sigma_{\psi,\phi}: \psi,\phi \in K\}.$$

When ϕ is a B-coordinate system of A, there is a natural way of generating a corresponding *set* of coordinate systems of A: Compose ϕ with every automorphism of A. Following Weyl (39), the resulting set of coordinate systems, $\{\phi\}_A$,[32] will be called a *distinguished class* of coordinate systems of A, and is defined by

$$\{\phi\}_A = \text{df. } \{\phi\circ\sigma: \sigma \in \text{Auto}(A)\}.$$

It is a simple matter to show that these distinguished classes are mutally exclusive, and since every B-coordinate system ϕ belongs to some distinguished class ($\{\phi\}_A$), the set of all B-coordinate systems is hereby partitioned into sets of distinguished classes.

Suppose now that $\{\phi\}_A$ is a distinguished class of coordinate systems of a structure A and that R is an invariant of A. The ϕ-image of R, $\phi[R]$, will now represent R as a structor of the set of coordinates B. Now suppose that ψ is another coordinate system in this same distinguished class. Then, for some $\sigma_i \in \text{Auto}(A)$, we have $\psi = \phi\circ\sigma_i$. Hence, $\psi[R] = (\phi\circ\sigma_I)[R] = \phi[\sigma_I[R]]$, and, since $\sigma_I[R] = R$, $\psi[R] = \phi[R]$. In other words: *If* R *is an invariant of A, then all coordinate systems in any distinguished class* $\{\phi\}_A$ *represent* R *in exactly the same way.*

Corresponding to each distinguished class of coordinate systems $\{\phi\}_A$, there will be a set of coordinate transformations $\Sigma(\{\phi\}_A)$ relating the systems in that class. As the following proposition asserts, these coordinate transformations are just the automorphisms of $\phi[A]$—the isomorphic representative of A on the coordinate universe B. (Notice that since A is an invariant of A, $\psi[A] = \phi[A]$; hence, all choices of a coordinate system in $\{\phi\}_A$ yield the same representation of A on B.)

PROPOSITION 2.25. *Let* $\{\phi\}_A$ *be a distinguished class of coordinate systems of A. Then,* $\Sigma(\{\phi\}_A)$, *the set of coordinate transformations of* $\{\phi\}_A$, *is just the set of automorphisms of* $\phi[A] = \psi[A]$, *for all* $\psi \in \{\phi\}_A$.

Proof. From the preceding comments, it suffices to show that $\Sigma(\{\phi\}_A) = \text{Auto}(\phi[A])$.

First, it is easy to show (exercise) that $\Gamma \in \Sigma(\{\phi\}_A)$ iff $\Gamma = \phi\sigma_i\phi^{-1}$, for some $\sigma_i \in$ Auto (A). Hence, at once, $\Gamma[\phi[A]] = \phi[A]$. So $\Gamma \in \text{Auto}(\phi[A])$.

Next, suppose $\tau \in \text{Auto}(\phi[A])$. Then, $\tau[\phi[A]] = \phi[A]$, so $\phi^{-1}\tau\phi[A] = A$. Hence, $\phi^{-1}\tau\phi = \sigma_i$, where $\sigma_i \in \text{Auto}(A)$. So $\tau\phi = \phi\sigma_i$, that is, $\tau = \phi\sigma_i\phi^{-1}$. So from above, $\tau \in \Sigma(\{\phi\}_A)$.

Since A and $\phi[A]$ are isomorphic, they have isomorphic symmetry groups. Hence, from the above, $\text{Sym}(A) = (\text{Auto}(A),\circ) \simeq (\Sigma(\{\phi\}_A),\circ) \simeq \text{Sym}(\phi[A])$. Since the above proposition holds for *any* distinguished class $\{\phi\}_A$, it follows that any such transformation group is isomorphic to the symmetry group of A:

COROLLARY 2.26. *For any distinguished class of coordinate systems* $\{\phi\}_A$ *of A, the group of transformations of* $\{\phi\}_A$ *is isomorphic to the symmetry group of A, that is* $(\Sigma(\{\phi\}_A),\circ) \simeq (\text{Auto}(A),\circ)$.

This result is important, since, even though the choice of a distinguished class of coordinate systems is arbitrary, the *structure* of its group of transformations is not. This structure is *always* that of the automorphism group of the coordinatized structure A.

These results suggest that we might invert the development thus far by *beginning* with a set of coordinate systems whose transformations are a group and regard these transformations as representing the automorphisms of an underlying structure A. Then we could define an invariant of A to be a structor on A that is represented by a special kind of structor on the coordinates B, namely, one that is left untouched by every coordinate transformation in the group. The following argument shows that such a strategy is justified.

We have already seen that the invariants of a structure are represented in the same way by the various coordinate systems of any distinguished class. It turns out the invariants of *A are* just those structors of A which every system of a distinguished class represents in exactly the same way:

PROPOSITION 2.27. *If* $\{\phi\}_A$ *is any distinguished class of B- coordinate systems of structure A, then* S *is an invariant of A iff for any* $\psi \in \{\phi\}_A$, $\psi[\text{S}] = \phi[\text{S}]$.

Proof. ($\Rightarrow$) has been established in the discussion above. Suppose $\psi[S] = \phi[S]$, all $\psi \in \{\phi\}_A$. Let σ_i be in Auto(A). Then, $\phi \circ \sigma_i \in \{\phi\}_A$. Thus, $\phi\sigma_I[S] = \phi[S]$, that is, $\sigma_I[S] = S$. Hence, S is an invariant of A.

Notice that this result applies to *any* distinguished class of coordinate systems of A. Different classes will represent the same invariant structor in different ways, yet *within* each class that structor obtains the same representation. This point does not seem to be well known—or at least well appreciated—by many who discuss invariants, especially in connection with relativity theory. For example, in special relativity an invariant is (rightly) required to have the same representation in all *Lorentz* coordinate systems, from which it is often concluded that these systems are *uniquely* distinguished by that theory. Curvilinear coordinates, for example, are *thus* held suspect. But, as we have seen, each member of a distinguished class of curvilinear coordinates represents any given invariant in exactly the same way, so *this* feature of Lorentz coordinates is not the source of their distinction.

Using the result of Proposition 2.27, we can now show that the inverse strategy discussed above succeeds: Those structors of the coordinates B that are preserved by every coordinate transformation in $\Sigma(\{\phi\}_A)$ *are* just the representatives—in $\{\phi\}_A$—of the invariants of structure A:

PROPOSITION 2.28. *If* $\{\phi\}_A$ *is any distinguished class of B-coordinate systems of A, and* S* *is a structor of* B, *then for any* $\psi \in \{\phi\}_A$, $\psi^{-1}[S^*]$ *is an invariant of A iff for any* $\Gamma \in \Sigma(\{\phi\}_A)$, $\Gamma[S^*] = S^*$.

Proof. Since isomorphisms preserve invariants (Proposition 2.11 (iii)), $A \mapsto \psi^{-1}[S^*]$ iff $\psi[A] \mapsto S^*$, and since $\psi[A] = \phi[A]$, iff $\phi[A] \mapsto S^*$. Furthermore, from Proposition 2.25, $\Sigma(\{\phi\}_A) = \text{Auto}(\phi[A])$. Thus, $A \mapsto \psi^{-1}[S^*]$ iff $\phi[A] \mapsto S^*$ iff for all $\Gamma \in \Sigma(\{\phi\}_A)$, $\Gamma[S^*] = S^*$.

Aside from showing the equivalence of the coordinate and invariant approaches to the definition of structure, the last result has useful technical applications. Often, the coordinate transformations of a distinguished class of coordinate systems are mathematically well understood and a wide variety of analytical techniques are available for their investigation. (The Poincaré transformations of the Lorentz coordinate systems, for example.) The above result shows that we may determine whether or not a structor is an invariant of A by first respresenting that structor in any coordinate system of a distinguished class (say, in some Lorentz coordinate system) and then determining whether or not the *coordinate structor* is left untouched by every coordinate transformation of that class (say, every Poincaré transformation). This technique will be used in a later section to show that the standard simultaneity relation ($\epsilon = 1/2$) is an invariant of Minkowski space-time, while the nonstandard simulaneity relations ($\epsilon \neq 1/2$) fail to be invariants—providing another proof of a similar result first obtained in a different way by David Malament (21).

A final result. When the coordinates B are pure sets, say, elements of R^n, then any distinguished class is itself an invariant of *A* in its own right. For if $\{\phi\}_A = \{\phi\sigma: \sigma \in \text{Auto}(A)\}$ is such a class and τ is an automorphism of *A*, then $\tau[\{\phi\}_A]$ will be just $\{\tau[\phi\sigma]: \sigma \in \text{Auto}(A)\}$. A simple computation shows that $\tau[\phi\sigma]$ is just $\phi\sigma\tau^{-1}$. Since σ and τ^{-1} are both automorphisms of *A*, so is $\sigma\tau^{-1}$. Hence, $\tau[\{\phi\}_A] = \{\phi\}_A$, and so $\{\phi\}_A$ is an invariant of *A*.

More than this, *any such distinguished class can serve to generate A,* that is,

PROPOSITION 2.29. *Let A be a structure, (ϕ,B) a coordinate system of A, with* B *a pure set. Then, Auto(A)* = Auto((A, $\{\phi\}_A$)), *that is A and* (A,$\{\phi\}_A$) *have the same invariants, and thus are invariantly equivalent.*

Proof. a. Assume $\tau \in$ Auto (*A*). Then, $\tau[\{\phi\}_A] = \tau[\{\phi\sigma: \sigma \in \text{Auto}(A)\}] = \{\phi\sigma\tau^{-1}: \sigma \in \text{Auto}(A)\}$. *Since* $\tau^{-1} \in \text{Auto}(A)$, $\{\sigma\tau^{-1}: \sigma \in \text{Auto}(A)\} = \text{Auto}(A)$. Hence, $\tau[\{\phi\}_A] = \{\phi\}_A$.

b. Assume $\tau \in \text{Auto}(A')$. Then, $\tau[\phi\sigma_i] = \phi\sigma_j$, some $\sigma_j \in \text{Auto}(A)$, that is, $\phi\sigma_i\tau^{-1} = \phi\sigma_j$, that is $\sigma_i\tau^{-1} = \sigma_j$, in other words, $\tau^{-1} = \sigma_i^{-1}\sigma_j$. Hence, $\tau^{-1} \in \text{Auto}(A)$, and the same for τ.

Notice that *any* distinguished class will do. For example, if *A* = (A,d) is a Euclidean plane, then the set of Cartesian coordinate systems will serve to determine *A*. However, if we choose any coordinate system ϕ: A $\mapsto$ R^2, however curvilinear or otherwise "odd," then from Proposition 2.29 the structure $A' = (A,\{\phi\}_A)$ also determines precisely the same invariants—including, of course, the metric d. For structural purposes, all distinguished classes are on a par. In practice, however, some classes have advantages over others in that many invariant structors of *A* may have especially simple representations in these coordinate systems. As we shall see in section 4 of this essay, this point will be useful in connection with the conventionality of simultaneity in special relativity.

Incidentally, these results serve to clarify the oft-debated merits of coordinate *versus* invariant approaches to structure. When coordinate systems map a structure *A* into a pure structure—the usual situation—such systems are structors of *A;* but unless *A* is rigid, they will not be invariant structors of *A*. Individually, coordinate systems are arbitrary. Distinguished classes of such are not, however. As we have shown, such classes are always structural invariants, and any such may be used to define that structure up to invariant equivalence. Thus, a definition of Minkowski space-time, say, in terms of (Lorentz) coordinate systems is as "invariant" an approach to this structure as any (using the space-time interval, say).

While there is no purely structural or "logical" difference between the two methods, it is true, however, that a distinguished class of coordinate systems can be intuitively less than illuminating. As with a theory's axioms, the choice of the distinguished structors used to generate a structure is not unique and is probably best left to the claims of epistemic or intuitive convenience.

2.4. *Codeterminate Structures and Natural Isomorphisms*

Invariantly equivalent structures, as we have seen, are just those structures with isomorphic automorphism groups. Thus, invariantly equivalent structures, even those with the same universe, need not have the *same* automorphisms, nor the same invariants. For example, if $A = (\{a,b,c\}, \{a\})$ and $A' = (\{a,b,c\}, \{b\})$, then A and A' are invariantly equivalent—indeed isomorphic—while their automorphisms, clearly, differ.

Consider, however, the structures $B = (\{a,b,c\}, \{a\})$ and $B' = (\{a,b,c\}, \{b,c\})$. A little thought shows that B and B', while they are set-theoretically distinct, have exactly the same automorphisms, and so, invariants. The "sameness" of B and B' is stronger than the "sameness" of A and A' above. Intuitively and roughly: while A and A' *have the same* "structure," B and B' *are* the same "structure."

We shall call prestructures like B and B' *codeterminate* structures, and write "$B \approx B'$." As a definition, we take:

DEFINITION 2.30. *Where (pre)structures A and A' have the same universe, A is codeterminate with A' ($A \approx A'$) just in case $A \mapsto$ and $A' \mapsto A$, that is, A' is an invariant of A, and conversely.*

As an immediate consequence we have:

PROPOSITION 2.31.

(i) $\approx$ is an equivalence relation on (pre) structures, and
(ii) A and A' are codeterminate just in case:
  (a) Auto*(A)* = Auto*(A')*,
  (b) A and A' have the same invariants.

Proof. Exercise.

Thus codeterminate structures differ merely in their choice of the distinguished structors used to generate the same set of invariants, much as the same formulation of a deductive theory may be generated from distinct sets of axioms.

Notice that we might have introduced codetermination at the outset and later defined invariant equivalence, because of the following:

PROPOSTION 2.32. *$A \simeq$i B just in case A is isomorphic to some structure B' codeterminate with B.*

It should now be clear that, with the notions of invariantly equivalent and codeterminate structures, the arbitrariness that attaches to the choice of distinguished structors within a structor has been overcome. Not so, however, for the choice of a structure's *universe.* Invariantly equivalent structures must have universes of the same cardinality, while codeterminate structures must share the *same* universe.

It is this "despotism of universes" that will be eliminated in the next few sections. Motivation comes from geometry, especially projective spaces. There we can take our universe to consist of points and then go on to define lines as subsets of these, or, on the other hand, begin with lines, defining points as subsets of these—intuitively, a point becomes all lines through that "point." Either way, the resulting structure is "essentially the same." It is this last sense of structural sameness whose explication is our goal. As a step on the way, a notion of considerable importance in its own right will be explicated and developed, the notion of *naturally isomorphic* structures.

A standard example involves vector spaces. If V is an n-dimensional real vector space, its dual V^* is defined as the set of real-valued linear functions on V, with addition and scalar multiplication defined in the obvious way. V^* then turns out to be an n-dimensional real vector space and (hence) isomorphic to V. This process may now be repeated again on V^*, obtaining still another vector space V^{**}—the second dual of V. Although all of these spaces are isomorphic, the choice of an isomorphism between V and V^* involves the arbitrary selection of a basis of V, whereas an isomorphism between V and its *second* dual V^{**} may be defined in a basis-independent way. Intuitively, it is this freedom from arbitrary invocation of a basis that makes this last isomorphism "natural."

A much simpler example shows what is involved quite clearly. Let A be an arbitrary set and $A' = A \times \{\phi\}$. Then we may define a bijection—an isomorphism of sets—between A and A′ by:

$$\text{For all } x \in A,\ \eta(x) = \langle x,\phi \rangle.$$

This pairing η is, of course, only one of many bijections between A and A′ (unless A is a unit set). but *this* pairing seems "natural," the others "arbitrary." (And this still seems correct even when sets A and A′ are finite and we can thus easily list these other pairings.)

Category theory nowadays provides the standard account of the "naturality" of natural isomorphisms; indeed, category theory was invented for just this purpose (see 11, §1). Here, another approach will be adopted; its relation to the category-theoretic account will be explored in the next section, where arguments showing the inadequacy of the category-theoretic approach will be provided.

The basic idea of the approach adopted here is quite simple: a "natural" isomorphism is just an *invariant isomorphism.* For example, any isomorphism from a vector space V to either V^* or V^{**} is a structor of V, and, hence, either is or is not an invariant of V. We can now show that no isomorphism of V and V^* is an invariant of V, whereas the standard "natural" isomorphism of V and V^{**} is indeed an invariant of V.

DEFINITION 2.33. *Where A and B are structures, η is an invariant isomorphism from A to B (η: $A \simeq$n B) just in case:*
(*i*) *η is an isomorphism of A and B (η: $A \simeq B$), and*
(*ii*) *η is an invariant of A ($A \mapsto \eta$).*

Consider the earlier example of set A and A′ = A × {ϕ}, when $\eta(x) = <x,\phi>$. Here we let $A = (A;S_A)$ and $A' = (A';S_{A'})$, that is, A has all permutations of A as its automorphisms, and similarly for A'. We can now easily show that η is an invariant of A, since

$$\eta = \{(x,<x,\phi>):\ x \in A\},$$

so if σ is any permutation of A,

$$\sigma[\eta] = \{(\sigma(x),\ <\sigma(x),\phi>):\ x \in A\}.$$

Since σ is just a permutation of A, we can then easily see that $\sigma[\eta]$ is just η; hence, η is an invariant of A.

But now let A = {a,b,c}, A′ = {<a,ϕ>, <b,ϕ>, <c,ϕ>}, and consider the "unnatural" isomorphism

$$\mu = \begin{pmatrix} a & b & c \\ <a,\phi> & <c,\phi> & <b,\phi> \end{pmatrix},$$

If we now consider the automorphism of A given by

$$\sigma = \begin{pmatrix} a & b & c \\ b & a & c \end{pmatrix}.$$

we obtain

$$\sigma[\mu] = \begin{pmatrix} \sigma(a) & \sigma(b) & \sigma(c) \\ \sigma[<a,\phi>] & \sigma[<c,\phi>] & \sigma[<b,\phi>] \end{pmatrix} = \begin{pmatrix} b & a & c \\ <,b,\phi> & <c,\phi> & <a,\phi> \end{pmatrix},$$

which is *not* just μ back again. In other words, μ is not an invariant of A.

We begin exploring invariant isomorphisms by establishing a proposition that provides them with an alternative characterization.

PROPOSITION. 2.34. *Where η is an isomorphism from A to B, the following conditions are equivalent:*
(*i*) η: $A \simeq$n B,

(*ii*) *for all* $a \in A$, $\sigma \in \text{Auto}(A)$, $\sigma[\eta(a)] = \eta\sigma(a)$,
(*iii*) *for all* $b \in B$, $\sigma[b] = \eta[\sigma](b)$.

Proof. (i) iff (ii). Assume $\eta: A \simeq n\ B$. Then, $A \mapsto \eta$ iff for all $a \in A$ and $\sigma^{-1} \in \text{Auto}(A)$, $\sigma^{-1}[\eta] = \eta$, that is $\sigma^{-1}[\eta](a) = \eta(a)$, that is by Lemma 2.9 (i), $\sigma^{-1}[\eta\sigma(a)] = \eta(a)$, in other words, $\eta\sigma(a) = \sigma[\eta(a)]$.

(i) iff (iii). At once from (i) iff (ii).

As a corollary we have:

COROLLARY 2.35. *If* η is an invariant isomorphism from A to B, then B is an invariant of A.

Proof. Assume $\eta: A \simeq n\ B$. Then $B = \eta[A]$. Defining σ^* as in Lemma 2.7, $\sigma^*(b) = \eta[\sigma](b)$, by (ii) above. Hence, $\sigma^*[B] = \eta[\sigma][B]$, and so by Lemma 2.7, $\sigma[B] = \eta[\sigma][B]$, all $\sigma \in \text{Auto}(A)$. By Lemma 2.10 (iii), $\sigma[B] = \eta\sigma\eta^{-1}[B] = \eta\sigma[A] = \eta[A] = B$. Hence, $A \mapsto B$.

The results of Proposition 2.34 are not merely of computational interest. For example, suppose the structure A is homogeneous, in other words, for any two items in its universe, there is an automorphism taking one to the other. Now, suppose that η is an invariant isomorphism from A to some structure B. Choose an arbitrary a_0 in A and let b_0 be $\eta(a_0)$. Substituting (ii) in the result of the above proposition we obtain:

$$\eta(\sigma(a_0)) = \sigma[b_0].$$

Now choose any $a_i \in A$ and let $\sigma_i(a_0) = a_i$. (Since A is homogeneous, recall, we know that such a σ_i must exist.) Substituting again,

$$\eta(a_i) = \sigma_i[b_0].$$

Hence, the value of η on a_i is completely determined by its value on a_0, that is, in homogeneous or transitive structures, natural isomorphisms are determined by their values on a *single member* of the universe of A. This result easily generalizes to nonhomogeneous structures, after a preliminary definition.

Where x is in the universe of a structure A, the *orbit of* x (in A) is just the set of those elements of A that are taken to x by some automorphism of A. Equivalently, $\text{Orb}_A(x) = \{\sigma(x): \sigma \in \text{Auto}(A)\}$. It is well known, and easy to show, that the orbits exhaustively partition the universe A into mutually disjoint (equivalence) classes, simply called "orbits." Thus, A can always be written as $A_0 \dot{\cup} A_1 \dot{\cup} \ldots \dot{\cup} A_\mu \dot{\cup} \ldots$, with each A_ν an orbit. Homogeneous structures are clearly those having exactly one orbit, namely, the entire universe A.

The generalization of the preceding result should now be clear. Since A may have more than a single orbit, it no longer suffices to choose a single pair $\langle a_0, \eta(a_0)\rangle$ to determine η; instead, we need to choose a pair from *each* orbit of *A*. We then take images of each pair for every automorphism of *A*, and we have recovered η in its entirety. This is established by the proof of the following proposition.

PROPOSITION 2.36. *Let* η: *A* $\simeq$n *B and let* $\bar{a}_0, \bar{a}, \ldots, \bar{a}_\mu, \ldots$ *be elements of the orbits of A*: $A_0, A_1, \ldots, A_\mu, \ldots$, *respectively. Then:*

$$\eta = \bigcup_\mu \{\sigma[<\bar{a}_\mu, \eta(\bar{a}_\mu)>]: \sigma \in \text{Auto}(A)\}.$$

Proof. In general, from Proposition 2.34 (ii), $\eta\sigma(a) = \sigma[\eta(a)]$, all $a \in A$, $\sigma \in \text{Auto}(A)$. So, in particular, $\eta(\sigma(\bar{a}_\mu)) = \sigma[\eta(\bar{a}_\mu)]$. Thus, $\{<\sigma(\bar{a}_\mu), \sigma[\eta(\bar{a}_\mu)]>: \sigma \in \text{Auto}(A)\} \subseteq \eta$, for all μ. That is, $\{\sigma[<\bar{a}_\mu, \eta(\bar{a}_\mu)>]: \sigma \in \text{Auto}(A)\} \subseteq \eta$, all μ. Now, $A = \bigcup_\mu \{\sigma(\bar{a}_\mu): \sigma \in \text{Auto}(A)\}$.

Hence, $\bigcup_\mu \{\sigma[<\bar{a}_\mu, \eta(\bar{a}_\mu)>]: \sigma \in \text{Auto}(A)\} = \eta$.

Later, we shall show how to construct invariant isomorphisms using such pairs as generators when we are not provided with η at the outset. First, however, some results on the uniqueness of invariant isomorphisms.

The invariant isomorphism η between the structures *A* and *A'* in the earlier example was unique. This need not always be the case, however. In general, the uniqueness of a natural isomorphism between structures *A* and *B* depends solely upon whether or not there is nontrivial automorphism of *A*, which is itself an invariant of *A*. If so, an invariant isomorphism is not unique; otherwise, it is. In other words:

PROPOSITION 2.37. *Let* η_1: *A* $\simeq$n *B. Then* η_1 *is the only such invariant isomorphism just in case there is no* $\tau \in$ Auto (*A*), $\tau \neq 1_A$, *that is an invariant of* $A(A \mapsto \tau)$.

The basic idea of the proof (in the appendix) of this result is that, were such a τ to exist, then $\eta_2 = \eta_1 \circ \tau^{-1}$ would be still another invariant isomorphism between *A* and *B*, and conversely, were η_2 to exist, $\tau = \eta_2{}^{-1} \circ \eta_1$ would then be an invariant automorphism of *A*. Group theorists call the subgroup Z(G) of a group defined by

$$Z(G) = \{g \in G: xgx^{-1}, \text{ all } x \in G\}$$

the *center* of G. By Lemma 2.10 (iii), xgx^{-1} is just x[g] when we are dealing with automorphisms, so what we have called the invariant automorphisms of *A* are just the members of Z(Auto(*A*))—the center of Auto(*A*).[33] Furthermore, the center of a group is called *trivial* when it contains the group's identity element as its sole member. In these terms, the above proposition becomes:

COROLLARY 2.38. *If A* $\simeq$n *B, then there is exactly one invariant isomorphism from A to B if and only if the center of the automorphism group of A is trivial.*

Applications are immediately forthcoming. For example, the center of the group of real vector space automorphisms (the general linear group) is not trivial and consists of those automorphisms that dilate each vector by the same constant factor. Hence, the standard natural isomorphism, between a real vector space V and its second dual V^{**} is only one of infinitely many such, obtainable, as we have seen, by composition with automorphisms in the center of Auto (V). In fact, as the proof of Proposition 2.34 reveals, all such invariant automorphism are so obtained. For structures $A = (A;S_A)$, whose automorphisms are just all the permutations of A—trivial structures—the situation is different. When the cardinality of A is greater thas two, S_A will have a trivial center (exercise). Hence, as in the earlier example, any invariant isomorphism η: $A \simeq$n B will be unique.

Unique or not, any invariant isomorphism between two structures A and B yields the same image on any automorphism of A. In other words, for any automorphism σ of A, $\eta_1[\sigma] = \eta_2[\sigma]$, where η_1 and η_2 are (possibly distinct) invariant automorphisms from A to B. Furthermore, the automorphism of B thus induced by η_1 (or η_2)—recall that *all* isomorphisms preserve automorphisms—is just the automorphism of B naturally induced by σ, that is, for all $b \in B$, $\eta_i[\sigma]\,(b) = \sigma[b]$. Notice that this is just Proposition 2.34 (iii) above. Furthermore, not only do invariant isomorphisms thus induce the natural mapping from Auto(A) to Auto(B), but any isomorphism that so behaves is an invariant isomorphism. We obtain thereby another criterion for an isomorphism's being "natural" or invariant. These results are summarized in the following propositions.

PROPOSITION 2.39. *If* η_1,η_2: $A \simeq$n B, *then for all* $\sigma \in$ Auto(A), $\eta_1[\sigma] = \eta_2[\sigma] = \sigma^*$, *the automorphism of B induced by* σ.

Proof. At once from Proposition 2.34 (i)-(iii).

PROPOSITION 2.40. *If* η: A $\simeq$ B, *and for all* $\sigma \in$ Auto(A), $\eta[\sigma] = \sigma^*$, *then* η: $A \simeq$n B.

Proof. At once from Proposition 2.34 (i)-(iii).

While we are discussing induced mappings, it turns out that invariant isomorphisms are preserved by "ordinary" isomorphisms in an especially strong way. If η is an invariant isomorphism from A to B and ϕ any isomorphism from A to some other structure C, then $\phi[\eta]$—the ϕ-image of η—is an invariant of C. Furthermore, $\phi[\eta]$ is independent of the choice of isomorphism from A to C. Hence, given η: $A \simeq$n B, and $A \simeq C$, a unique structure D is determined that is invariantly isomorphic to C. Of course, D is only uniquely specified with respect to the original invariant isomorphism η. In many cases, however, η will be unique—as we have seen—and thus D will be uniquely fixed, given $A \simeq$n B and

$A \simeq C$. This result will be useful in the next section where comparisons are made with category-theoretic approaches to natural isomorphisms. The following proposition sums up the situation. (Those unfamiliar with category theory might skip this and the following section upon first reading.)

PROPOSITION 2.41. *Let* $\eta: A \simeq n\ B$ *and* $\phi: A \simeq C$. *Then letting* $\phi^*(b) \equiv \phi[b]$, *all* $b \in B$, *the following diagram commutes. (Here,* $D \equiv \phi[\eta][C]$.)

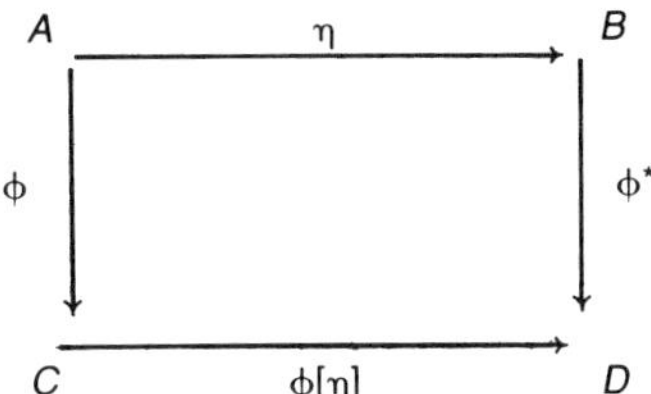

That is, for all $a \in A$, $(\phi^* \circ \eta)(a) = (\phi[\eta] \circ \phi)(a)$.

Furthermore, $\phi[\eta]$, *and, hence,* D, *is independent of the choice of isomorphism* ϕ, *that is for any* $\psi: A \simeq C$, $\phi[\eta] = \psi[\eta]$.

Finally, the mapping $\phi[\eta]$ *is an invariant isomorphism from C to D, that is,* $\phi[\eta]: C \simeq n\ D$.

While the invariant isomorphism $\simeq n$ is reflexive and transitive (exercise), it is clearly not symmetrical. This is because $A \simeq n\ B$ implies that B is a structor of A, but, in general, the reverse is not the case. (Attempting to symmetrize by going to the relation "$A \simeq n\ B$ or $B \simeq n\ A$" does not succeed since—for reasons we need not discuss here—transitivity then fails.) Even when naturally isomorphic, A and B (i.e., A and B) will not in general have the same structors, so it is thus too much to expect them to have the same invariants. The most that we could hope for is that any structor of *both* A and B is an invariant of A if and only if it is also an invariant of B. And *this hope is fulfilled,* as the following theorem reveals.

PROPOSITION 2.42. *If* $A \simeq n\ B$ *and* S *is a structor of both* A *and* B, *then* $A \mapsto S$ *iff* $B \mapsto S$, that is, S *is an invariant of* A *iff* S *is an invariant of* B.

Proof. ($\Leftarrow$) Since $A \simeq n\ B$, by Corollary 2.35, $A \mapsto B$. But $B \mapsto S$, and, hence (as in the proof of Proposition 2.8), $A \mapsto S$.
($\Rightarrow$) Assume $A \mapsto S$, with $\eta: A \simeq n\ B$. Suppose $\tau \in$ Auto (B). Then by Proposition 2.39, there is a $\sigma \in$ Auto (A) that induces τ on B. Hence, by Lemma 2.7, since S is a structor of B, $\tau[S] = \sigma[S] = S$.

Putting it picturesquely, perhaps: All that keeps invariantly isomorphic structures from having the *same* invariants is their possibly having different structors. When this is ruled out by specializing to a common universe, sameness of invariants is reinstated. (This follows at once from the last proposition.) Indeed, we can assert:

COROLLARY 2.43. *Where A and A′ have the same universe* A, *then if A is invariantly isomorphic to A′, A and A′ are codeterminate structures; that is, if A* $\simeq$n *A′, then A* $\approx$ *A′.*

The converse of this result fails, since invariantly isomorphic structures are, after all, isomorphic. As the example cited earlier in this section shows, this need not be the case for codeterminate structures. As we shall see in section 2.6, however, invariant isomorphisms may be generalized in a natural way that reduces to codetermination when the structures involved have the same universe.

Before going deeper into the structure of invariant isomorphisms, let us first notice an important feature of rigid structures. Recall that a rigid structure is one that totally individuates its universe: its only automorphism is the identity mapping. As a result, *every* structor of a rigid structor is an invariant, and, as we have seen (Proposition 2.21), any two rigid structures of the same cardinality are invariantly equivalent. We are now in a position to show that all rigid structures function as pure sets. Rigid structures, pure or no, are all invariantly isomorphic to pure structures.

PROPOSITION 2.44. *If A* $\simeq$n *B and B is pure, then A is rigid.*

Proof. Let η: *A* $\simeq$n *B*. Then $\eta = \{<a,\eta(a)>: a \in A\}$. Hence, since $\eta(a)$ is pure, for all $\sigma \in S_A$, $\sigma[\eta] = \{<\sigma(a),\eta(a)>: a \in A\}$. Hence, if $\sigma \neq 1_A$, $\sigma[\eta] \neq \eta$. Thus, since $A \mapsto \eta$, Auto(*A*) $= \{1_A\}$, that is, *A* is rigid.

COROLLARY 2.45. *A is rigid iff there is a pure structure B such that A* $\simeq$n *B.*

Proof. ($\Rightarrow$) Suppose *A* is rigid. Let ϕ be a well-ordering of *A*. Then $\phi^{-1}[A]$ is pure and ϕ^{-1}: $A \simeq \phi^{-1}[A]$. Since *A* is rigid, 1_A is the only automorphism of *A*. Hence, $A \mapsto \phi^{-1}$.

($\Leftarrow$) At once from the preceding proposition.

Later developments will show that it is necessary to go still deeper into the structure of invariant isomorphisms; in particular, we shall need to see how it is possible to build structures that are invariantly isomorphic to *A*, starting merely with structors of *A*. Proposition 2.36 and the discussion preceding it provide important clues.

For simplicity's sake, let us consider a homogeneous (i.e., single-orbit) structure *A*. We have already seen that if we are given an invariant isomorphism η from *A* to a structure *B*, then η can be recovered from any of its pairs $<a_0,b_0>$, $b_0 = \eta(a_0)$, by closing under automorphic images. Now why not construct invariant isomorphisms at the outset by this procedure? In other words, let S_0 by any structor of A, a_0 any element of A, and begin with the pair $<a_0,S_0>$. Then, define η as just $\{\sigma[<a_0,S_0>]: \sigma \in \text{Auto}(A)\}$. Finally, let *B* be just the η-image of *A*, η [*A*].

(Notice that the universe B of B is just $\{\sigma[S_0]: \sigma \in \text{Auto}(A)\}$.) Will η then always be an invariant isomorphism from A to $\eta[A] = B$?

No, not always. To see why not, let us look at the process more carefully.

First, we have let η be defined by

$$\eta = \{\sigma[<a_0,S_0>]: \sigma \in \text{Auto}(A)\}.$$

Since $<a_0,S_0>$ is a structor of A, and η is just its automorphic closure, it is easy to see that η, so defined, must be an invariant of A, that is, $A \mapsto \eta$. So far, then, so good. The problem, however, is that we have no guarantee that η, so defined, is one-one, or indeed, even a function. This, in fact, will only be the case when the chosen structor S_0 is "well behaved." Suppose, for example, that distinct automorphisms σ_1 and σ_2 were to yield:

$$\sigma_1[<a_0,S_0>] = <\sigma_1(a_0),\sigma_1[S_0]>, \tag{2.2}$$

$$\sigma_2[<a_0,S_0>] = <\sigma_2(a_0),\sigma_2[S_0]>, \tag{2.3}$$

and it turned out that

$$\sigma_1(a_0) = \sigma_2(a_0), \tag{2.4}$$

but

$$\sigma_1[S_0] \neq \sigma_2[S_0]. \tag{2.5}$$

Then η, as defined above, would assign distinct values $(\sigma_1[S_0],\sigma_2[S_0])$ to the same argument $(\sigma_1(a_0) = \sigma_2(a_0))$, and thus fail to be a function. Such a structor S_0 would have been a poor choice; what we need is a structor S_0 that satisfies the condition:

(C_1) If $\sigma_i[a_0] = \sigma_j[a_0]$, then $\sigma_i[S_0] = \sigma_j[S_0]$, for all $\sigma_i,\sigma_j \in \text{Auto}(A)$.

Let us call the set of all automorphisms of A that take a structor S of A to itself, the *stability group in A of* S, and write "$\text{Stab}_A(S)$." Since in condition (C_1), $\sigma_i(a_0) = \sigma_j(a_0)$ iff $\sigma_j^{-1}\sigma_i(a_0) = a_0$ iff $\sigma_j^{-1}\sigma_i \in \text{Stab}_A(a_0)$, and similarly for S_0, it is easy to see that (C_1) becomes:

(C_1)′ $\qquad \text{Stab}_A(a_0) \subseteq \text{Stab}_A(S_0)$.

Similarly, it is easy to show in the same way that if η is to be a *one-one* function, the converse of (C_1)′ must hold. Hence, for the above construction to succeed, we require simply

(C) $\qquad \text{Stab}_A(a_0) = \text{Stab}_A(S_0)$.

In other words, given our choice of a_0 from A, we must now choose a structor S_0 of A which is left intact by all and only those automorphisms that leave a_0 intact. Furthermore, as the proof of the following proposition shows, this is all that is required; η was always an invariant, now η will be a one-one function; hence, η: $A \simeq n\ \eta[A]$.

When A is not homogeneous, then we merely choose representatives $\bar{a}_0,\bar{a}_1, \ldots$ from each orbit of A and corresponding distinct structors S_0, $S_1, \ldots$ so that each pair $<\bar{a}_\mu, S_\mu>$ satisfies condition (C) and no automorphism of A takes any S_μ to a distinct S_ν. Closing under all automorphisms of A now yields an invariant isomorphism η as before. Hence, the following proposition (whose complete proof is in the appendix).

PROPOSITION 2.46. *Let A be any structure with* $A = A_0 \dot{\cup} A_1 \dot{\cup} \ldots \dot{\cup} A_\mu \dot{\cup} \ldots$, *each* A_μ *an orbit of A. Let* $\bar{a}_0,\bar{a}_1, \ldots, \bar{a}_\mu, \ldots$ *be such that* $\bar{a}_\mu \in A_\mu$. *Furthermore, let* $S_0,S_1, \ldots ,S_\mu, \ldots$ *be distinct structors of A such that, for all* $\sigma \in$ Auto (A):

(*i*) $\text{Stab}_A(\bar{a}_\mu) = \text{Stab}_A(S_\mu)$,
(*ii*) $\sigma[S_\mu] = S_\nu$ *implies* $\mu = \nu$.

If we now define:

$$\eta = \bigcup_{\mu}\{\sigma[<\bar{a}_\mu,S_\mu>]: \sigma \in \text{Auto}\ (A)\},$$

then

$$\eta: A \simeq n\ \eta[A].$$

Finally, one more reduction. The above proposition requires finding suitable structors S_μ for each orbit-representative $\bar{a}_\mu$. We can, however, do much better: a single structor S of A that is altered by all automorphisms of A except the identity permutation 1_A of A will do. First, we index such an S by the orbits of A, obtaining $<S,0>,<S,1>, \ldots,$ $<S,\mu>$, and so forth. Clearly, these are distinct. Next, with each representative $\bar{a}_\mu$, we pair $S_\mu = \{\sigma[<S,\mu>]: \sigma \in \text{Stab}_A(\bar{a}_\mu)\ \sigma \in \text{Stab}_A(\bar{a}_\mu)\}$. The S_μ are also distinct, and, further, the construction guarantees that $\bar{a}_\mu$ and S_μ have the same stability group (condition (C) above). Using these S_μ in the previous proposition, we may now obtain an invariant isomorphism η as before. The details are provided by the proof of the following lemma.

LEMMA 2.47. *Let* S *be a structor of A and assume that for all* $\sigma \in$ Auto(A), $\sigma[S] = S$ *implies* $\sigma = 1_A$. *If* $\bar{a}_0,a_1, \ldots ,\bar{a}_\mu, \ldots$ *are representatives of each orbit of A, and we define:*

$$S_\mu = \{\sigma[<S,\mu>]: \sigma \in \text{Stab}_A(\bar{a}_\mu)\},$$

then:

(*i*) *the* S_μ *are distinct, and for all* $\sigma \in$ Auto (A):

(*ii*) $\sigma[S_\mu] = S_\nu$ *implies* $\mu = \nu$,
(*iii*) $\text{Stab}_A(\bar{a}_\mu) = \text{Stab}_A(S_\mu)$.

The next step is to generalize the notion of invariant isomorphism to structures that are not isomorphic, in fact to structures that may have universes differing in cardinality. The above results have perhaps been rough going, but they will help to smooth out these later developments.

2.5. *Critique of the Category Theory Account of Natural Isomorphisms*[34]

The standard contemporary account of natural isomorphisms is given by category theory. Indeed, in their original paper founding this discipline, Eilenberg and MacLane begin: "The subject matter of this paper is best explained by an example, such as that of the relation between a vector space L and its 'dual' or 'conjugate' space T(L)" (ll, pp. 231–32). And later in this section they go on to write: "The study of functors also provides a technical background for the intuitive notion of naturality and makes it possible to verify . . . the naturality of an isomorphism or equivalences in all those cases where it has been intuitively recognized that the isomorphisms are indeed 'natural' " (p. 236).

In this section, the claim that category theory accurately explicates the intuitive notion of a "natural" isomorphism will be critically examined. As we shall see, while there are many similarities between the approach adopted here and the categorical account, there are also important differences. Moreover, these differences give reason to doubt the adequacy of the category-theoretic approach to this issue.[35]

The essence of the category theory account of structure is its deliberate neglect of the "internal" properties of a mathematical object and the use of the mappings or "morphisms" between such objects as the sole determinants of their structure. Thus, for example, the category of topological spaces consists of these spaces considered as objects—logicians would say "individuals"—together with all continuous mappings between them. The category of sets takes sets as objects and all mappings between sets as morphisms; the category of vector spaces takes all linear mappings as its morphisms. The general definition of a category abstracts from such examples in the expected way. Thus, a category may be defined as a quintuple

$$K = (O,M,\text{dom},\text{cod},\circ),$$

where the elements of O are its K *objects* (Ob(K)), the elements of M its *morphisms,* dom and cod are functions from M to O yielding (intuitively) the *domain* and *codomain* (range) of each morphism, and $\circ$ is a partial function from M $\times$ M to M which yields the result of composing two morphisms whenever the domain of the first is equal to the codomain of the second. The axioms defining K are now chosen so as to insure that morphisms compose the way functions do.[36]

It is well known (see 18, pp. 21–22) that a category may also be defined as simply a pair

$$K = (\mathrm{M},\circ),$$

where the previous objects are now identified with the identity morphisms.[37] The previous functions, dom and cod, may now be defined in the obvious way (e.g., where e is an identity morphism, dom(ϕ) = e just in case $\phi \circ e$ is in M).

Clearly, using this definition, a category turns out to be a structure in the sense used here: M is just a set[38] and $\circ$ a function from M $\times$ M to M, and so a structor of M.[39]

A *functor* is a mapping from one category to another that preserves composition and identity morphisms in the expected way (see 18, §9). When the first definition of a category is used, a functor is regarded as a mapping that takes objects to objects as well as morphisms to morphisms, again preserving relations between the two. Thus, if a functor F takes object A in category K_1 to object A′ in category K_2, then it is required to take any morphism in K_1 with domain A to a morphism in K_2 with domain A′, and so forth. Intuitively, a functor is a homomorphism of categories.

We are now in a position to illustrate the category-theoretic approach to natural isomorphisms. As a simple example, consider the "natural" isomorphism η_A that exists between any set A and A $\times$ {a*} where η_A is given by

$$\eta_A(x) = \text{df.} <x,a^*>, \text{ for all x in A,} \tag{2.6}$$

and a* is some fixed individual. The naturality of η is now taken to consist in the fact that, for any sets A and B and any morphism ϕ from A to B, the following diagram commutes:

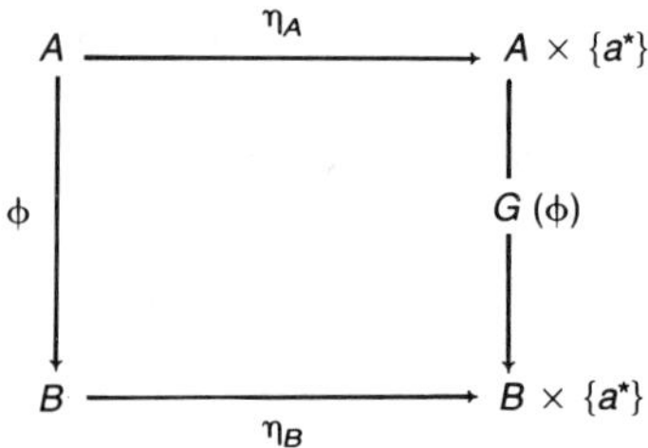

Here, the functor G takes ϕ to the morphism "induced by" ϕ and is defined by:

$$G(\phi)(<x,a^*>) = <\phi(x),a^*>, \text{ all } x \in A; \tag{2.7}$$

and to say that the above diagram *commutes* is just to say that:

$$\eta_B \circ \phi = G(\phi) \circ \eta_A. \tag{2.8}$$

Notice that η_A is more properly written as $\eta(A)$, since η is a function defined on the objects of the category of sets which yield an isomorphism (η_A say) for each such object. Category theorists therefore call the function η a natural isomorphism rather than so calling *each* η_x, $x \in \text{Ob}(K)$, as we have. This, however, is an unimportant difference in terminology. What *is* important is that category theory regards the generality of η so considered, together with commutativity with *all* morphisms as getting at what is behind the "naturality" of natural isomorphisms.[40]

The above example now generalizes as follows:

DEFINITION. *Let K be any category,* I: $K \mapsto K$ *be the identity functor on K, and* G: $K \mapsto K$ *be a functor. Then,* (I,η,G) *is a natural isomorphism from* I *to* G *just in case:*

(*i*) For each $A \in \text{Ob}(K)$, $\eta(A)$ is an isomorphism from A to G(A), and
(*ii*) For each *K*-morphism ϕ, the diagram than follows commutes.[41]

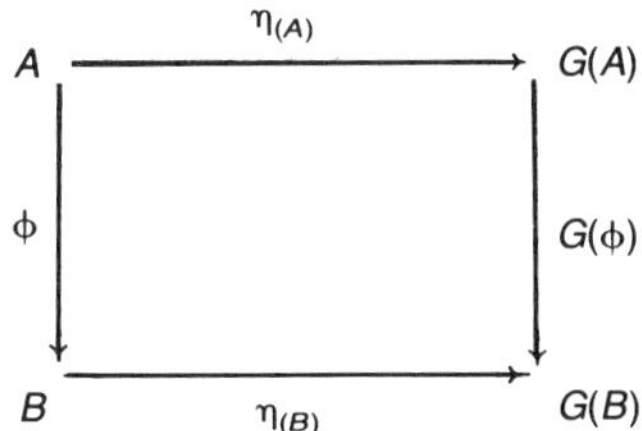

In the earlier example, the functor G was defined by (2.7) together with the obvious requirement that G take any set A to $A \times \{a^*\}$. It is now routine to check that (I,η,G) is indeed "natural" according to the above definition.

With the above definition on hand, we are now in a position to compare the two accounts.

To begin with, suppose that A^* is a structure and η_{A^*} is an invariant isomorphism from A^* to some structure B^*. It is now possible to construct a category A^*_c in such a way that η_{A^*} becomes one of the values of a categorical natural isomorphism η, with the remaining values being in some sense "appropriate." We proceed as follows.

First, let the objects of A^*_c be just those structures isomorphic to A^*, that is,

$$\text{Ob}(A^*_c) = \{D: A^* \simeq D\}. \tag{2.9}$$

Let the morphisms of A^*_c be just the isomorphisms of these structures, in other words,

$$\text{Mor}(A^*_c) = \{\phi:(\phi:C \simeq D),\ C, D \in \text{Ob}(A^*_c)\}. \tag{2.10}$$

The functions dom, cod, and $\circ$ are now defined in the standard set-theoretical way. Clearly, since $\eta_{A^*}: A^* \simeq B^*$, B^* is in Ob(A^*_c), and it is easy to show that A^*_c so defined is a category in the first sense defined earlier.

Next, we define the natural isomorphism η. Suppose A is an object in the category A^*_c. Then $A^* \simeq A$, and so we simply let η (A) be the ϕ-image of η_{A^*}, where ϕ is *any* isomorphism from A^* to A. In other words, for all $A \in$ Obj(A^*_c),

$$\eta(A) = \text{df. } \phi[\eta_{A^*}], \tag{2.11}$$

where ϕ: $A^* \simeq A$.

Notice that when $A = A^*$, $\eta[A^*] = \phi[\eta_{A^*}] = \eta_{A^*}$, since ϕ is then an automorphism of A^* and η_{A^*} is an invariant of A^*. Clearly, η is well defined if and only if the choice of the isomorphism ϕ: $A^* \mapsto A$ is irrelevant; but this is exactly what Proposition 2.41 of the previous section guarantees.

Finally, we define the "inducing" functor G. Let A be an object in A^*_c; then we let

$$G(A) \equiv \eta_A[A], \tag{2.12}$$

the η_A-image of A. And if ϕ: $A \simeq B$, then, for all x in G(A),

$$G(\phi)(x) = \text{df. } \phi[x], \tag{2.13}$$

that is, G(ϕ) is just the ϕ-image of the members of G(A).

For this definition to succeed we must show that: (a) G(A) is a structor of A, for all objects A; and (b), G so defined is indeed a functor. Then we may go on to show that (I,η,G) is a natural isomorphism with respect to the category A^*_c. These facts are established in the proof of the following proposition:

> PROPOSITION 2.48. *Let A^* be a structure, and A^*_c, η, and* G *be defined as above. Then* (I,η,G) *is a natural isomorphism of* I *and* G *in the category A^*_c. Furthermore, for all A in the objects of A^*_c, η (A) is an invariant of A.*

Thus, a single isomorphism $\eta_{A^*} \simeq$n B may be used to generate an entire set of such (namely, rng (η)) on the isomorphism category A^*_c of A^*, yielding a natural isomorphism η in the category-theoretical sense.[42] Hence, a natural isomorphism in the invariant sense will yield in a "natural" way a categorical natural isomorphism of which it is an instance.

A partial converse of this result is also true. Given a category of set-theoretical structures in the sense defined here, and where G is the "inducing" map (implying that G(A) is always a structor of A), then η_A will always be an invariant of A and, hence, a natural isomorphism in our sense. To see this, notice that commutativity implies that when we take ϕ to be an *automorphism* of structure A,

$$\eta_A \circ \phi = G(\phi) \circ \eta_A,$$

for all ϕ in Auto (A). But since $G(\phi)$ is just the ϕ-induced mapping, we obtain at once:

$$\eta_A \circ \phi(x) = \phi[\eta_A(x)],$$

for all $x \in A$ and ϕ in Auto(A). And by Proposition 2.34, this condition is necessary and sufficient for η_A to be an invariant isomorphism.

So invariant isomorphisms can be made to yield categorical natural isomorphisms, and categorical natural isomorphisms defined on categories of structures are also invariant, that is, each instance η_A of η is an invariant isomorphism.

Nevertheless, the harmony of the two approaches is only partial. For recall that, in both situations, the function G was taken to be the "inducing" functor and defined in terms of taking images in the set-theoretical way. Now, in some of the classical examples, such as the natural isomorphism between a vector space V and its second dual V^{**}, the functor G, indeed, delivers on any homomorphism ϕ: $V_1 \to V_2$ the set-theoretically induced mapping ϕ^{**}: $V_1^{**} \to V_2^{**}$ (proof: exercise [tedious]), but even in the case of the earlier example in the category of sets, this will not always be the case. For recall that according to (2) above, $G(\phi)(\langle x,a^*\rangle)$ is just $\langle \phi(x),a^*\rangle$. But the induced mapping would have the result be $\langle \phi(x),\phi[a^*]\rangle$, and when a* is an arbitrary structure of the set A—say, a member of A—$\phi[a^*]$ will not in general be equal to a*. Only when a* is taken to be a *pure* set based on A will G so defined yield the set-theoretically induced mapping. In fact, a glance at the mapping G as defined by (2.7) shows that it, in effect, *treats* a* as a pure set, whether it is or is not.

While at first sight it might seem that this is a mere quibble over a single example, the issue raised here is fundamental: How *is* G to be defined "correctly" in order that η in (I,η,G) correspond accurately to a "natural" isomorphism in the preanalytic sense? The problem is especially acute in the light of the following fact:

PROPOSITION 2.49. *Let K be a category and* η *be any function from K-objects to K-isomorphisms. Then there is a functor* G *which makes* (I,η,G) *a natural isomorphism in K.*

Proof. On objects K in *K*, define G(K) as just rng(η(K)). Where ϕ: K $\mapsto$ J is a morphism, define $G(\phi) = \eta_J \circ \phi \circ \eta_K{}^{-1}$. It follows at once from the definition that (I,η,G) is a natural isomorphism in *K*.

In other words, *any* set of isomorphisms is "natural" for *some* functor G or other. For example, if on the category of sets, say, we define η so that for A = {a,b,c}, η_A = {<a,a>,<b,c>,<c,b>}, and let η_x be the identity on set X when X is not equal to set A, a functor G can be "rigged" so as to make (I,η,G) a natural isomorphism; yet, clearly, η_A is not an invariant of A (i.e., of $A = (A;S_A)$).

Herein lies the basic difficulty with the category-theoretic explication of "naturality." When the example considered does as a matter of *set-theoretical* fact have G(ϕ) always equal to the mapping *set-theoretically induced* by ϕ, then, as we have seen, η will be invariantly natural. But category theory also allows *any* η to be considered natural with respect to some factor G or other.

Now, if a functor G could be shown on *category-theoretic grounds alone* to be "the inducement functor," there would be no problem. But it is just such a demonstration that the categorical approach relinquishes by its studied refusal to consider the internal structures of its objects and confine attention to their morphisms.

To see this clearly, suppose that structures *A* and *B* are objects in a category *K,* and that *B* is an invariant structor of *A*. It now follows (see the discussion in section 2.6 following Proposition 2.51) that any automorphism of *A* will induce a unique automorphism of *B,* thereby enabling us to define a homomorphism from the automorphisms of *A* to those of *B*. Suppose, for simplicity's sake, that this homomorphism is an isomorphism, and thus we have obtained from *A* and *B,* not only their various morphisms, but a natural mapping of their automorphisms as well. It is just this additional information that category theory deliberately excludes by its exclusive attention to the morphisms of its objects.

More precisely, let A_1 be any structure, and B_1 another structure distinct from A_1, but invariantly isomorphic to A_1. Let category K_1 have $\{A_1, B_1\}$ as its set of objects, and all homomorphisms of these as its morphisms. Now let B_2 be a structure that is isomorphic, but not *invariantly* isomorphic, to A_1, and let the category K_2 have A_1 and B_2 for its objects and, as before, homomorphisms of these as its morphisms. Then it is easy to see that there is a functor F: $K_1 \mapsto K_2$ that is a bijection from the morphisms of K_1 to the morphisms of K_2 and is, thus, an isomorphism of these two categories. Hence, the induced isomorphism described above between the automorphisms of A_1 and B_1 cannot be categorically defined, since it is not preserved by category isomorphisms.[43]

In summary, categorical natural isomorphisms (I,η,G) will determine invariant isomorphisms when G *happens* to be the set-theoretically definable functor which assigns to any morphism ϕ the corresponding induced function. For this reason, the category-theoretic definition does, indeed, capture the standard examples of natural isomorphisms. In general, however, for *any* function η that assigns *any* isomorphism to each structure (or object) in a category, there will always exist a functor G making (I,η,G) a natural isomorphism in the categorical sense. Furthermore, the categorical approach, by restricting consideration to object morphisms exclusively, necessarily fails to distinguish functors G that yield induced mappings from those that do not. Hence, while category theory may be successful in providing general characterizations of many constructions frequently encountered in various areas of mathematics, it does not provide an accurate account of natural isomorphisms. While isomorphisms that are "natural" in the intuitive sense are categorically natural, categorically "natural" isomorphisms are often intuitively perverse. Invariant isomorphisms (and, certainly, freely invariant isomorphisms) do not seem to suffer from this defect.

2.6. Naturally Generated Structures and Structural Equivalence

In this section, we at last come to a solution to the problem of defining a nontrivial equivalence relation which may equate structures even if their universes differ in cardinality. Just as the choice of distinguished structors on a universe is not unique, so, too, we shall see, for the choice of its universe. As motivation for the developments to follow, a simple example will be presented in some detail.

Consider the four points a,b,c,d arranged in a unit square in figure 2. (Ignore the lines for now.)

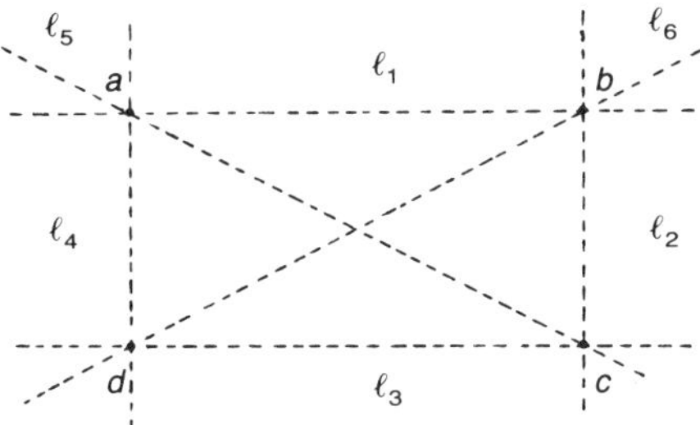

Figure 2

We may now consider these points as the elements of a universe A = {a,b,c,d} of a structure *A*, whose automorphisms are just the group G of all symmetries of the square, that is, all permutations of A which leave the geometrical configuration of the square intact. Intuitively, these are the four rotations of the figure about its center (thus, 1_A is just the 360° rotation), the two reflections about each diagonal pair, and the two reflections about the midpoints of its opposite sides—eight permutations in all. Thus, G = Auto(*A*) is just $\{\sigma_0,\sigma_1, \ldots ,\sigma_7\}$, where

$$\sigma_0 = 1_A \qquad \sigma_4 = \begin{pmatrix} a,b,c,d \\ a,d,c,b \end{pmatrix}$$

$$\sigma_1 = \begin{pmatrix} a,b,c,d \\ b,c,d,a \end{pmatrix} \qquad \sigma_5 = \begin{pmatrix} a,b,c,d \\ c,b,a,d \end{pmatrix}$$

$$\sigma_2 = \begin{pmatrix} a,b,c,d \\ c,d,a,b \end{pmatrix} \qquad \sigma_6 = \begin{pmatrix} a,b,c,d \\ b,a,d,c \end{pmatrix}$$

$$\sigma_3 = \begin{pmatrix} a,b,c,d \\ d,a,b,c \end{pmatrix} \qquad \sigma_7 = \begin{pmatrix} a,b,c,d \\ d,c,b,a \end{pmatrix}$$

Here, $\sigma_0, \ldots, \sigma_3$ are the rotations, σ_4 and σ_5 the diagonal reflections, while σ_6 and σ_7 are the midline reflections. The group G is known as the *dihedral group* D_4 (or the dihedral group of order 8 of a square). Using the abbreviated notation introduced in section 2.1, the structure *A* is equal to (A;G). In particular, we shall take $A = (A, \{\sigma[<a,b,c,d>]: \sigma \in G\})$ which, by the proof of Proposition 2.11, will have G as its automorphisms.

Using figure 2 as a guide, we now see that *A* determines six "lines," each corresponding to a pair of distinct points. We label them as follows.

$$\ell_1 = \{a,b\},\ \ell_2 = \{b,c\},\ \ell_3 = \{c,d\},$$
$$\ell_4 = \{a,d\},\ \ell_5 = \{a,c\},\ \ell_6 = \{b,d\}.$$

For obvious reasons, the pairs $\ell_1, \ldots, \ell_4$ will be called "sides," the pairs ℓ_5 and ℓ_6, "diagonals." Since these lines are structures of A, the set $B = \{\ell_1, \ell_2, \ldots, \ell_6\}$ of all such is also a structor of A, and even more—as running through the group G reveals—B is an invariant of *A*.

Furthermore, notice that each automorphism σ_i of *A* induces a unique permutation of B. For example, σ_2 induces

$$\sigma_2^* = \begin{pmatrix} \ell_1,\ell_2,\ell_3,\ell_4,\ell_5,\ell_6 \\ \ell_3,\ell_4,\ell_1,\ell_2,\ell_5,\ell_6 \end{pmatrix},$$

as can easily be checked using σ_2 and the definitions of the lines ℓ_I given above. (Geometrically, σ_2 rotates our points 180° clockwise; σ_2^*rotates the "lines" in the figure in the same way.) Similar computations show that distinct σ_i yield distinct σ_i^*.

We now may form a new structure $B = (B;G^*)$, where G* is just the set of all such induced permutations of B. In particular, let $B = (B, \{\sigma^*[<\ell_1, \ldots, \ell_6>]: \sigma^* \in G^*\})$. Since $\sigma^*[<\ell_1, \ldots, \ell_6>]$ is just $\sigma[<\ell_1, \ldots, \ell_6>]$, we can now easily show that *B* is an invariant of *A;* intuitively, *A* determines *B*.

But we can now also show that, in a sense, *B* "determines" *A*. Of course, since *A* is not a structor of *B,* we cannot hope to show that *A* is an invariant of *B,* so we do the next best thing—we show that *there is an invariant A′ of B which is naturally isomorphic to B*.

There are, in fact, many such candidates for *A′*, but perhaps the most obvious choice is to let our new "points" just be the set of all lines through these points. In other words, we let:

$$a' = \{\ell_1,\ell_4,\ell_5\}, \qquad b' = \{\ell_1,\ell_2,\ell_6\},$$

$$c' = \{\ell_2,\ell_3,\ell_5\}, \qquad d' = \{\ell_3,\ell_4,\ell_6\}.$$

The universe of our new structure A' is then defined by $A' = \{a',b',c',d'\}$.

Checking the automorphisms G^* of B, we now find that each induces a permutation σ^{**} on A', and, in fact, for all $x \in A$,

$$\sigma_i(x) = \sigma_i^{**}(x').$$

Calling the set of all such permutations G^{**}, we may now let $A' = (A';G^{**})$; in particular, let $A' = (A',\{\sigma^{**}[<a',b',c',d'>]: \sigma^{**} \in G^{**})$. As before, $B \mapsto A'$, that is, A' is an invariant of B, and furthermore

$$\eta = \{<x,x'>: x \in A\}$$

is an invariant isomorphism form A to A'. (Again, this is easily checked.)

Finally, if we take the η-image of B, and call this B', then it turns out that η thus supplies another invariant isomorphism η^* from B to B'. Here, when $b \in B$, $\eta^*(b)$ is defined as just $\eta[b]$—the η-image of B.

All in all, we have an alternating chain of natural copies, shown in figure 3, which can be extended indefinitely. (For the sake of clarifying his intuitions, the reader might well compute η^* and so construct B^* in the above example.)

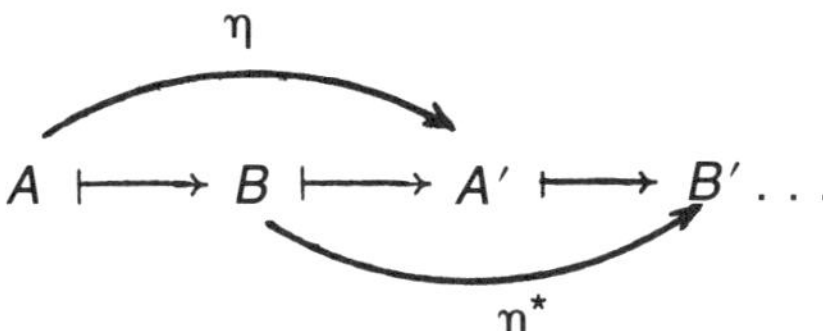

Figure 3

To summarize: Beginning with a structure A, we proceeded to find a structure B which was an invariant of A. Then from B we constructed a structure A', which was, in turn, an invariant of B; in addition, however, there existed a natural isomorphism η from A to A', which also could be used to construct a natural copy of B, which was an invariant of A'. Notice that it follows from this last point that it makes little difference whether we begin the process with A—as we have done here—or use lines B (and G^*) as our starting point. The chain would still remain essentially the same—a structure yielding another which generates a natural copy of the first, and so forth.[44]

In such situations we shall say that the structure A *naturally generates* B and write $A \overset{n}{\mapsto} B$. This notion generalizes natural (invariant) isomorphisms by allowing A and B to have universes of different cardinality, and is intended as a precise explication of a situation that is common in mathematical practice, especially in geometry. There, it is well known that the choice of what is "basic" in a geometry is conventional. We may begin with points, then define lines as subsets of these, or on the other hand, begin with lines and treat points as subsets of lines. The result is "the same" in either case. Similarly, Whitehead (41), taking extended spatial regions as primitive, constructs points as certain sets of overlapping regions. And a category (as we have seen in the last section) may be alternatively considered as having a universe containing two distinct subsets, objects and morphisms, or as having only morphisms in its universe, with objects identified with identity morphisms. With these examples in mind, let us begin the formal development of these ideas, starting with the definition of natural generation.

DEFINITION 2.50. *Let A and B be structures. Then A naturally generates B* ($A \overset{n}{\mapsto} B$) *just in case:* (*i*) *B is an invariant of A* ($A \mapsto B$) and (*ii*) *there is a structure A′ and a function* η *such that:* (a)η *is an invariant isomorphism from A to A′* (η: $A \simeq n\ A'$), *A′ is an invariant of B* ($B \mapsto A'$), *and* η^* *is an invariant of B* ($B \mapsto \eta^*$), *where* η^* *is the* η*-image function on* B *defined by:* $\eta^*(\mathrm{b}) = \eta[\mathrm{b}]$, *all* $\mathrm{b} \in \mathrm{B}$.

Suppose that A is invariantly isomorphic to B. Then, where η: $A \simeq n\ B$, we simply choose B as A' in the above definition and show that $B \mapsto \eta^*$ to obtain:

PROPOSITION 2.51. *If* $A \simeq n\ B$, *then* $A \overset{n}{\mapsto} B$.

An important feature of the earlier example of naturally generated structures was the fact that the automorphisms of B were just those permutations induced on its universe B by the automorphisms of A. Let us be quite clear about what is involved here, since induced automorphisms are crucial to later developments.

Suppose that set B is an invariant of a structor A, that is, $A \mapsto B$. Then, if σ is an automorphism of A, $\sigma[\mathrm{B}]$ is just B, and so σ must take every member of B to some other member as its σ-image; furthermore, if $\mathrm{b} \in \mathrm{B}$, since σ^{-1} is also an automorphism of A, $\sigma[\sigma^{-1}[\mathrm{b}]]$ is just b, so every member of B is in the range of the induced mapping. Finally, if $\sigma[\mathrm{b}_1] = \sigma[\mathrm{b}_2]$, then by Lemma 2.9 $\mathrm{b}_1 = \mathrm{b}_2$; hence, the induced mapping is a bijection from B to B, that is, the induced mapping is in S_{B}—the set of all permutations of B. Hence, if we define $\rho_{A,B}$ as the mapping that takes each automorphism of B to the permutation it induces of B, in other words,

DEFINITION 2.52. *Where* B *is an invariant of* A, *and* $\sigma \in \mathrm{Auto}(A)$, *then* $\rho_{A,B}$ *is defined as:*

$$\rho_{A,B}(\sigma) = \{<b,\sigma[b]>: b \in B\},$$

that is, for all b $\in$ B,

$$\rho_{A,B}(\sigma)\ (b) = \sigma[b],$$

then we have just shown that $\rho_{A,B}$: Auto(A) $\rightarrow$ S_B, *that is,* ρ *is a function from* Auto(A) *to* S_B.

Next, suppose that B is the universe of a structure B that is an invariant of A. Then, since B is an invariant of A, if σ is an automorphism of A, $\sigma[B]$ is just B, and so $\rho_{A,B}(\sigma)$—the induced permutation of B—will also be an automorphism of B (from Lemma 2.7, $\sigma[B] = \rho_{A,B}(\sigma)\ [B]$). But the automorphisms of B need not all be so induced (take A not trivial and B = (A;S_A), for example), or, even if they are, the same automorphism of B may be induced by distinct automorphisms of A. (In other words, $\rho_{A,B}$ is in general, a *homomorphism* of Auto(A) and Auto(B); incidentally, it is easy to show that it is a *natural,* in other words, invariant, *homomorphism,* that is $A \mapsto \rho_{A,B}$.) When, however, $\rho_{A,B}$ is a bijection from Auto(A) to Auto(B), we shall say that the automorphisms of B are bi-induced (short for bijectively induced) by A. As in the previous extended example, it turns out that if A naturally generates B, then the automorphisms of B are *just* the A-induced automorphisms, that is, we have:

PROPOSITION 2.53. *If* $A \overset{n}{\mapsto} B$, *then* $\rho_{A,B}$: Auto(A) $\rightarrow$ Auto(B) *is a bijection, and is also an invariant isomorphism of Sym*(A) and Sym(B), *the symmetry groups of A and B, respectively.*

From this result—which requires some labor (see the appendix)—a number of basic properties of $\overset{n}{\mapsto}$ follow easily. Thus, $\overset{n}{\mapsto}$, like $\simeq$n, is reflexive, but clearly not symmetrical. (Transitivity follows from Proposition 2.56 below.) Furthermore, when $A \overset{n}{\mapsto} B$, a structor of both A and B is an invariant of A just in case it is an invariant of B. These results are summarized in the following proposition.

PROPOSITION 2.54. *Where A, B, and C are structures,*
(*i*) $A \overset{n}{\mapsto} A$,
(*ii*) *When* $A \overset{n}{\mapsto} B$ *and* S *is a structor of both A and B,* S *is an invariant of A iff* S *is an invariant of B.*

Furthermore, counterparts of Proposition 2.44 and Corollary 2.45 of the last section are easy to establish (exercise); for example, A is rigid just in case A naturally generates some pure structure B.

We are now in a position to generalize the relation of invariant equivalence of section 2.2 in the promised way. The basic idea is simple. Since a structure A is here regarded as "the same" structure as those that it naturally generates, we say that B has the same structure as A just in case B is invariantly equivalent to some structure naturally generated by A.

DEFINITION 2.55. *A is structurally equivalent to B* ($A \approx s\ B$) *if and only if there is a structure B′ such that A naturally generates B′* ($A \overset{n}{\mapsto} B'$) *and B is invariantly equivalent to B′* ($B \simeq i\ B'$).

The natural thing to do at this point is to at once show that $\approx s$ is indeed an equivalence relation on structures. However, this will be easier after we have answered a crucial question: Is the converse of Proposition 2.53 above true? In other words, does the fact that the automorphisms of *B* are bi-induced by *A suffice* to show that *A* naturally generates *B*? If so, we are spared the task of finding a suitable A' naturally isomorphic to *A*.

The earlier results on natural isomorphisms (Proposition 2.46 and Lemma 2.47) were preparations for a positive answer to this question. For suppose *B* is a structor of *A* and the automorphisms of *B* are just those bi-induced by *A*. Let B^{ν} be a well-ordering of B. Then, if $\sigma \in \text{Auto}(A)$, $\sigma[B^{\nu}] = B^{\nu}$ implies that $\sigma = 1_A$. Hence, B^{ν}—being a structor of A (as well as a structor of B)—qualifies as S in Lemma 2.47, and an application of Proposition 2.46 yields η: $A \simeq n[A']$, where $A' = \eta[A]$. It only remains then to show that $B \mapsto A'$ and $B \mapsto \eta^*$. The details are given in the proof of the following proposition.

PROPOSITION 2.56. $A \overset{n}{\mapsto} B$ *iff B is a structor of A and the automorphisms of B are bi-induced by A, that is,* $\rho_{A,B}$ *is a bijection from* Auto(*A*) *to* Auto(*B*).

Of course, this result shows that we could have adopted a much simpler definition of $\overset{n}{\mapsto}$ to begin with. But then we would not have made the ability of *B* to generate a naturally isomorphic copy of *A* explicit. Somewhere this fact should emerge, either in the definition or—in this case—as a theorem.

Earlier, we defined codeterminate structures as those having the same universe and, in effect, the same invariants. We also saw (Corollary 2.43 and the discussion that followed) that, while naturally isomorphic structures having the same universe were codeterminate, the converse of this result failed, and why. Naturally generated structures with the same universe, however, *are* just codeterminate structures, as the proof of the following proposition shows.

PROPOSITION 2.57. *When structures A and B have the same universe, A is codeterminate with B iff A naturally generates B (and iff B naturally generates A).*

Proof. By Proposition 2.31 (ii) and Proposition 2.56, when A = B, $A \approx B$ iff $\text{Auto}(A) = \text{Auto}(B)$ iff $\rho_{A,B} = 1_{\text{Auto}(A)}$ iff $A \overset{n}{\mapsto} B$. (Similarly for $B \overset{n}{\mapsto} A$.)

Finally, we ask: Is there a simple test for determining just when *A* and *B* are structurally equivalent? As a start, suppose they are, and that $A \overset{n}{\mapsto} B'$ and $B \simeq i\ B'$. Since $\rho_{A,B'}$ is a group isomorphism of Auto (*A*) and Auto(B'), we have at once that Sym(A) = <Auto(A), $\circ$> is isomorphic to Sym(B') = <Auto(B'), $\circ$>. Since B and B' are invariantly equivalent, by Proposition 2.22 we also have Sym(B') $\simeq$ Sym(B). Hence, Sym(A) $\simeq$ Sym(B), that is, A and B have isomorphic symmetry groups.

Remarkably, the converse is true as well, but the proof is somewhat more difficult. What needs to be shown is that if Sym(A) and Sym(B) are isomorphic, then there exists a structure B', naturally generated by A, and invariantly equivalent to B. Since the proof is tedious, and the basic construction somewhat complicated, what is involved will only be illustrated by continuing with our point-line example.

As before, let A = (A;G) where A = {a,b,c,d} and G = $\{\sigma_0, \ldots, \sigma_7\}$ as illustrated by figure 2. Now consider the structure B = $(\{\ell_1, \ldots, \ell_6\};G^*)$, where we now consider the elements ℓ_i of B as individuals, and G* = $\{\tau_0, \ldots, \tau_7\}$ given by:

$$\tau_0 = 1_B \qquad \tau_4 = \begin{pmatrix} \ell_1,\ell_2,\ell_3,\ell_4,\ell_5,\ell_6 \\ \ell_4,\ell_3,\ell_2,\ell_1,\ell_5,\ell_6 \end{pmatrix}$$

$$\tau_1 = \begin{pmatrix} \ell_1,\ell_2,\ell_3,\ell_4,\ell_5,\ell_6 \\ \ell_2,\ell_3,\ell_4,\ell_1,\ell_6,\ell_5 \end{pmatrix} \qquad \tau_5 = \begin{pmatrix} \ell_1,\ell_2,\ell_3,\ell_4,\ell_5,\ell_6 \\ \ell_2,\ell_1,\ell_4,\ell_3,\ell_5,\ell_6 \end{pmatrix}$$

$$\tau_2 = \begin{pmatrix} \ell_1,\ell_2,\ell_3,\ell_4,\ell_5,\ell_6 \\ \ell_3,\ell_4,\ell_1,\ell_2,\ell_5,\ell_6 \end{pmatrix} \qquad \tau_6 = \begin{pmatrix} \ell_1,\ell_2,\ell_3,\ell_4,\ell_5,\ell_6 \\ \ell_1,\ell_4,\ell_3,\ell_2,\ell_6,\ell_5 \end{pmatrix}$$

$$\tau_3 = \begin{pmatrix} \ell_1,\ell_2,\ell_3,\ell_4,\ell_5,\ell_6 \\ \ell_4,\ell_1,\ell_2,\ell_3,\ell_6,\ell_5 \end{pmatrix} \qquad \tau_7 = \begin{pmatrix} \ell_1,\ell_2,\ell_3,\ell_4,\ell_5,\ell_6 \\ \ell_3,\ell_2,\ell_1,\ell_4,\ell_6,\ell_5 \end{pmatrix}$$

It is easy to check that $\phi(\sigma_i) \equiv \tau_i$ is an isomorphism of <Auto(A),$\circ$> and <Auto(B),$\circ$>.

Next, we forget about intuitive interpretations of A and B, and relying only on the fact that ϕ: Sym(A) $\simeq$ Sym(B), construct a suitable B' from A so that $A \overset{n}{\mapsto} B'$, and <B$'$, Auto($B'$)> is isomorphic to <B, Auto(B)>. It will then follow by Proposition 2.33 (ii) that $B \simeq$i B', and so A and B are structurally equivalent.

The actual construction involves choosing representatives of each orbit of B, finding the left cosets of the stability group of each representative, pulling these back to Auto(A) by using an isomorphism of Sym(A) and Sym(B), and finally, using an arbitrary well-ordering of A together with the retrieved cosets to define counterparts to the elements of B that are structors of A. The details are given in the proof of the following proposition.

PROPOSITION 2.58. *Let* Sym(A) *and* Sym(B) *be isomorphic. Then there is a structure B' such that $A \overset{n}{\mapsto} B'$ and $B \simeq$i B'.*

Hence, immediately:

COROLLARY 2.59. *A is structurally equivalent to B* ($A \approx_s B$) *iff A and B have isomorphic symmetry groups* Sym(*A*) Sym(*B*).

In the end, then, we see that it is only the structure of a (pre)structure's symmetry group that is essential. Again, we could have begun here, using isomorphism of symmetry groups as the criterion of structural equivalence at the outset. But we would still want to know what such an equivalence implied: the ability of each structure to naturally generate an invariantly equivalent copy of the other.

Incidentally, we have not established thus far that $\approx$s is an equivalence relation. Now is a good time to do so, since the last result makes the proof an easy exercise.

PROPOSITION 2.60. $\approx$s *is reflexive, symmetrical, and transitive, in short, an equivalence relation.*

Proof. At once from Corollary 2.59.

Just as natural generation generalizes the notion of codeterminate structures (Proposition 2.57), so too for structural equivalence and invariant equivalence.

PROPOSITION. 2.61. *If* $A \simeq$i B, *then* $A \approx$s B.

Proof. By Proposition 2.22, if $A \simeq$i B, then Sym(A) $\simeq$ Sym(B), and so by Corollary 2.59, $A \approx$s B.

Do structurally equivalent structures fail to be invariantly equivalent merely by virtue of having universes of different cardinalities? If so, we should have the partial converse to the above: If $A \approx$s B and $|A| = |B|$, then $A \simeq$i B. This statement, however, is false. A counterexample is provided by the structures A and A' following Proposition 2.22. There, it was shown that A and A' are not invariantly equivalent, while they have the same universe and isomorphic symmetry groups. This last fact shows, by Corollary 2.59, that A and A' *are,* however, structurally equivalent.

One moral should be clear: universes should not be taken too seriously, at least for structure's sake. Hence, philosophical arguments against scientific realism, such as those recently put forward by Nelson Goodman (15), which point to the conventionality involved in such choices of a universe, merely show the inadequacy of isomorphism as the criterion of structural equivalence.

3. ALTERNATIVE GEOMETRIES

3.1. Isomorphic Geometries and Multiple Concretion

Any permutation of a structure's universe will yield as its image a structure isomorphic to the original and having the same universe. This fact, although trivial from a set-theoretical standpoint, has important implications when specific physically realized structures are considered. In this section, the physical geometry of space will be chosen as an illustration; many of the results, however, are applicable to all physically realized structures.

For reasons that will soon be clear, the above principle will be called the principle of *multiple concretion;* more precisely, it asserts:

Multiple Concretion. Where A is a structure and τ is a permutation of A, then the τ-image of A, namely $A' = \tau[A]$, is isomorphic to A, and hence invariantly equivalent to A.

The truth of this assertion follows at once from the definitions of the notions involved. It also immediately follows that a structure with universe A will have distinct isomorphic copies with universe A when that universe has permutations that are not automorphisms. Or, put a bit differently, fixing the universe A of a structure of a given type $A = (\mathrm{A}, \mathrm{R}^\beta)$ still permits the realization of that structure in, roughly, as many ways as there are nonautomorphic permutations of A. If the universe is then a physically interpreted set (spatial points or events, say), the structors on that universe that will yield a structure of a given (isomorphism) type may be physically realized—or "concretized"—in a wide variety of ways. Hence, the choice of "multiple concretion" as the name of this principle.[45]

Now consider a set P of physical spatial points, that is, a physical space. What are the implications of the claim that the geometry of this space of Euclidean? At the very least, this would seem to imply the existence of a physically significant metric d, realized by a set of "measurement procedures" in some broad sense of that term, and such that the structure $E = (\mathrm{E}, \mathrm{d})$ is a Euclidean space. Thus, d must be a metric on P, externally convex, and so forth. (An axiomatic characterization of Euclidean spaces as metric spaces may be found in Blumenthal (6, §7.6). A definition of Euclidean spaces that will do for our purposes is the following:

DEFINITION 3.1. $E = (\mathrm{P}, \mathrm{d})$ *is an* n*-dimensional Euclidean space if and only if:*

(*i*) P *is a nonempty set,*

(*ii*) d *is a function from* $\mathrm{P} \times \mathrm{P}$ *to the nonnegative real numbers, and*

(*iii*) *there is a bijection* X *from* P *to* R^n *such that for any elements* p_1, p_2 *of* P, $d(p_1,p_2) = [(x^1(p_1) - x^1(p_2))^2 + \ldots + (x^n(p_1) - x^n(p_2))^2]^{1/2}$, *where* $x^i(p) = \text{df. } u^i(X(p))$, *the* i*-component of the* n*-tuple* X(p).

Any bijection Y from P to R^n will be called a *coordinate system* of E, and when Y satisfies condition (iii) above, Y is said to be a *Cartesian* coordinate system.

Suppose now that d_1, empirically interpreted, results in $E_1 = (\mathrm{P}, d_1)$ being a Euclidean space, say, a plane. Knowing *just this* about d_1, what empirical predictions are we now entitled to make concerning the actual outcome of applying d_1 to specific points of P? In particular, can we now assert that any particular point-pairs (p_1,p_2), (p_3,p_4) are—or fail to be—congruent in *E?*

As Grünbaum has pointed out (17, p. 98), Reichenbach apparently believed that the congruence of point-pairs in P was determined by the existence of a physically specifiable d_1 which yielded $E = (P,d_1)$ Euclidean. But a simple argument, based upon the "truism" stated above, shows that this is not so. Let $E_1 = (P,d_1)$ be a (physically interpreted) Euclidean plane. Suppose that the pairs:

(A) $$(p_1,p_2),\ (p_3,p_4)$$

are congruent in E_1 (i.e., $d_1(p_1,p_2) = d_1(p_3,p_4)$), whereas the pairs:

(B) $$(q_1,q_2),\ (q_3,q_4)$$

fail to be congruent in E_1. (Assume throughout that no pair is degenerate.) Next, let τ be one of the many nonautomorphic permutations of P such that

(C) $$\tau(p_i) = q_i,\quad i = 1,2,3,4.$$

Finally, we let $E_2 = (P,d_2)$ be simply the τ-image of E_1, that is, $E_2 = (P, \tau[d_1])$.

By multiple concretion, E_2 is isomorphic to E_1 and so is also a Euclidean space. However, the pair (B), which failed to be congruent in E_1, is now congruent in E_2. To see this, notice that in general,

$$d_2(n_1,n_2) = d_1(\tau^{-1}(n_1), \tau^{-1}(n_2)). \tag{3.1}$$

Hence, at once from (C),

$$d_2(q_1,q_2) = d_2(q_3,q_4).$$

Recall that d_2 as well as d_1 yields a Euclidean geometry on P. Hence, since d_1 and d_2 differ in their congruence verdicts, the mere fact that a function d yields a Euclidean geometry on P fails to imply the congruence of *any particular* point-pairs of P.

It is a simple matter to illustrate the above situation graphically. In figure 4, x^1 and x^2 represent Cartesian coordinates for the Euclidean space $E_1 = (P,d_1)$, whereas Y^1 and Y^2 represent the E^{-1}-images of x^1 and X^2, where we have—generously—let τ^{-1} be a nonautomorphic bicontinuous (in E_1) transformation of P.

The first thing to notice is that Y, although clearly not a Cartesian coordinate system of E_1, *is* a Cartesian coordinate system of E_2. To see this, recall that

$$y^i = \tau[x^i] = x^i \circ \tau^{-1},$$

and so

$$x^i = y^i \circ \tau. \tag{3.2}$$

From (3.1) and Definition 3.1 (iii),

$$d_2(r_1,r_2) = [(\Sigma(x^i(\tau^{-1}(r_1)) - x^i(\tau^{-1}(r_2)))^2]^{1/2},$$

and now substituting, using (3.2), we obtain

$$d_2(n_1,n_2) = (\Sigma((y^i(r_1) - y^i(r_2))^2)^{1/2},$$

showing that Y is a Cartesian coordinate system of E_2.

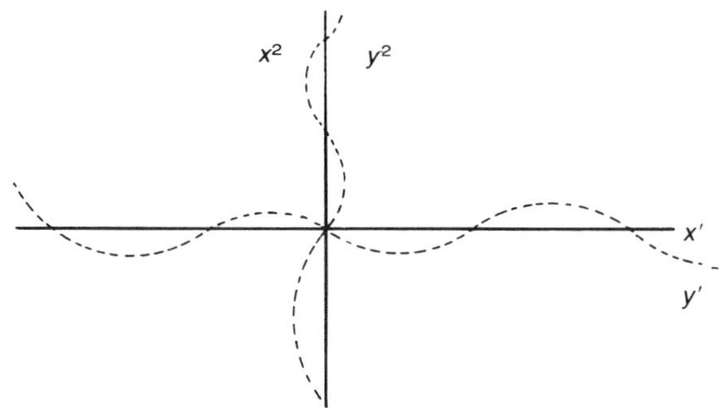

Figure 4

Having shown this, it is easy to see from figure 4 that each of the two Euclidean metrics d_1 and d_2 "squirms" with respect to the other. Our choice of a continuous transformation τ results in the two spaces having the same Euclidean topology,[46] but differing in their verdicts on linearity, parallelism, and congruence.

The principle of multiple concretion also comes in a coordinate version, at first sight not quite so obvious. Let P, as before, be a fixed set, and let the structors of P be represented as $\overline{R^\beta}$ in some coordinate system ϕ: P → B. Then if ψ: P → B is any other coordinate system that represents the structors S^β as $\overline{R^\beta}$ also, the two structures (P,R^β) and (P,S^β) must be isomorphic. More precisely:

Multiple Concretion (coordinate version). Let ϕ: P→ B and ψ: P→ B be coordinate systems of the structures $A = (P,R^\beta)$ and $A' = (P,S^\beta)$, respectively. Then if $\phi[R^\beta] = \psi[S^\beta]$, A and A' are isomorphic (and so invariantly equivalent).

This result follows at once from the fact that, since $\psi[S^\beta] = \phi[R^\beta]$, $S^\beta = (\psi^{-1} \circ \phi)[R^\beta]$. Hence, since $\psi^{-1} \circ \phi$ is a permutation of P, multiple concretion (the direct version) yields the result at once.

For example, let $E = (P,d)$ be a Euclidean plane with X a Cartesian coordinate system of E. If we now transform X to a coordinate system Y, where

$$\begin{aligned} x^1(p) &= ky^1(p) \quad (k \neq 0,1) \\ x^2(p) &= y^2(p), \end{aligned} \tag{3.3}$$

and now substitute in Definition 3.1 (iii), we find that *the same metric* d is now given in terms of the y-coordinates by

$$d^2(p,q) = k^2(\Delta y^1(p,q))^2 + (\Delta y^2(p,q))^2. \tag{3.4}$$

So, clearly, the existence of a function Y wherein a structor d^2 of P takes form (4) above, yields a Euclidean plane—in fact, the *same* plane E = (P,d) with which we began.

But now let us consider a metric d_* which is represented *in coordinate system* X in exactly the way that d is represented *in coordinate system* Y; that is, d_*^2 is given by:

$$d_*^2(p,q) = k^2 (\Delta x^1(p,q))^2 + (\Delta x^2(p,q))^2. \tag{3.5}$$

First, notice that the metrics d_* and d are not identical, and even differ on their congruence verdicts. For example, if X(p) = (0,0), X(q) = (0,1), and X(r) = (1,0), then the pairs (p,q) and (p,r) are congruent according to metric d, while $d_*(p,q) = 1$ and $d_*(p,n) = k \neq 1$. Nevertheless, the coordinate version of multiple concretion assures us that E_* = (P,d_*) is isomorphic to E = (p,d) and, hence, is also a Euclidean plane. This follows immediately from the fact that Y represents d in exactly the same way that X represents d_*. This is not *merely* to say that (3.4) and (3.5) exhibit the same functional form of their respective coordinates, but that *because* of this, they represent their metrics in *exactly the same way:* The images of d and d_* (in $R^2 \times R^2 \times R$) are *identical.* Thus, by working through the definitions of the various notions involved, it is easy to show that

$$Y[d] = \{<(a_1,a_2), (b_1,b_2)m>: a_i,b_i,m \in R, \text{ and } m^2 = k^2(a_1-b_1)^2 + (a_2-b_2)^2\},$$

and the same is true for X[d_*]. Hence, the coordinate version of multiple concretion is applicable, and the result that E_* is Euclidean follows immediately.

Since Y[d] = X[d_*], we have at once $d_* = \tau[d]$, where $\tau \equiv X^{-1} \circ Y$ is the permutation of P represented (by either X or Y) in R^2 by $X^{-1} \circ Y$.[47] Hence, as should now be clear, the specification of the new metric d_* by the method of coordinate transformations is the exact counterpart of defining such a metric as the image of the original under a suitable permutation of the underlying set of points P.

As far as I know, the first to use the coordinate transformation method to demonstrate the existence of alternative Euclidean metrics was Grünbaum (17, pp. 98–100). In the context of Riemannian structures, he considers a structure E_R = (P,*D*,g), where (P,*D*) is a two-dimensional differentiable manifold having a globally Euclidean topology, and the metric tensor g is represented in some region by a local coordinate system X by

$$ds^2 = dx^2 + dy^2.$$

By a suitable coordinate transformation, the same metric ds^2 may be represented in a coordinate system Y by

$$ds^2 = k^2(dx)^2 + (dy)^2.$$

Returning now to system X, Grünbaum asks that we consider the metric

$$ds_1{}^2 = kdx^2 + dy^2,$$

observes that $ds_1{}^2$ and ds^2 yield incompatible congruence verdicts on the region of P under consideration and then goes on to argue that $ds_1{}^2$, like ds^2, yields a Euclidean geometry on this region.

As we have seen, this claim is indeed correct, and we have also seen why: $ds_1{}^2$ is just the τ-image of the metric ds^2, where τ is a (local) permutation of that region represented by the transformation from coordinate system X to Y. In this instance, the transformation is smooth (i.e., c^∞ differentiable), and so the differentiable structure *D* is left untouched.

While Grünbaum's claim is correct, his supporting reasons are not, and are in addition a potential source of philosophical confusion. The metric $ds_1{}^2$ is Euclidean, Grünbaum asserts, because the sameness of the functional form of ds^2 in coordinate system Y and $ds_1{}^2$ in system X implies that the Gaussian curvature must be the same (zero) in both cases, since this curvature depends only upon the form of the g_{ik}'s.

There are, however, two things "wrong" with this argument. The first is that its invocation of curvature obscures the reason for the success of the general strategy: the same device works for any coordinated structure, geometry or no, so the reference to Gaussian curvature is gratuitous.

The second problem with the argument is that it creates the impression that, while congruence patterns are susceptible to multiple realizations on a set P, the curvature distribution is immune to such manipulations. However, just as congruence is multiply imaged by any permutation of P, so, too, for curvature. This fact is obscured in the example considered, since the space is Euclidean and so its curvature is zero throughout. But if we were to consider a two-dimensional manifold, like a Euclidean plane save for having a single "bump" somewhere, say in some small region around a point p_o, then the permutation represented by $Y \circ X^{-1}$ could well have the effect of transferring the bump to some region far from p_o, leaving p_o and its immediate surroundings flat. In a sense, of course, the curvature of the space is not "changed," for the alternative curvature distribution is just an isomorphic copy of the original. But *in just this sense,* neither has the congruence of point-pairs changed. Clearly, congruence and curvature fare alike in the general case, and it is only for spaces of constant curvature that the above shift from a ds^2 to a ds_1^2 metric will leave the curvature distribution intact. Hence, the appeal to curvature unnecessarily undermines the generality of Grünbaum's claim, which does not require the restriction to spaces of constant curvature, apparently implied by the use of "each of the" when he generalizes on his Euclidean example: "It is clear that the multiplicity of metrizations which we have proven for Euclidean geometry obtains as well for each of the non-Euclidean geometries" (17, p. 100).

Having established the existence of alternative realizations on a given manifold P of one and the same kind of geometry, we now consider the problem of the choice of one metric out of the many competing alternatives yielding the same geometry.

First, an elementary point. From the purely structural standpoint, there is no way to distinguish a standard—say, Euclidean—metric from one of its nonstandard Euclidean alternatives. And the fact that a metric takes the Pythagorean form in a coordinate system X is of no avail here.[48] For, as we have seen, the coordinate systems X and Y of the incompatible Euclidean metrics (3.4) and (3.5) considered earlier which "squirm" with respect to one another (see figure 4), are each *Cartesian* coordinates of their respective spaces E_1 and E_2. Talk of the "standard" metric thus presupposes the adoption of some generally accepted class of measurement procedures that realize the verdict of that metric. Only with the specifications of some such set do other metrics on P become "nonstandard."

Next, it should be realized that there is no a priori reason to exclude the possibility of multiple "standard" metrics, in the sense that each metric is realized by a distinct set of physically significant measurement procedures. If this were the case, then space would be monogeometric but *multiply* metric.

However, what has aroused most interest is not this version of the possibility of multiple geometries, but the problems posed by the various versions of conventionalism espoused by Poincaré, Reichenbach, and Grünbaum (see Grünbaum 17, chap. 1, for a survey of these views.) Briefly, the issue is this. Suppose that E = (P,d) is Euclidean, with P being a set of physical spatial points and d realized by a set of standard physical measurement procedures. The conventionalist now poses the following question: Are we not also at liberty to remetrize P according to some nonstandard (Euclidean) metric d, with the result that the rigid rods (or devices) that realize the standard metric d are now deemed to "squirm," as judged by the nonstandard metric d?[49]

It is important to be quite clear about what is being asked here. As I shall construe it, the question—when loose but suggestive talk like "we are at liberty" is put aside—asks: *Given the physical significance of the metric* d which yields a Euclidean geometry on P, is not any other metric d′ that also yields a Euclidean geometry on P equally entitled to physical significance? And this last question may be posed even more precisely as: Is the alternative (Euclidean) metric d′ objectively equivalent to the metric d of $E = (\mathrm{P},\mathrm{d})$?

All structures obtained by permuting the universe of a given structure are, of course, isomorphic to the original structure and so are invariantly equivalent to that structure. In other words, all structures obtained by multiple concretion *have the same structure,* but this is not to say that they *are the same structure* in the sense discussed in section 2.4 and 2.6 above.[50]

Since the two structures in question have the same universe, then by Proposition 2.57, they will *naturally* generate each other just in case they are codeterminate ($\approx$) structures, that is, just in case each structure is an invariant of the other.

Let us now suppose that the structure $E = (\mathrm{E},\mathrm{d})$ is a Euclidean plane consisting of a set E of physical points and a metric d delivered by a set of measurement procedures that yield a physical geometry on E. We have seen above that if τ is any permutation of E, then $E' = (\mathrm{E}, \tau[\mathrm{d}])$ is structurally equivalent to E, and so also a Euclidean plane. We now ask, however, the stronger question: Under what conditions will E' be the *same* structure as E, or, more precisely, when will E' be *codeterminate* with E? The existence of Euclidean geometries codeterminate with E, yet differing from E in a nontrivial way, would provide decisive support for geometric conventionalism. If no such interesting alternative geometries are possible, then the existence of alternative isomorphic Euclidean geometries that differ in their congruence verdicts becomes of no more *geometrical* interest than the very general *set-theoretical* fact of *multiple concretion.*

Since the question of when a structure A and $A' = \tau[A]$ are codeterminate is not peculiar to geometric structures, we may first obtain a general solution and then apply it to the Euclidean case later. The answer, in short, is that A and A' will be codeterminate just when the permutation τ is in the normalizer of the group of automorphisms of A. (For a definition of the normalizer of a group, see the discussion preceding Proposition 2.13.) This is what is established in the proof of the following:

PROPOSITION 3.2. *Let* $A' = \tau[A]$, *where* $\tau \in \mathrm{S_A}$. *Then* A *is codeterminate with* A' ($A \approx A'$) *iff* $\tau \in \mathrm{Norm}(\mathrm{Auto}(A))$.

Proof. ($\Rightarrow$) Assume $A \approx A'$. Then by Definition 2.30 (i), $A \mapsto \tau[A]$. Let σ be in Auto (A). Then, from (i), $\sigma[\tau[A]] = \tau[A]$, that is, $\tau^{-1}\sigma\tau[A] = A$. Hence, $\tau^{-1}\sigma\tau \in \mathrm{Auto}(A)$. Thus τ^{-1} commutes with Auto(A).

Since $\tau^{-1}[A'] = A$, $\sigma(\tau^{-1}[A']) = \sigma[A] = A$. Hence, $\tau\sigma\tau^{-1}[A'] =$

$\tau[A] = A'$. Thus, $\tau\sigma\tau^{-1} \in \text{Auto}(A') = \text{Auto}(A)$, by Proposition 2.31. Hence, τ commutes with Auto(A).

Since both τ,τ^{-1}commute with Auto(A), $\tau \in$ Norm(Auto(A)).

($\Leftarrow$) Assume $\tau \in$ Norm(A). Then so is τ^{-1}. Let σ be in Auto(A). Then $\tau^{-1}\sigma\tau \in$ Auto(A), and so $\tau^{-1}\sigma\tau[A] = A$, that is, $\sigma[\tau[A]] = \tau[A]$, in other words, $\sigma[A'] = A'$. Hence, $\sigma \in$ Auto(A'), and so Auto(A) $\leqslant$ Auto(A').

Next, let σ be in Auto(A'). Then $\sigma[A'] = A'$, that is, $\sigma\tau[A] = \tau[A]$, in other words, $\tau^{-1}\sigma\tau \in$ Auto(A). Hence, τ commutes with $\tau^{-1}\sigma\tau$ and so $\tau\tau^{-1}\sigma\tau\tau^{-1} = \sigma \in$ Auto(A). Hence, Auto(A') $\leqslant$ Auto(A). Thus, Auto(A) = Auto(A'), and so, by Proposition 2.31, A $\approx$ A$'$.

Now what does the result amount to, geometrically? In a Euclidean metric geometry (and in all freely mobile geometries), the normalizer of the isometry (automorphism) is just the similarity group of the geometry and may be obtained by merely adding the dilations to the isometries and closing under composition.[51] Hence, if τ in the Euclidean case is not an automorphism, it is at least a similarity and so $\tau[E]$ and E differ at most by a scale factor (i.e., τ[d] and d so differ). As a result, both metrics yield the same congruence verdicts on the point set E. Multiple concretion does not, therefore, furnish us with alternative metrics that differ nontrivially.

In short, for the classical geometries, the *only isomorphic geometries that are codeterminate with the geometry E* = (P,d) *are those whose metrics differ from* d *by a scale factor.* Hence, we are here provided with an excellent reason for not regarding all geometries that are isomorphic to a *given* physical geometry E = (P,d) as thereby equally entitled to physical significance. However, as we shall now see, the restriction to *isomorphic* alternative metrics in the above discussion is crucial, and there are indeed alternative "Euclidean" metrics that differ from a given d on congruence verdicts yet are equally entitled to be considered physically significant geometries.

3.2. Invariantly Equivalent Geometries

Although multiple concretion generates isomorphic codeterminate structures, not all codeterminate structures are isomorphic, even when they share the same universe. Consider a Euclidean plane E = (P,d) and the structure E^* = (P,d*) having the same universe of points P with the new "metric" d* defined for all p_1, p_2 in P by

$$d^*(p_1,p_2) = \ell n(d(p_1,p_2) + 1). \qquad (3.6)$$

Since d* is explicitly definable in terms of d, it follows at once that E^* is an invariant of E. Furthermore, since d is explicitly definable in terms of d* (d = exp(d*) − 1), the structure E is an invariant of E^* as well. Hence, E and E^* are codeterminate structures and, from the present point of view, essentially the same structure.

Now it is easy to show that the structure E^* satisfies the axioms of a *metric* space, that is, $d^*(p,q) = 0$ iff $p = q$, $d^*(p,q) = d^*(q,p)$, and $d^*(p,q) + d^*(q,r) \geqslant d^*(p,r)$. A little calculation, however, shows that the equality part of the last equation *never* holds. (I am indebted to Eugene Pidzarko for pointing this out.) In a Euclidean plane, collinear points do satisfy the equality condition; hence, E^* is not Euclidean. Since all Euclidean planes are isomorphic, it follows at once that E and E^* are not isomorphic. In short, E and E^* *are codeterminate, but not isomorphic, structures.*

So much the worse, then, for axiomatic characterizations of geometries and structures. In this case, since we are dealing with invariants that are interdefinable, it is easy to feel comfortable about the situation. For we need only replace "d" by "exp(d*) − 1" in the usual axioms and *translate* the axiomatic conditions accordingly. Thus, the triangle inequality

$$d(p,q) + d(q,r) \geqslant d(p,r) \tag{3.7}$$

of some axiomatic approaches (say, Blumenthal 5) becomes

$$\exp(d^*(p,q)) + \exp(d^*(q,r)) \geqslant \exp(d^*(p,r)) + 1. \tag{3.8}$$

In the case of Definition 3.1 of the Euclidean plane, the Pythagorean formula for distances changes similarly. The result, of course, is a pair of intertranslatable axiomatizations.

From the invariant viewpoint these axiomatizations differ only in that they characterize different but codeterminate generators of the same structure. Since these generators have quite different structural properties, it is not appropriate to say that the two *axiomatizations* assert the same thing; rather, they specify the same structure by singling out different, but codeterminate generators of that structure.

When codeterminate structures are not interdefinable but merely mutually invariant, then intertranslatability (in some language or other) may well fail. But this should be of little concern at this point. Intertranslatability was merely used here as a linguistic crutch to make the equivalence of E and E^* plausible. Once it becomes clear that d and d* generate the same invariants (since the automorphisms of E and E^* are obviously identical), the failure of E^* to satisfy some standard set of axioms for a Euclidean plane may be regarded as having little structural significance.

The codeterminate structures E and E^*, while not isomorphic, nevertheless agree on their congruence verdicts: $d(p,q) = d(r,s)$ implies $d^*(p,q) = d^*(r,s)$, and conversely. Agreement on congruence, however, is not required by codetermination, as the following (somewhat perverse) example reveals.

Let E = (P,d) be, as before, a Euclidean plane. Let d′ be defined by:

$$d'(p,q) = 1 \text{ iff } d(p,q) = 1, \quad \text{otherwise } d'(p,q) = k_o, \tag{3.9}$$

where k_o is any fixed positive real number not equal to 1. Let E' = (P,d′).

Remarkably, it turns out that E and E' have the same automorphisms and so are codeterminate structures. The important fact behind this result is that any permutation of the Euclidean plane that preserves all unit intervals or, for that matter, all intervals of any fixed (nonzero) length, is an isometry.[52] Clearly, E and E' radically differ in their congruence verdicts, since they agree only on intervals of length k_o and point-intervals. Clearly, then, they fail to be isomorphic.

Suppose, now, that the structure E is physically interpreted, with its d-verdicts realized by the empirical outcomes of a set of measurement procedures which we shall call "d-rods." Then, clearly, we may obtain the metric d′ from these d-rod verdicts by assigning unit d′-lengths when d-rods yield unit intervals, and assigning length k_o otherwise. Conversely, if, as a matter of awkward empirical fact, a set of d′-rods were to exist, we could obtain the unit intervals of the d-metric by direct d′-measurement and then extend these results to arbitrary intervals by the methods of the proof of the theorem on the automorphisms of E referred to in the previous footnote.

In this sense, the choice of metrics d or d′ is a matter of convention: each objectively determines the other. From the structural standpoint, however, this way of describing the situation is not especially apt. It would be more accurate to say that *neither* E nor E' (nor E^*) *is* a Euclidean space, but each of these structures generates the same structure by distinguishing different, but codeterminate aspects of that structure. Linguistically, E is easiest to handle, but, structurally and physically, these systems are on a par. For recall that the very same *physical* procedures that provide a realization of any one of these structures will do the same for the others.

But do not the d-rods "shrink" and "expand" with respect to the d′-rods, as Grünbaum has described such situations? The "aspect" approach outlined above shows, I believe, how such talk can be misleading. The d-rods physically realize the metric d, while d′-rods physically realize metric d′. But d and d′ are not *competing* metrics on P; instead, they *coexist* in the structure generated by E or E'. In general, *given* an acceptable physical realization of a structure A = (A,R^β) we thereby also physically realize the codeterminate structure A' = (A,S^γ). The realization of the invariant structors S^γ—and all other invariant structors—is thus indirectly provided by realizing A, that is, by realizing the structors R^β of A which suffice to generate the structure determined by A.

It is important to note, however, that we do not thereby resolve problems relating to what constitutes an "acceptable" physical realization of A in the first place. The above argument is conditional in form and answers the question: given a physically realized structure A, what other structors are *thereby* colegitimate with A? We do not, however, thereby restrict the scope of these proposals too severely. For it is just within the context of alternatives to an *initially acceptable geometry* that most conventionalistic issues arise. Thus, Grünbaum (17, pp. 18–22) invites us to consider an alternative metrization of (the upper half of) a blackboard which issues in a hyperbolic geometry, and the challenge is to show why such a remetrization is not every bit as physically admissible as its *standard* Euclidean alternative. Thus we are asked to consider a Euclidean plane $K = (\mathrm{P},\mathrm{g})$, where g is a metric tensor field on P expressed in some coordinate system X of P by:

$$\mathrm{g} = \mathrm{dx}^2 + \mathrm{dy}^2. \tag{3.10}$$

Assume further that K is physically interpreted, P is some set of physical points, and the coordinate system X identified as one of many such systems determined by the verdicts of rigid rods used to construct "Cartesian" coordinates in the standard way. So realized, K becomes a physical Euclidean plane.

At this point, Grünbaum now asks that we consider the alternative metrization of the upper half of this plane ("blackboard") given by:

$$\mathrm{g}^* = (\mathrm{dx}^2 + \mathrm{dy}^2)/\mathrm{y}^2,\ \mathrm{y}^2 > 0. \tag{3.11}$$

This metric, he points out, endows the upper half of the plane with a hyperbolic geometry and results in congruence verdicts incompatible with those of g in this region. But the new metric, he claims, does not *empirically* contravene the old in these regions; it is just that, unlike g, it makes the lengths of our original rigid rods a function of their position and orientation in the (upper half of the) plane.

Now, at first sight, the structures $K = (\mathrm{P},\mathrm{g})$ and $K^* = (\mathrm{P},\mathrm{g}^*)$ appear to be in the same relation as E and E^* discussed earlier. Since $\mathrm{g}^* = \mathrm{g}/\mathrm{y}^2$, the two metrics might be considered as interdefinable, hence the structures K and K^* codeterminate.

But this would be incorrect. The relation (3.11) above does not *define* g* in terms of K, since it invokes a *particular* coordinate system X of K. For, recall (see sec. 2.3) that unless a structure happens to be rigid—which K is not—no *single* coordinate system of that structure will be one of its invariants. It is only the entire distinguished class of such systems that will be an invariant—indeed a generator—of that structure. As a result, the metric g* is not an invariant of K, and is even locally altered by the rotations of K.

Nor will it do to consider (3.11) as defining g* in *any* Cartesian coordinate system X, because in that case g* will not be unique and, hence, not well defined.

We can be more precise about this. An examination of the "definition" (3.11) of g* shows that we have, in the guise of the function y, smuggled in an additional scalar field on P. In effect, we have defined a structure $K^+ = (P,d,\phi)$, where ϕ: P → Reals, and in the Cartesian coordinate system (x,y),

$$\phi(p) = (y(p))^2 \text{ when } y(p) > 0, \text{ otherwise } \phi(p) = 1.$$

It is this scalar field ϕ, presumably, which is held to "adjust" the lengths of the rigid rods as a function of their position in the plane. But as Grünbaum has repeatedly emphasized, this "adjustment" is not to be taken as the literal action of a physically significant field of some sort (as some have read Poincaré and Reichenbach), but rather the introduction of such fields or forces is to be considered merely as a metaphorical way of referring to the possibility of using the above "definition" of g* as an alternate "coordinative definition" of spatial congruence. We have already seen that (3.11) is *not* an explicit definition of g* with respect to the Euclidean space *K*—or for that matter, with respect to any linguistic formulation of the theory of *K*. But could (3.11) be considered an alternative *coordinative* definition of g* on P? In order to answer this question, a few remarks on coordinate definitions are in order.

First of all, a coordinative definition, for Reichenbach, Salmon, and others, is merely an assignment of physical meaning to a previously unintepreted term in some formal calculus. In our terms, this would amount to providing a structure—say $K = (P,g)$ or $K^* = (P,g^*)$—with a physical realization by identifying (in the case of *K*, say) the sets P and g with physical objects, relations, or operations. In our example, the set P is coordinated with the points of a blackboard, and Cartesian coordinates are interpreted as those constructed in the usual way, using a specific set of measurement procedures—our customary "rigid rods."

Now with coordinative definitions so construed, admittedly, (3.11) provides a metric g* that is not identical to g and, furthermore, differs from g in geometrically nontrivial ways. But this is not to admit anything of *geometrical* importance. For let the books on a shelf be arranged in the left-to-right order a,b,c,d, yielding the strict ordering $B = (\{a,b,c,d\},\ll)$, where $\ll$ is the order in which I have listed the set. Now label the books arbitrarily from one to four, say, by

$$X = \begin{pmatrix} a & b & c & d \\ 3 & 1 & 4 & 2 \end{pmatrix},$$

and "define" $\ll'$ by

$$\ll'(p,q) \text{ iff } X(p) < X(q).^{53} \qquad (3.12)$$

The result, clearly, is another strict ordering of the same books, and an ordering that differs from the original in nontrivial ways. To say this, however, is not to say that the way in which the new ordering was obtained is not trivial. Having fixed the universe B = {a,b,c,d}, the arbitrary choice of the coordinate assignment X allowed us to pull back a strict ordering onto our books from the standard strict ordering of the first four integers.

And, of course, the device always works. Coordinatize a physical domain of a suitable cardinality and pull back standard mathematical structures to that domain, and the result is inevitably a physical realization of that structure. Structure, as always, is preserved by isomorphisms.

But what of the fact that in the Euclidean case and other examples cited by conventionalists, the topology and manifold structure of our points is left intact? Does this not show that the existence of alternative metrics is not trivial after all?

Consider again the structure $K = (P,D,g)$, a Euclidean plane, say, with D the manifold structure on P. Let us suppose, first of all, that there is *no* alternative metric g^* such that $K^* = (P,D,g^*)$ is also a Euclidean plane. (Notice the manifold structure D, and, hence, the topology, is the same in both cases.) What would now follow about the structure K?

To see, let σ be an automorphism of the reduced structure (P,D) which is *not* an automorphism of the metric g. By multiple concretion then, $\phi[K] = (P,D,\phi[g])$ is also a Euclidean plane, but $\phi[g] \neq g$, contradicting our original assumption that no such alternatives to the Euclidean plane K existed. In other words, the failure of K to admit such alternatives would be tantamount to the geometry of a space being an invariant of its smooth structure. It is thus not surprising, at least nowadays, that topology and smoothness fail to determine a metric geometry.[54]

Similarly, it is not surprising that smoothness alone fails to determine the *class* of all Euclidean metrics. In other words, there are automorphisms of (P,D) that will take some Euclidean metric to a non-Euclidean metric. Again, this is an interesting, but hardly deep, result of contemporary geometry, and it fails to support a geometric conventionalism of physical or philosophical significance.

But now let us return to our shelf of books. There we saw an alternative ordering $\ll'$ and how we obtained it. That ordering was not an invariant of the original order, as can be easily checked, and the two structures $B = (B,\ll)$ and $B' = (B,\ll')$ are not codeterminate. Our coordinative definition (3.12) was a "definition" only in the sense that it singled out one of the many orderings of our books isomorphic to the original. On the other hand, the explicit definition

$$\gg(p,q) \text{ iff not } \ll(p,q) \qquad (3.13)$$

also singles out another ordering of these same books. However, this ordering, the converse of our original, *is* an invariant of the original ordering, and the structures B = (B,<<) and B' = (B,>>) *are* codeterminate. Hence, I suggest, it is legitimate to regard the *same* structure B as *also* ordered by >>, and the two relations << and >> as physically colegitimate. But this is not to say that we are forced to make a choice between *conflicting* orderings. Rather, the shelf of books is at once *both* ordered from left to right *and* from right to left.

So too for the codeterminate geometries. The fact that the Euclidean metric and the "metric" d* (= ℓn(d + 1)) differ in their "grammar" does not imply that we must choose one over the other as being exhibited by physical space. Both metrics, if you will, are "there"; both metrics are colegitimate.

On the other hand, alternatives to a given structure obtained by multiple concretion or "coordinative" definitions will in general *not* be codeterminate with the original structure. That such alternatives are available in the first place, is a matter of general set-theoretical fact and has little to do with properties of special geometric structures such as their curvature or continuity. The hidden truth of conventionalism, I suggest, is its appreciation that a domain characterized merely *structurally* is not thus characterized empirically at all, and in proceeding to do so, we have, within our structural guidelines, many permissible empirical alternatives. I have argued here that the initial choice of one of these alternatives yields a class of others, and these are best regarded as "completing," rather than competing, aspects of the same physical structure.

4. RELATIVIZATION AND REALITY

4.1. Relations and Relativity

In a Galilean space-time (see Geroch 13), the spatial distance between nonsimultaneous events is not an objective matter, but depends upon the frame of reference adopted. Successive events that occur at the same place in one reference frame can be far apart in another frame moving rapidly with respect to the first. Sameness of spatial location, in short, is not an invariant relation in a Galilean space-time, but varies in a systematic way from inertial frame to inertial frame.

This dependence of spatial distance on the frame of reference, while showing the nonobjectivity of spatial distances, also hints that objectivity may be reinstated. Instead of regarding, say, the relation $S(e_1,e_2)$ of sameness of spatial location as a *two-placed* relation between events, we switch over to the *three-placed* relation $S(e_1,e_2,F)$ between event-pairs (e_1,e_2) and (inertial) reference frames (F). We may now regard the expanded relation as objective just in case every automorphism that preserves the frame F also preserves the set of pairs e_1,e_2 in such a way that $S(e_1,e_2,F)$. Objectivity is thus "restored" by regarding the previous two-placed relation as "incomplete," holding or not with respect to previously unmentioned factors that its replacement now sets forth explicitly. What was earlier—say in an Aristotelian space-time—taken to be objective *simpliciter* is relativised *and* objectified by the transition to a relation of higher degree. As Weyl puts it.

> Our knowledge stands under the norm of *objectivity*. He who believes in Euclidean geometry will say that all points in space are objectively alike, and that so are all possible directions. However, Newton seems to have thought that space has an absolute center. Epicurus certainly thought that the vertical is objectively distinguishable from all other directions. He gives as his reason that all bodies when left to themselves move in one and the same direction. Hence the statement that the line is vertical is elliptic or incomplete, the complete statement being something like this: the line has the direction of gravity at the point P. Thus the gravitational field, which we know to depend on the material content of the world, enters into the complete proposition as a contingent factor. (39 p. 71)

Our task here is to clarify and develop these ideas.

Let *A* be a structure and T be a structor of *A* that is not, let us say, an invariant of *A*. (To fix your ideas, you might let *A* be a Galilean space-time (globally flat) and T be a two-placed relation holding between events just in case they are at the same spatial location in some particular inertial frame.) Next, let us suppose that there is a structor S of *A* (the inertial frame in the example) which is such that, even though S is not an invariant of *A* either, every automorphism of *A* that *does* preserve S is an automorphism of T.[55] In this case, we shall say that T *is objective in A relative to* S and write $(A,S) \mapsto T$. Notice that it follows at once that T is objective with respect to *A* itself just in case T is an invariant of *A* ($(A,A) \mapsto T$ iff $A \mapsto T$), and if T is objective with respect to an *invariant* S, then T is an invariant also ($A \mapsto S$ and $(A,S) \mapsto T$ implies $A \mapsto T$).

But there is more to relative objectivity than a relation between specific structors S and T. We say in special relativity theory, for example, that "time" is relative to "an inertial frame," not merely "this time" is objective with respect to this or that frame. To capture this idea, we consider those structors *congruent* (in *A*) to S and congruent to T.

Recall that two structors of *A* are said to be *congruent* or *indiscernible* just in case there is an automorphism of *A* that takes one to the other. Congruent figures in Euclidean geometry are examples of congruent structors in this sense. So if we now take the image of our "same-place" structor T under an automorphism σ of *A*, we obtain a congruent structor σ[T] of the same "type," that is, also a "same-place" structor of *A*. All such structors, of course, are obtained by closing T under all automorphisms of *A*. We call the result $[T]_A = \{\sigma[T]: \sigma \in \text{Auto}(A)\}$ the *type* of T (in *A*).

Now let us suppose that T is objective relative to S (in *A*). Let S′ be in the type of S and in particular, let S′ = τ[S], where τ is some automorphism of *A*. Then T′ = τ[T] will also be in the type of T, and we now claim that T′ will also be objective relative to S′, as illustrated in figure 5.

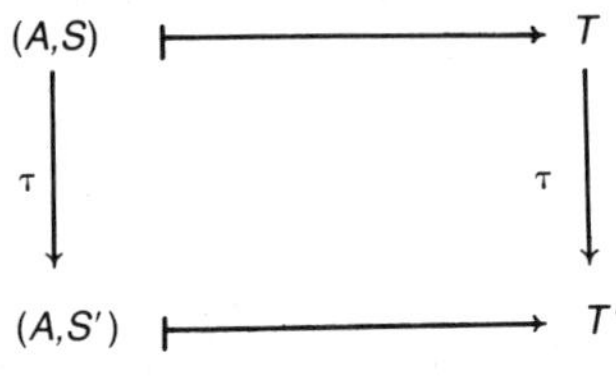

Figure 5

To see this, let σ be an automorphism of *A* that preserves S′. Then $\sigma[S'] = \sigma[\tau[S]] = \tau[S]$, and so $\tau^{-1}\sigma\tau[S] = S$. Hence, $\tau^{-1}\sigma\tau[T] = T$, since T is objective in *A* relative to S. But then $\sigma\tau[T] = \tau[T]$, that is, $\sigma[T'] = T'$, as claimed.

So we see that when T is objective in *A* relative to S, the pair (S,T) generates an entire set of pairs $[(S,T)]_A = \{(\sigma[S],\sigma[T]) : \sigma \in \text{Auto}(A)\}$, and in *each* of these pairs, (S′,T′), the structor T′ is objective in *A* with respect to S′. The result is thus a *relation* $[(S,T)]_A$ which, since it is the closure of a structor (S,T) of *A* under all automorphisms of *A*, is obviously always an invariant of *A*. What is not quite so obvious, although easy to prove, is that this relation is a *function* just in case S is objective relative to T in *A*.

Proposition 4.1. *Where* S *and* T *are structors of the structure A:* (*A*,S) $\mapsto$ T *iff* $[(S,T)]_A$ *is a function.*

Proof. Suppose $(\sigma_1[S], \sigma_1[T])$ and $(\sigma_2[S], \sigma_2[T])$ are both in $[(S,T)]_A$ and $\sigma_1[S] = \sigma_2[S]$. Then, $\sigma_2^{-1}\sigma_1[S] = S$, and so, since $(A,S) \mapsto T$, $\sigma_2^{-1}\sigma_1[T] = T$. Hence, $\sigma_1[T] = \sigma_2[T]$.

($\Leftarrow$) Assume σ is in Auto(*A*) and let $\sigma[S] = S$. Now both (S,T) and $(\sigma[S],\sigma[T]) = (S,\sigma[T])$ are in $[(S,T)]_A$. So, since $[(S,T)]_A$ is a function, $\sigma[T] = T$. Thus, $(A,S) \mapsto T$.

So far, we have seen that the relative objectivity (in *A*) of the pair S,T yields the same result for any pair σ[S], σ[T] when σ is an automorphism of *A*. The pair S,T and the pair σ[S],σ[T] both yield the same invariant relation $[(S,T)]_A = [(\sigma[S],\sigma[T])]_A$. Notice that the pair (S,T) itself will not generally be an invariant of *A*, nor will any of its congruent counterparts (S′,T′). It is, rather, the set of *all* such pairs, $[(S,T)]_A$, which is the invariant of *A* in itself. Is $[(S,T)]_A$ then the structor of *A* whose objectivity is necessary and sufficient for the relative objectivity of T to S? Not quite.

First, of all, notice that $[(S,T)]_A$ is *always* an invariant of *A*, whether or not T is objective relative to S. (This is simply because it is an automorphic closure of a structor—(S,T)—of A). We have seen from Proposition 4.1 that it is not the invariance of $[(S,T)]_A$ that coincides with the relative objectivity of T to S, but the *functionality* of this relation. Hence, our problem remains, and is very similar to the following situation in elementary logic.

Suppose that a proposition q, not logically implied by the theory T alone, follows from the set {T,p}. By the deduction theorem, this will be so just when the theory T yields the implication "if p then q" on its own. Our problem is to find the structor composed from S and T that behaves like the conditionals of linguistic theories.

The following construction yields the desired structor. First, we define the automorphisms $(\bar{S} \vee T)_A$ as follows:

DEFINITION 4.2. $(\bar{S} \vee T)_A$ = df. $[\text{Auto}(A) - \text{Auto}((A,S))] \cup \text{Auto}((A,T))$.

(These automorphisms. it should be noted, do not necessarily form a group.)

As a straightforward set-theoretical fact we have at once:

LEMMA 4.3. $(\bar{S} \vee T)_A = \text{Auto}(A)$ *iff* $(A,S) \mapsto T$.

The structor we want is now obtained by closing the pair (S,T) under the automorphisms in $(\bar{S} \vee T)_A$. We write the result $(S \rightarrow T)_A$, and are now in a position to prove the crucial result.

PROPOSITION 4.4. $(A,S) \mapsto T$ *iff* $A \mapsto (S \rightarrow T)_A$.

Proof. ($\Rightarrow$) Since $(A,S) \mapsto T$, by Lemma 4.3, $(\bar{S} \vee T)_A = \text{Auto}(A)$. Hence, $(S \rightarrow T)_A = [(S,T)]_A$ and so, being the automorphic closure of a structor of A, is an invariant of A.

($\Leftarrow$) We proceed by showing that $A \mapsto (S \rightarrow T)_A$ implies that $(\bar{S} \vee T)_A = \text{Auto}(A)$. The result then follows from the preceding lemma. Obviously, we need only show that Auto(A) is a subset of $(\bar{S} \vee T)_A$.

Suppose σ is in Auto(A). Then, since (S,T) is in $(S \rightarrow T)_A$, so is $\sigma[(S,T)]$, that is, $(\sigma[S],\sigma[T]) = (\tau[S],\tau[T])$, for some τ in $(\bar{S} \vee T)_A$. Hence, $\sigma[S] = \tau[S]$, and $\sigma[T] = \tau[T]$, and either: (i) $\tau \in \text{Auto}(A) - \text{Auto}((A,S))$, or (ii) $\tau \in \text{Auto}((A,T))$.

Suppose (i). Then, $\tau \in \text{Auto}((A,S))$. But $\sigma \in \text{Auto}((A,S))$, otherwise $\tau[S] = \sigma[S] = S$, and so $\tau \in \text{Auto}((A,S))$ after all. Thus, $\sigma \in \text{Auto}(A) - \text{Auto}(A,S)$, and so $\sigma \in (\bar{S} \vee T)_A$.

Suppose (ii). Then $\tau \in \text{Auto}(A,T)$ and so $\sigma[T] = \tau[T] = T$. Hence, $\sigma \in \text{Auto}((A,T))$ and so again $\sigma \in (\bar{S} \vee T)_A$.

Thus, $\text{Auto}(A) \subseteq (\bar{S} \vee T)_A$, and we have the result at once from Lemma 4.3.

An examination of the above proof shows that $A \mapsto (S \rightarrow T)_A$ if and only if $[(S,T)]_{Auto(A)} = [(S,T)]_{(\bar{S} \vee T)_A}$, and so we have the following result, which summarizes the situation up to now:

COROLLARY 4.5. *Where* S *and* T *are structures of the structure A, the following conditions are equivalent:*
(*i*) T *is objective (in A) relative to* S,
(*ii*) $(A,S) \mapsto T$,
(*iii*) $A \mapsto (S \rightarrow T)_A$,
(*iv*) $[(S,T)]_{Auto(A)} = [(S,T)]_{(\bar{S} \vee T)_A}$,
(*v*) $[(S,T)]_{Auto(A)}$ *is a function.*

With these results, we have vindicated Weyl's claim in the passage quoted earlier, and more. We have shown that relative objectivity is a special case of invariance—the invariance of the relation $(S \rightarrow T)_A$. Hence, when someone asks, as Nelson Goodman (in effect) does (15, p. 3): "When T is relativized to S, what is it that is not so relativized and remains objective in its own right?" we have a precise reply: "The invariant relation $(S \rightarrow T)_A$."

4.2. *Invariance and Simultaneity*

An inertial reference frame and its adopted simultaneity relation yield a corresponding geometrical object in Minkowski space-time. Thus, a single inertial particle will travel on a straight timelike line (call these "inertial lines"), and an inertial reference frame—an idealized collection of such particles—fills space-time with a set of parallel inertial lines. The simultaneity relation of an inertial frame stratifies events into spacelike slices; geometrically, these slices will be linear, spacelike, hypersurfaces (see figure 6). The standard ($\varepsilon = ½$) simultaneity relation produces hypersurfaces orthogonal to the frame's inertial lines; nonstandard choices of ε yield nonorthogonal hypersurfaces.

Defenders of the thesis of the *conventionality* of simultaneity within the special theory of relativity (STR) maintain that, while the choice of orthogonal supersurfaces is customary, it is not required. Those hypersurfaces corresponding to the choices of ε other than one-half, it is held, are physically colegitimate with the $\varepsilon = ½$ choices. The two sorts of simultaniety stratifications differ in their *descriptive* simplicity, but not in their physical significance.[56]

The discussion and results of the preceding section allow the issue to be put quite precisely. For a given inertial frame, *F*, we have a number of candidates for a simultaneity relation with respect to *F*, namely, one for each value of ε between zero and one. Call these relations "Sim *F*(ε)." (Sim *F*(½) is then the standard simultaneity relation in *F*.) The issue now becomes: Which, if any, of these relations are objective with respect to *F*? Conventionalists hold that, at the very least, the various simultaneity

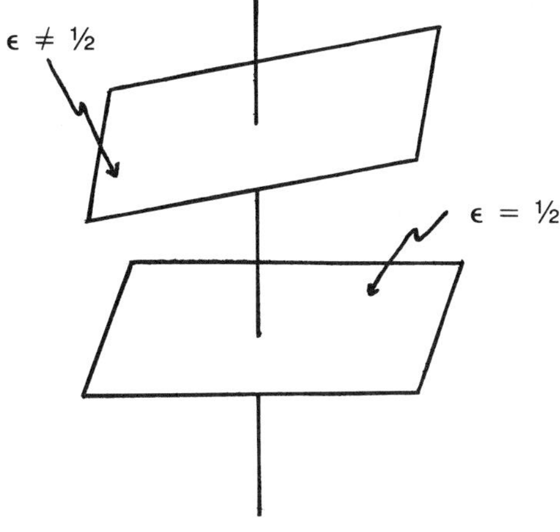

Figure 6

relations are objectively on a par: either none are objective or they all are. Nonconventionalists have held that the standard ε = ½ simultaneity slicings are physically significant, the other slicings, arbitrary.

Recent discussions of these issues have raised what appear to be decisive objections to the conventionalist view. As Winnie (in 43) has emphasized, A. A. Robb, early in this century, was able to define space-time congruence and standard simultaneity solely in terms of an asymmetrical relation of causal precedence. In the same paper, Winnie defines these notions in terms of the symmetrical relation of causal connectability. Since causal connectability is often taken by conventionalists themselves as a paradigm of a *nonconventional* physical relation, these results seriously impugn those versions of the conventionalist thesis that refuse to grant objective status to the standard simultaneity relation.

The *coup de grace,* however, was administered by David Malament. For Malament showed (in 21) that the standard simultaneity relation was the *only* causally definable simultaneity relation in a Minkowski space-time. Essentially, this was done by showing that for any nonstandard simultaneity slicing that was transitive, there exists a causality-preserving automorphism of Minkowski space-time that, while preserving the frame *F*, would fail to preserve the nonstandard slicing.

The same idea can also be used to prove an analogous result, namely, that the standard simultaneity relation is the only ε-simultaneity relation that is an *invariant* of Minkowski space-time. Relativization of simultaneity to an inertial frame does, in the sense made precise in the previous section, result in a corresponding invariant relation between frames and simultaneity stratifications in a Minkowski space-time.

Let *F* be an inertial reference frame, and $X = (x^0,x^1)$ a Lorentz (and so $\varepsilon = ½$) coordinate system adapted to *F*. For simplicity's sake, we shall consider a two-dimensional Minkowski space-time, and comment on the general case as we proceed.) Since x^0 is the $\varepsilon = ½$ time coordinate function, and, for all values of ε between zero and one, $t_\varepsilon = t_{1/2} + x(2\varepsilon - 1)$, we may write ε-coordinates in the form

$$X_\varepsilon = (x^0 + x^1(2\varepsilon - 1),\ x^1). \tag{4.1}$$

Note that since a change from Lorentz to ε-coordinates leaves the spatial coordinate function(s) intact, the ε-coordinate systems are also adapted to the frame *F*.

It is now a simple matter to express ε-simultaneity relative to the frame *F* in terms of the *Lorentz* coordinates X, for, as an examination of (4.1) reveals, a pair of events e_1, e_2 will be ε-simultaneous just in case

$$x^0(e_1) + x^1(e_1)(2\varepsilon - 1) = x^0(e_2) + x^1(e_2)\,(2\varepsilon - 1). \tag{4.2}$$

Using "$\mathrm{Sim}_\varepsilon(F)$" for the ε-simultaneity relation with respect to frame *F*, it is easy to see that

$$\mathrm{Sim}_\varepsilon(F) = \{(e_1,e_2)\colon \Delta x^0(e_1,e_2) + \Delta x^1(e_1,e_2)\,(2\varepsilon - 1) = 0\}.^{57} \tag{4.3}$$

When $\varepsilon = ½$ in $\mathrm{Sim}_\varepsilon(F)$, then, as we would expect, events are simultaneous in *F* just in case $\Delta x^0 = 0$.

The results and discussion of section 4.1 now allow a succinct formulation of the conventionality problem. For what values of ε (if any) is the relation $\mathrm{Sim}_\varepsilon(F)$ objective with respect to frame *F* in Minkowski space-time? And as we have seen, this amounts to determining whether or not those automorphisms of Minkowski space-time that preserve frame *F* also preserve the associated simultaneity relation $\mathrm{Sim}_\varepsilon(F)$.

First of all, notice that in answering this question, we may ignore the typically *relativistic* automorphisms, the boosts that take one inertial frame to another. For it is only the automorphisms that preserve frame *F* that matter here.

One type of these automorphisms takes an especially simple form in Lorentz coordinates: the temporal inversions. Consider, in particular, the inversion τ about the $x^0 = 0$ hypersurface, and represented in Lorentz coordinates simply by:

$$\tilde{\tau}(a_0,a_1) = (-a_0,a_1). \tag{4.4}$$

Clearly, τ preserves the frame *F;* when applied to the $\mathrm{Sim}_\varepsilon(F)$, however, the result is just:

$$\tau[\mathrm{Sim}_\varepsilon(F)] = \{(e_1,e_2)\colon -\Delta x^0(e_1,e_2) + \Delta x^1(e_1,e_2)(2\varepsilon - 1)\} = 0. \tag{4.5}$$

Since relative objectivity requires that $\mathrm{Sim}_\varepsilon(F) = \tau[\mathrm{Sim}_\varepsilon(F)]$, a comparison of (4.3) and (4.5) shows this in turn requires that

$$\Delta x^0(e_1,e_2) + \Delta x^1(e_1,e_2)(2\varepsilon - 1) = 0 \tag{4.6}$$

just in case

$$-\Delta x^0(e_1,e_2) + \Delta x^1(e_1,e_2)(2\varepsilon - 1) = 0, \tag{4.7}$$

for all events e_1,e_2.

But now choose an event pair e_1,e_2 which are spatially distant but ε-simultaneous. (It is clear that many such pairs exist, for each choice of ε.) Then relative objectivity of $\mathrm{Sim}_\varepsilon(F)$ requires that *both* (4.6) and (4.7) hold for e_1,e_2. Hence, at once, $\Delta x^1(e_1,e_2)(2\varepsilon - 1) = 0$, and since $\Delta x^1(e_1,e_2) \neq 0$ (e_1 and e_2, recall, were chosen to be spatially distant) we must have $2\varepsilon - 1 = 0$, that is, $\varepsilon = \frac{1}{2}$. Hence, we have proved the following proposition.

PROPOSITION 4.6. *If* $\mathrm{Sim}_\varepsilon(F)$ *is objective with respect to F in a Minkowski space-time, then* $\varepsilon = \frac{1}{2}$*; that is, the* $\varepsilon = \frac{1}{2}$ *simultaneity relation, if it is an invariant, is the only invariant ε-simultaneity relation.*

To show that the $\varepsilon = \frac{1}{2}$ simultaneity relation is, indeed an invariant of a Minkowski space-time, we need only now show that any automorphism that preserves the frame *F* also preserves the relation:

$$\mathrm{Sim}_{1/2}(F) = \{(e_1,e_2)\colon \Delta x^0(e_1,e_2) = 0\}. \tag{4.8}$$

Now just what automorphisms are relevant here? First of all, there are the temporal inversions, and these take standard simultaneity slices of *F* to other such slices, and so preserve $\mathrm{Sim}_{1/2}(F)$. Then there are the spatial translations, rotations, and reflections. But just because these are *spatial,* they are all represented by the same temporal coordinate function x^0, and so, once again, $\mathrm{Sim}_{1/2}(F)$ is left intact by any such automorphism. Finally, there are those frame-preserving automorphisms that, since they are composed from those above, must also preserve $\varepsilon = \frac{1}{2}$ simultaneity. Thus we have shown:

PROPOSITION 4.7. *In Minkowski space-time, the only ε-simultaneity relation that is objective with respect to an inertial frame F is the* $\varepsilon = \frac{1}{2}$ *simultaneity relation.*[58]

Although we have considered only a single frame *F* and $\varepsilon = \frac{1}{2}$ with respect to *F,* the results of Corollary 4.5 show that we have also proved the invariance *simpliciter* of the relation.

$$(F \rightarrow \mathrm{Sim}_{1/2}(F))_M,$$

in which *each* frame of *F* is paired with its associated $\varepsilon = \frac{1}{2}$ simultaneity relation. This, of course, is where those automorphisms that are boosts—frame transformations—come in. Each frame's $\varepsilon = \frac{1}{2}$ simultaneity relation is defined by the condition $\Delta x^0 = 0$, expressed in

Lorentz coordinates adapted to that frame. But the boosts of a Minkowski space-time are represented by the Lorentz transformation which, by definition, take Lorentz coordinates to Lorentz coordinates, and the Δx^0 of one frame to the Δx^0 (or Δy^0) of another. As a result, the relation

$$(F \rightarrow \mathrm{Sim}_{½}(F))_M$$

is the same, regardless of the choice of frame F. It is, in other words, *the* standard simultaneity relation between inertial frames and their $\varepsilon = ½$ simultaneity slicings.

The above result complements Malament's theorem on space-time structure by showing the unique *invariance* of standard simultaneity in Minkowski space-time, rather than its unique *causal definability*.[59] For those who would ground the nonconventional basis of the special theory in causal connectibility, the result is weaker. On the other hand, if, as seems reasonable, one holds that the structure of a Minkowski space-time may be generated by a variety of conventions, none of which are infected with *simultaneity* conventions, then this result is just what the nonconventionalist requires. An example may help to clarify the issue here.

Reichenbach was, in some sense, dimly aware of the causal definability of the structure of Minkowski space-time, but appears to have rejected a causal theory of space-time congruence on operational grounds. Thus, in his discussion (27, §27) of the possibility of distinguishing inertial from accelerated frames by using signal relations, Reichenbach rejects any method that would not enable the distinction to be made *in a finite region.* He describes such a distinction as "not fruitful," since the correct application of the concept ('inertial' vs. 'accelerated' frames) cannot be guaranteed in some finite region (for more technical details, see Reichenbach 27, pp. 187–92). A property or relation is then physically inadmissible if, for any finite region, there could be entities within that region that fail to have the property of relation in question only by virtue of their relations to entities outside that region.

For the moment, let us set aside the question of the value of this condition as a test of a concept's physical admissibility, and consider how the causal theory of Minkowski space-time fares in this respect. Assume a causal theory that uses a symmetrical relation of causal connectibility as its only nonlogical primitive (as in Winnie 43). Then, in Minkowski space-time, this relation will satisfy Reichenbach's finite admissibility condition. This is almost trivial, since if in any finite region of Minkowski space-time, two events are causally connectible in that region, they are *a fortiori* causally connectible. *Failure* of causal connectibility—let us call it *causal simultaneity*—is more interesting. The finite regions of Minkowski space-time illustrated in figure 7 (a) show that a pair of events may *fail* to

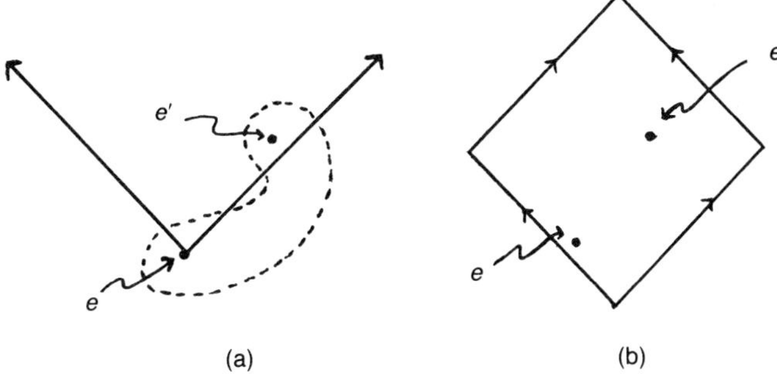

Figure 7

be causally connectible within a region, but causally connectible outside that region, and so causally connectible after all.

Still, causal simultaneity passes Reichenbach's test. For recall that a concept's inadmissibility required there being *no* decisive finite regions within which to test its applicability. But if there is *some* finite region in which a decisive answer may be given wholly in terms of the situation in that region, then the concept is physically admissible, or, at least, passes on this score.

Now such turns out to be the case for causal simultaneity in a Minkowski space-time. The situation is illustrated in figure 7 (b). Given any two events e, e′, there will always be a finite region—like that in the figure—which contains both events. Such regions are called *Alexandrov intervals,* and may be defined as the set of all events causally between some pair of events having timelike separation. (For the definition of causal betweenness in terms of causal connectibility, see Winnie 43, p. 148.) Alexandrov intervals are finite regions, and for any events e and e′ in such an interval, e and e′ are or fail to be causally connectible just in case they are causally connectible within the interval. Hence, we may establish the causal simultaneity of any event pair by "surveying" some finite region—any Alexandrov interval—containing the pair.

Causal connectibility satisfied Reichenbach's admissibility criterion, and causal simultaneity, explicitly definable in terms of causal connectibility, also turns out to satisfy this criterion. So it might be thought that the same holds for *any* relation explicitly definable in terms of causal connectibility. Perhaps a concept explicitly definable in terms of an admissible concept will always be admissible as well.

It turns out, however, that this is not so. An example is Latzer's definition of space-like collinearity used in the causal theory (Latzer 20, Winnie 43). Given three events, e, e′, and d, under what conditions will they all lie on the same spacelike line? To begin with, of course, we require that no two of these events be causally connectible, This is the first clause in the definition. The second is not so obvious: we require that the light cones of the three events *fail* to have a common point of intersection. (This constuction fails in two-dimensional Minkowski space-time, so the picture should be drawn in three dimensions to get a feel for this condition.) Now the three-placed relation of spacelike collinearity, explicitly defined in terms of admissible concepts, fails to pass Reichenbach's test. The situation is illustrated in figure 8.

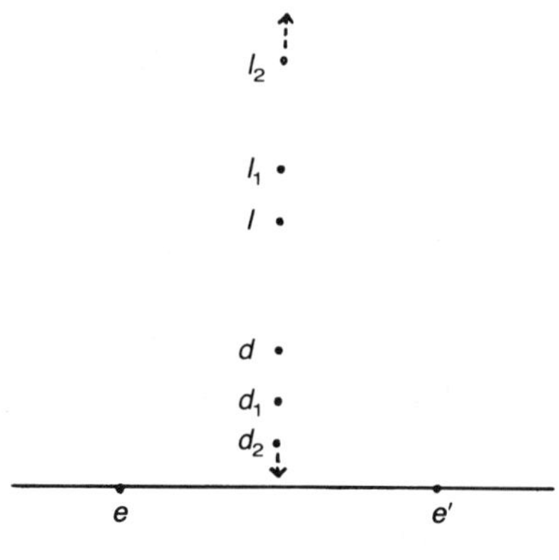

Figure 8

Assume that events e, e′, and d fail to be causally connectible. Then (using the figure in a rough way) since event d does not lie on the spacelike straight line connecting e and e′, the light cones of e, e′, and d will have a common point of intersection, say I. Now as we consider events d_1, d_2, and so on ever closer to the line ee′, the common intersections—I_1, I_2, and so forth—lie farther and farther away from e and e′, until—when we have the d-event *on* line ee′—the intersection is the "point at infinity," that is, the cones fail to intersect. (Remember, this only becomes geometrically obvious when a more complicated, three-dimensional Minkowski space-time is depicted. The figure above is merely used to show *what* happens, not make it plausible that it *does* happen.)

So in any finite region of space-time containing events e and e′, the choice of an event d sufficiently close to the line ee′ will result in the common intersection point lying outside the finite region. Thus, even though the notion of spacelike collinearity of three events is explicitly definable in terms of causal connectibility—and causal connectibility is locally ascertainable in Reichenbach's sense—spacelike collinearity is *not* locally ascertainable. Local ascertainability is not preserved by explicit definitions.

The causal construction of the linear structure of Minkowski space-time is crucial for the success of the causal theory. Latzer's definition does, of course, explicitly define the spacelike linear structure, but as we have seen, that definition provides no means for determining in practice whether or not three events *are* spatically collinear: a failure to observe a common intersection of the events' light cones might merely be the result of our having failed to extend our observations over an only slightly larger region.

There is, I submit, something *to* Reichenbach's requirement, and to recognize this is not necessarily to espouse the verifiability theory of meaning or operationalism. Consider again our three events e_1, e_2, and d. Suppose now that if d *is* spatically collinear with e_1 and e_2, then the sun will explode tomorrow, otherwise it will continue as is for another billion or so years. Suppose further that you must judge the collinearity issue solely on the basis of observable causal structure. Finally, suppose that you have surveyed a large, but (naturally) finite region of space-time and could find no intersection of the three events' light cones. What do you now predict—will the sun explode or no?

It appears that, in such a situation, either prediction is evidentially ill founded, and the same would hold for any physical concept linked with empirical prediction in the manner of this example. Of course, the example is a fantasy, for at least two reasons. The first is that nothing so decisive as the sun's exploding is likely to follow from errors in judgment about spacelike collinearity, and, the second, more important reason, is that in practice the spacelike collinearity of events is not measured causally; if measured at all, standard simultaneity coupled with spatial linearity would be the more usual method.

While these considerations raise legitimate concerns as to the empirical or epistemological adequacy of the causal theory of space-time, they should afford little comfort to most defenders of the thesis of the conventionality of simultaneity, namely, those who approach the conventionality of simultaneity *assuming that the inertial frames of the special theory are on hand at the outset.*[60] According to these conventionalists, the problem becomes: given the *prior* availability of the class of inertial frames of reference, and perhaps any attendant conventions involved in their specification, what *other* conventions are now required in order to associate a simultaneity slicing, with each such frame? Conventionalists now go on to argue, as we have seen, that an additional arbitrary choice of ε is required.

Given causal connectibility *and* inertial frames, however, a "local" definition of simultaneity is no longer a problem: it is just the standard $\varepsilon = \frac{1}{2}$ signal synchrony definition! (The light cone, as shown in Winnie 43, is explicitly definable in terms of causal connectibility.) The other, $\varepsilon \neq \frac{1}{2}$, signal definitions will still fail, however, and for the same reason: they will require the arbitrary introduction of a temporal or a spatial orientation, neither of which is objectively available in Minkowski space-time. Furthermore, since these constructions of a standard simultaneity slicing may all be carried out within a finite region of space-time, it is clear that for these purposes we need only assume the availability of reference frames that are inertial within such finite regions. In short, the availability of locally inertial frames, together with the causal connectibility relation of Minkowski space-time allows the unique explicit definability of the standard simultaneity relation in these frames by means of criteria that apply within finite regions of the space-time.

4.3. Simultaneity and Translation

Thus far, we have shown that the $\varepsilon = \frac{1}{2}$, and only the $\varepsilon = \frac{1}{2}$, simultaneity relation is an invariant of Minkowski space-time. As section 4.2 reveals, this result is not difficult to establish; it (or, rather, Malament's result) took so long in coming, I believe because it apparently conflicted with another commonly held[61] belief, here to be called "the translation theses"; roughly stated, it runs as follows:

Translation Thesis. Let $T(\varepsilon)$ be the kinematics of the STR formulated in terms of ε-simultaneity relations, where $0 < \varepsilon < 1$. Then, for all such values of ε, the theories $T(\varepsilon)$ are mutually intertranslatable.

To see the basis of the conflict, suppose the translation thesis is true, and let $T_1(\mathrm{Sim}(\frac{1}{2}))$ and $T_2(\mathrm{Sim}(\varepsilon_0))$ be formulations of the STR in terms of $\varepsilon = \frac{1}{2}$ and some $\varepsilon_0 \neq \frac{1}{2}$ simultaneity relation, respectively. (Of course, we allow here for the presence of other primitive predicates as well.) Then at once, and trivially, each model of the first version will display the three-placed relation of $\varepsilon = \frac{1}{2}$ simultaneity as a distinguished structor; hence, that relation is an invariant in any such model. Clearly, the same is true for the second version also: the ε_0-simultaneity relation will be an invariant of each of that theory's models as well. But now suppose, in accord with the translation thesis, that the two versions are intertranslatable in the strict sense, that is, suppose they have a common definitional extension. Then the predicate "$\mathrm{Sim}(\varepsilon_0)$" must be definable in T_1 and, for the same reason, "$\mathrm{Sim}(\frac{1}{2})$" must also be definable in T_2. Definitions, however, preserve invariance; hence, in each theory, *both* relations must be invariant. Thus, the translation thesis is incompatible with the *unique* invariance of standard simultaneity. Clearly, unless we are converts to the recent philosophy of Nelson Goodman, restoration of some semblance of consistency seems called for.

Naturally, there is a problem here only for those who adhere to the translation thesis in some form or other. For the above argument, together with the previous results, shows conclusively that the translation thesis, at least in the strict version given here, is simply false. The puzzle now is that it has been so easy for defenders of the conventionality of simultaneity thesis to undercut in detail every interesting thought experiment attempting to fix the value of epsilon in a non-question-begging way.[62] In *some* sense, the predictions of standard formulations of special relativity can be "translated" into corresponding and compatible predictions of the nonstandard versions. But how is this possible, if the translation thesis is false?

In order to answer this question, let us begin by considering the following two definitions of a Minkowski space-time.[63]

DEFINITION A. *A structure M* = (E,ω^2) *is a (two-dimensional) Minkowski space-time just in case:*

(a_1) E *is a nonempty set and* ω^2 *is a function from* ExE *to the real numbers, and*

(a_2) *there is a bijection* $X: E \rightarrow R^2$ *such that for all* $e_1, e_2, \in E$,

$$\omega^2(e_1,e_2) = -(\Delta x^0(e_1,e_2))^2 + (\Delta x^1(e_1,e_2))^2,$$

where $x^i(e) \equiv u^i(X(e))$, *the* i-*th number in the pair* X(e).

DEFINITION B. *As in Definition* A, *except that we replace* (a_2) *by:*

(b_2) *there is a bijection* $X: E \rightarrow R^2$ *such that for all* $e_1,e_2 \in E$,

$$\begin{aligned}\omega^2(e_1,e_2) = & -(\Delta x^0(e_1,e_2))^2 \\ & + 2\Delta x^0(e_1,e_2)\Delta x^1(e_1,e_2)(2\varepsilon_0 - 1) \\ & + (\Delta x^1(e_1,e_2))^2(4\epsilon_0(1 - \varepsilon_0)),\end{aligned}$$

(a_2) *where* $x^i(e) \equiv u^i(X(e))$, *the* i-*th number in the pair* X(e), *and* ε_0 *is a constant between zero and one.*

Definition B is to be thought of as definition-schema; in each instance, ε_0 is to be a real number between 0 amd 1. (Definition A is then an instance of B when one-half is substituted for ε.)

First, note that structures that qualify as Minkowski space-times are of the same form, according to both definitions: $M = (E,\omega^2)$, as specified in clause (a_1). Furthermore, not only are these structures identical in form, they are also set-theoretically identical. More precisely, if $M = (E,\omega^2)$ is a Minkowski space-time according to Definition A, then it also so qualifies according to (each instance of) Definition B. This is not so obvious at first sight, but follows at once from the fact that clause (b_2) in the second definition is just the form that the interval ω^2 takes when we switch from the Lorentz coordinates of Definition A to the ε-coordinates of Definition B.[64]

Thus, Definitions A and B specify equivalent versions of Minkowski space-time in the strictest sense: they supply the *same* class of models. Yet they do so in different ways. The first version (A) specifies the interval using Lorentz coordinates, the second (B), using ε-coordinates. In *this* sense then, these are equivalent (and intertranslatable) versions of the STR which utilize different choices of epsilon. But in both cases, the only descriptive primitive is the function (symbol) for ω^2; reference to ε is by way of *coordinate* relations, and they are used only as a means of getting at ω^2 by "pulling back" from R^2 (or, in the general case, R^4).

A more interesting case, however, is provided by making use of the results of section 2.3. There, recall, we showed (Proposition 2.29) that for any structure A, if $\{\phi\}_A$ is a distinguished class of coordinate systems of A, then A and $A' = (A, \{\phi\}_A)$ are invariantly equivalent. Actually, we showed even more than that: A and A' have the same automorphisms and invariants, and so are what we have called "codeterminate" structures (section 2.4, Definition 2.30).

Now, a Lorentz coordinate system, by definition, is one in which the interval ω^2 is given by the equation in (a_2) of Definition A above. Similarly, an ε-coordinate system (for a fixed value of ε) is defined as one in which the interval is given as in Definition B, clause (b_2). Hence, the class of all such coordinate systems is defined as a class of all coordinate systems in which the generator (ω^2) of a structure takes the same form, that is, the generator is represented identically the same in all coordinate systems within the class. As the proof of the following proposition reveals, this condition is enough to show that such a class of coordinate systems is a distinguished class, that is, equal to the closure of any one of its members under the automorphism of A.

PROPOSITION 4.8. *Let A be a structure and* R *be a structor of* A *such that* Auto(A) = Auto((A,R)). *Let ϕ be a coordinate system of A and let $\{\phi\}_R$ be the class of all coordinate systems ψ of A*(rng(ψ) = rng(ϕ)) *which are such that* $\psi[R] = \phi[R]$. *Then,* $\{\phi\}_R$ *is a distinguished class of coordinate systems of A.*

Proof.
(a) Suppose $\phi\sigma \in \{\phi\}_A$, $\sigma \in$ Auto(A).
Then, $\phi\sigma[R] = \phi[R]$. Hence, $\phi\sigma \in \{\phi\}_R$. Thus, $\{\phi\}_A \subseteq \{\phi\}_R$.
(b) Suppose $\psi[R] = \phi[R]$, that is, $\psi \in \{\phi\}_R$.
Then, $\phi^{-1}\psi[R] = R$, and so $\phi^{-1}\psi = \sigma_*$, σ_* some permutation in Auto(A). Hence, $\psi = \phi\sigma_*$, and so $\psi \in \{\phi\}_A$. Thus, $\{\phi\}_R \subseteq \{\phi\}_A$.

It follows at once from this result and Proposition 2.29 that the structures $M = (E,\omega^2)$ and $M' = (E,\{\phi\}_{\omega_2})$ have exactly the same automorphisms and so are codeterminate structures.

If we now fix ε and let $\{\phi\}_\varepsilon$ be the class of all ε-coordinate systems, the above result, together with Proposition 2.29, implies that the structures $M = (E,\omega^2)$ and $M' = (E,\{\phi\}_\varepsilon)$ are codeterminate, that is, they have the same automorphisms. So, for example, the structures $M(\frac{1}{2}) = (E,\{\phi\}_{\frac{1}{2}})$ and $M(\frac{3}{4}) = (E,\{\phi\}_{\frac{3}{4}})$ both represent the same Minkowski space-time $M = (E,\omega^2)$ in the strict sense of having exactly the same automorphisms and invariants.

Now, let us consider the structures $M(\frac{1}{2}) = (E, \{\phi\}_{1/2})$ and $M(\frac{3}{4}) = (E, \{\phi\}_{3/4})$ from a linguistic standpoint, each as a model of its appropriate linguistic theory $T(\frac{1}{2})$ and $T(\frac{3}{4})$, respectively. Notice that, in each case, *a single primitive term is involved* and that term is to be interpreted as referring to a *class,* in particular, a class of *coordinate systems.* If we call these terms "$S(\frac{1}{2})$" and "$S(\frac{3}{4})$," respectively, our theories become $T(S(\frac{1}{2}))$ and $T(S(\frac{3}{4}))$. Each theory is now considered to be (in some appropriate higher order language) the set of all sentences true in the appropriate corresponding structure. Thus, $T(S(\frac{1}{2}))$ is just the set of all sentences true in $M(\frac{1}{2}) = (E, \{\phi\}_{1\frac{1}{2}})$ and similarly for $T(S(\frac{3}{4}))$.

Now these theories, in a sufficiently rich language, may be intertranslatable, but without paradox. For no term referring to either the $\varepsilon = \frac{1}{2}$ *simultaneity relation or* the $\varepsilon = \frac{3}{4}$ *simultaneity relation* is taken as primitive by either theory. Instead, our primitives refer to *classes* of *coordinate systems of each type.* And, as we have seen, *each* such *class* is an invariant of Minkowski space-time, and indeed both models have the same invariants. Since, as we have seen, $\varepsilon = \frac{3}{4}$ *simultaneity* is not objective with respect to an inertial frame, it follows that this *relation* is not definable in out theory $T(S(\frac{3}{4}))$—*even though that theory takes as its single primitive a term referring to just the* $\varepsilon = \frac{3}{4}$ *coordinate systems.* Thus, while it is permissible to formulate the theory of Minkowski space-time in terms of systems in which $\varepsilon = \frac{1}{2}$, it does not follow that simultaneity slicings other than the $\varepsilon = \frac{1}{2}$ slicings are thereby given equal physical legitimacy.

Traditionally, the structure of Minkowski space-time was given by Lorentz transformations between what we have called here Lorentz (or $\varepsilon = \frac{1}{2}$) coordinate systems. As a result, these coordinate systems were held by some to be "privileged." We have seen here that *all* distinguished classes of coordinates are equally "privileged" in the sense that they all encode the same space-time structure. Hence, those systems that use $\varepsilon \neq \frac{1}{2}$ are, as Reichenbach and others have emphasized, equally suitable for the formulation of the kinematics of special relativity. The mistake that conventionalists made, however, was in believing that the colegitimacy of $\varepsilon = \frac{1}{2}$ and $\varepsilon \neq \frac{1}{2}$ *classes* of coordinate systems implied that a similar colegitimacy attached to *coordinate simultaneity* with respect to the inertial *frames* of the special theory. Thus, if we choose an inertial frame F and let the simultaneity slices be given with respect to F by the condition that $x^0 = \text{const.}$ for some distinguished coordinate system adapted[65] to F, then we obtain a slicing that is objective with respect to frame F only when our distinguished coordinates are Lorentz coordinates.

In summary, while the translation thesis, when taken as stating that alternative versions of the STR adopt different $\varepsilon =$ *simultaneity relations* among their primitive terms, is certainly false, there are other reasonable senses in which versions of special relativity theory may involve the use of values of ε other than one-half. In one such version, the choice amounts to the use of a primitive term referring to a *nonstandard* ($\varepsilon \neq \frac{1}{2}$) but distinguished class of ε-Lorentz coordinate systems. The resulting structures are then codeterminate with their standardly specified counterparts; but no paradox is now forthcoming, since in neither case can a nonstandard simultaneity relation be obtained as an invariant. Standard simultaneity remains the unique causally definable invariant relative simultaneity relation in a Minkowski space-time.

5. FREE INVARIANCE AND FREE MOBILITY

5.1. Local Automorphisms and Free Invariants

There are good reasons, we have seen in Section 1, for regarding invariance as *necessary* for a structor's objectivity with respect to a given structure. But a structure's automorphism group may be "small," perhaps to the point of containing only the identity, and then to take invariance as also a *sufficient* condition of objectivity leads to results that violate many sensible intuitions about structural equivalence. In the case of rigid structures, for example, *all* of a set's structors turn out to be invariants, and any two rigid structures with universes of equal cardinalities will be invariantly equivalent (see Proposition 2.21 and the discussion that follows it).

Take, for example, the case of a plane that is Euclidean except that it has three (nonequidistant and noncollinear) "bumps" on it. This structure is rigid, since any automorphism must fix these bumps. Next, take another structure, just like the first, except that in addition to the three bumps mentioned earlier, there are infinitely many other such trios, all with different distances between their bumps, scattered about the surface. Obviously, this second structure is also rigid, and so, by the result mentioned earlier (Proposition 2.21), invariantly equivalent to the first.

But just as obviously, there are what seems to be structural differences between these two structures. These differences do not appear on the global level, of course, since both structures are globally rigid. But they do appear when we consider *subregions* of the plane and their automorphisms. For example, in the case of the first structure, if we enclose its sole triad of bumps with a circle and remove the interior of that circle from the surface, the remaining region will not be rigid, having (intuitively) rotations about the center of the removed circle as automorphisms. To so remove any such triad from the second structure, however, leaves a structure that is still rigid, there being infinitely many triads remaining to "nail it down." While the (global) automorphisms of these two structures are alike, the automorphism groups of their *substructures* are not. By going to the substructures of a given prestructure

$A = (A,R^{\beta})$, we may develop a finer measure of structural equivalence and a stronger version of invariance than those presented so far. Eventually, precise definitions will be given, but first, the general approach.

As before, we begin with (pre) structures $A = (A,R^{\beta})$. But now, if B is a subset of A, we define the *restriction* of A to B, $A_{|B}$, as the structure that results when we "restrict" each of the distinguished structors of A to the subset B. Thus, $A_{|B} = (B,R^{\beta}_{|B})$, where for any structor R^{α} in the sequence R^{β}, $R^{\alpha}_{|B}$ is, roughly, the intersection of R^{α} with the structors of B (see Definition 5.1). The result is clearly another structure ($A_{|B}$) with universe B together with a sequence $R^{\beta}_{|B}$ of distinguished structors of B.

Of course, the structure $A_{|B}$ will have automorphisms in its own right (Auto($A_{|B}$)), and these will be called the B-local automorphisms of A.[66] If we now consider a structor S of A, and restrict S to B, the resulting structor $S_{|B}$, may or may not be by an invariant of $A_{|B}$. If it is (i.e., if $A_{|B} \vdash\!\!\rightarrow S_{|B}$), then S is said to be a (B) *local invariant* of A.

Finally, if a structor S of A is a (B) local invariant of A, for *any* subset B of A, then we say that S is a *free invariant* of A ($A \models\!\!\Rightarrow$ S).[67] Clearly, any free invariant of A is also an invariant of A, but as we shall see in the example below, the converse fails in general.

To fix these ideas, consider the following structure (also used later) depicted in figure 9.

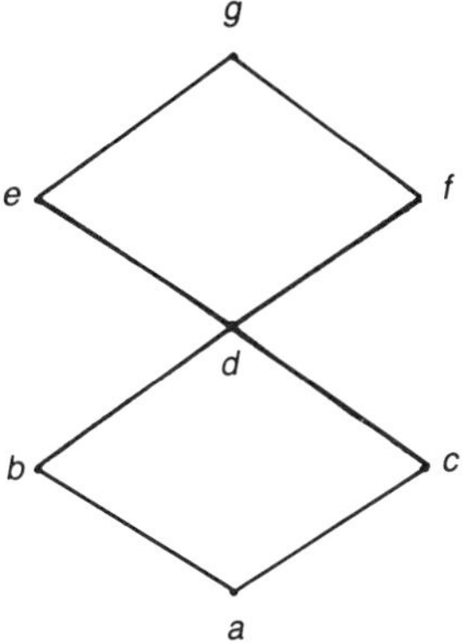

Figure 9

Let the universe A consist of the set {a,b,c,d,e,f,g} of points vertically ordered by the strict betweenness relation Btw. Thus, <a,c,d>, <a,c,e>, <a,c,f>, and <a,c,g> are in Btw because point c is (vertically) between point a and these other points, and similarly <d,c,a>, <e,c,a>, <f,c,a>, and <g,c,a> are in Btw, for the same reason. The other members of Btw should now be obvious from the diagram.

The automorphisms of the (pre)structure A = ({a,b,c,d,e,f,g},Btw) are also easy to see using the diagram, and, clearly, each automorphism of A maps point d to itself. Thus, the set {d} will be an invariant of A ($A \mapsto$ {d}).

But now restrict the structure A to the set B of points {a,b,c,d}, obtaining $A_{|B}$ = (B, $Btw_{|B}$). Again, using the (bottom portion of the) figure, it is easy to see that the mappings

$$\sigma_1 = \begin{pmatrix} a,b,c,d \\ d,b,c,a \end{pmatrix} \qquad \sigma_2 = \begin{pmatrix} a,b,c,d \\ d,c,b,a \end{pmatrix}$$

are automorphisms of $A_{|B}$, but they do not preserve $\{d\}_{|B}$ = {d}. Hence, the set {d}, althought an invariant of A, is not an invariant of $A_{|B}$, and so {d} is not a *free* invariant of A. We shall return to this example later, when the problem of justification of free invariance as a criterion of objectivity is discussed. But now it is time to make the above ideas precise.

Following the above rough outline, the restriction of a structor S of A to a subset of A is defined as follows:

DEFINITION 5.1 *Where* S *is a structor of the set* A *and* B $\subseteq$A, *the restriction* $S_{|B}$ *of* S *to* B *is defined as:*
(*i*) $S_{|B}$ = S, *when* S *is a structor of* B, *and*
(*ii*) $S_{|B}$ = S $\cap$ *Structors*(B), *otherwise.*[68]

The first clause of the definition will apply when the structor considered is an individual in the universe A. If an individual a is in B, then a restricted to B is just a itself; otherwise, clause (ii) applies and the result of restriction is the empty set. When S is not an individual, clause (ii) leads to the expected results. Notice that it follows at once that S restricted to B($S_{|B}$) is always a structor of B. As a result, if we begin with a structor A—$A_{|B}$ is well-defined, equal to (B,$R^{\beta}_{|B}$), and a (pre)structure in its own right. Hence, we are now in a position to define free invariants.

DEFINTION 5.2. *Let* A = (A,R^{β}) *be a (pre)structure and let* S *be a structor of* A. *Then* S *is a free invariant of* A ($A \Rrightarrow$ S) *iff for all subsets* B of A, $S_{|B}$ *is an invariant of* $A_{|B}$($A_{|B} \mapsto S_{|B}$).

Clearly, $A \Rrightarrow$ S implies $A \mapsto$ S, but as we have seen with {d} of the earlier example (figure 9), the converse fails.

At this point we might proceed to define an equivalence relation on (pre) structures using the approach adopted earlier for invariant equivalence (Definitions 2.14 and 2.15): Call two structures *locally invariantly equivalent* just when they have isomorphic *free invariant* extensions, with such extensions defined in the same style as invariant extensions (Definition 2.14). An equivalent approach, however, is to define local invariant equivalence in terms of the local automorphism groups directly; two structures are then held equivalent when their local automorphisms can be appropriately matched up. The definition we adopt then is:

DEFINITION 5.3. *Let A and B be structures. Then A and B are locally invariantly equivalent* ($A \simeq_{\ell i} B$) *iff there is a bijection* ϕ: A → B *such that for all subsets* D *of* A, ϕ *maps the automorphisms of* $A_{|D}$ *to the automorphisms of* $B_{|\phi[D]}$, *that is,* $\phi[\text{Auto}(A_{|D})] = \text{Auto}(B_{|\phi[D]})$.

To get a feel for this idea, imagine the structure *A* as given by a list of each of its subsets (A included, of course) together with a set of permutations of that subset. (These will be the automorphisms of *A* restricted to that subset.) From the former, globally invariant standpoint, two structures *A* and *B* were held to be equivalent if it was possible to find a bijection from A to B that paired the set A and its associated permutations (Auto(*A*)) with the set B and its permutations (see Corollary 2.17, i, iii). The new definition (5.3) requires that this be possible for *all* subsets of the universe of structure *A* if we are to have structural equivalence.

Earlier (in section 2.4), codeterminate structures were defined as those having the same universe and the same invariants. Now, codeterminate structures (*locally* codeterminate?) would be those having the same *free* invariants. Whereas natural isomorphisms previously were invariant isomorphisms (Definition 2.33), now we would have them be freely invariant isomorphisms, and the new definition of natural generation (Definition 2.50) would be modified accordingly. In short, the entire development of section 2 may now be reworked using free invariants instead of invariants. Exactly how the theorems of that section will be affected by these changes remains to be seen. But before (someone, someday) takes on such a job, are there good grounds for regarding free invariance as necessary and sufficient for objective determination?

Invariance, we argued in section 1, is necessary for objectivity. The reason given was simply this: automorphisms, preserving a structure's generators as they do, must also preserve any structure implicitly determined by these generators. But can we not argue now that, for the same reason, free invariance is also necessary for objectivity? After all, a free invariant is just a structor that is preserved by all local automorphisms, and since these local automorphisms also "preserve" a structure's generators, must they not also preserve any structor objectively determined by these generators?

To see what's wrong with this argument, let us return to the structure *A* = ({a,b,c,d,e,f,g}, Btw) illustrated by figure 9. As we saw, the set K = {d}, although an invariant of *A*, is *not* a free invariant of *A*. Hence, under the present proposal, K is not entitled to be regarded as objectively determined by *A*.

Recall the specific difficulty. For the restricted structure corresponding to the lower half of the diagram, there were automorphisms (σ_1,σ_2 listed earlier) which inverted the vertical order of the points {a,b,c,d}, thus mapping K = {d} onto the set {a}. The set K, which was globally fixed in the structure *A*, no longer plays this unique, "central" role in the substructure corresponding to the "lower half" of *A*, since it is interchangeable or congruent with the set {a}.

So, as the example shows, local automorphisms need not preserve a structure's generators—in this case, the single generator Btw. They do preserve, of course, the *restriction* of these generators to any chosen subset B, but the result of a local automorphism cannot always be "glued back" into the original structure. (Putting it more accurately: a local automorphism need not be extendible to a global automorphism.) Since local automorphisms need not preserve a structure's generators, at least not in any straightforward way, the above argument for the necessity of free invariance as a condition of objectivity collapses.

Another favorable feature of (global) invariance is that it is preserved by (first-order, say) definitions. Not so, however, for free invariants. Again, consider the example depicted by figure 9. Trivially, the betweenness relation Btw, being the generator of *A*, is a free invariant of *A*. The set of extremal points E = {a,g} is first-order definable in *A* by the condition (recalling that Btw is *strict* betweenness, so no point is between another and itself):

$$(C_1) \qquad \underline{E}_x \equiv (y)(z)[\sim \underline{Btw}_{yxz}],$$

which asserts that x is not between any two points. The set K = {d} is now definable by

$$(C_2) \qquad \underline{K}_x \equiv (\exists y)(\exists z)[\underline{Btw}_{yxz} \cdot \sim (\underline{E}_y \vee \underline{E}_z)],$$

which asserts that K is the set of all points that lie between two non-extreme points. The set K, therefore, *is first-order definable* in terms of the (freely invariant) generator of *A*, but *is not*, as we have seen, itself a *free invariant* of *A*. Free invariance, therefore, is *not* preserved by (first-order) definitions.[69]

Free invariance, while perhaps not necessary for objectivity, certainly seems to be sufficient. It would be hard to imagine a plausible, yet stronger, condition. Nevertheless, such proposals are not settled by a potpourri of examples and intuitions, but by following out their consequences in some detail. For example, in the case of relational systems, which free invariants are first-order definable or definable in $L_{\omega,\omega}$ or larger infinitary languages? These and other detailed questions about free invariance need answers before the objectivity problem can be given a confident solution.

Fortunately, without resolving these issues, we are in a position to say something of interest about the solution in a particular class of cases.

5.2. The Free Extendibility Theorem

If invariance is too weak an objectivity criterion, free invariance too strong, and the latter implies the former, then the "correct" criterion must be somewhere "in between," implied by free invariance and implying invariance. Hence, the question arises: Under what conditions do the two criteria—invariance and local invariance—coincide? For, if the premises of the above argument are correct, then—in these cases at least—simple invariance (or free invariance) may be taken as the criterion of objective determinability.

The answer to this question turns out to be remarkably simple. Restrict a structure A to an arbitrary set B of its universe, obtaining $A_{|B}$. We thus obtain, as we have seen, a structure with automorphisms in its own right. We can now show that the invariants of A will coincide with the free invariants of A just in case every automorphism of $A_{|B}$ can be extended to an automorphism of A, regardless of the choice of the subset B. In order to prove this result, we first need some results on how functions commute with structors under restriction.

Restriction of structors was defined previously (Definition 5.1). The customary notion of the restriction of *functions,* however, is a somewhat different idea. As is customary, where f: A → B is a function from A to B, and set C is a subset of A, the restriction of f to C ($f_{\restriction C}$) is defined as the set of pairs (x,f(x)) such that x is in C. ($f_{\restriction C} = \{<x,f(x>: x \varepsilon C\}$). The two types of slashes ("|" and "⌈") serve to distinguish the two types of "restriction." We are now ready for the lemma:

LEMMA 5.4. *Where* $B \subseteq A$, f: B → B *a bijection and* f^+: A → A *a bijection which is an extension of f (i.e.,* $f^+_{\restriction B} = f$), *then*

(*i*) *If* S *is a structor of* B, $f^+[S] = f[S]$, *and where S is a structor of* A:

(*ii*) S *is a structor of* B *iff* $f^+[S]$ *is a structor of* B, *and*

(*iii*) $(f^+[S])_{|B} = f[S_{|B}]$.

Proof of (i). By induction on the rank of S.

Proof of (ii). Suppose $f^+[S]$ is a structor of B. By (i), $f^+[S] = f[S]$, so $f(S) = T$ is a structor of B. Hence, $S = f^{-1}[T]$, and since f^{-1}: B → B is a bijection, by Lemma 2.6, S is a structor of B. The converse is proved similarly.

Proof of (iii). Case 1: S is a structor of B. The by Definition 5.1, $S_{|B} = S$, and by (i) above, $f^+[S] = f[S]$. Since S is a structor of B, by Lemma 2.6, so is f[S]. Putting these together: $(f^+[S])_{|B} = (f[S])_{|B} = f[S] = f[S_{|B}]$.

Case 2. S is not a structor of B. Then by Definition 5.1, $f^+[S] = \{f^+[x]: x \in S \cap Str(B)\} = \{f^+[x]: x \in S \ \& \ f^+[x] \in Str(B)\} =$ (by (ii)) $= \{f^+[x]: x \in S \text{ and } x \in Str(B)\} =$ (by (i)) $= \{f[x]: x \in S \text{ and } x \in Str(B)\} = f[\{x: x \in S \text{ and } x \in Str(B)\}] = f[S \cap Str(B)] = f[S_{|B}]$.

Clause (iii) is the important part of Lemma 5.4 for our purposes. It states that restriction of functions ($\upharpoonright$) commutes with restriction of structors ($|$); it figures importantly in the main result which now follows.

PROPOSITION 5.5. (*Free Extendibility Theorem*) *Let A be any (pre)structure. Then the invariants of A will coincide with the free invariants of A just in case: for every subset* B *of the universe* A *of A, every automorphism of A restricted to* B($A_{|B}$) *may be extended to an automorphism of A.*

Proof. Recall that every free invariant is always an invariant, so we need only show that the converse holds.

($\Leftarrow$) Assume that every automorphism of $A_{|B}$ is extendible to an automorphism of A. We now show that $A \mapsto S$ implies $A \Mapsto S$.

Suppose $A \mapsto S$. Let σ be an automorphism of $A_{|B}$. Then there is an extension σ^+ of σ that is an automorphism of A. Hence, $\sigma^+[S] = S$. But then $(\sigma^+[S])_{|B} = S_{|B}$. By Lemma 5.4(iii), $(\sigma^+[S])_{|B} = \sigma[S_{|B}]$. Hence, $\sigma[S_{|B}] = S_{|B}$. Thus, $A_{|B} \mapsto S_{|B}$, for all B, B $\subseteq$ A, and so $A \Mapsto S$.

($\Rightarrow$) Assume $A \mapsto S$ implies $A \Rightarrow S$.

Suppose τ is an automorphism of $A_{|B}$ that is not extendible to an automorphism of A. We now produce an invariant S* of A which is not a free invariant of A—contradicting our assumption.

First, well-order B, calling the result B^μ. Let $S^* = \{\sigma[B^\mu]: \sigma \in \text{Auto}(A)\}$. Since S* is the automorphic closure of a structor of A, $A \mapsto S^*$.

Suppose $\tau[S^*_{|B}] = S^*_{|B}$. Then, since $B^\mu \in S^*_{|B}$, $\tau[B^\mu] \in S^*_{|B}$ also. Hence, for some $\sigma^* \in \text{Auto}(A)$, $\tau[B^\mu] = \sigma^*[B^\mu]$. Since B^μ is a well-ordering of B, we must then have $\tau(b) = \sigma^*(b)$, for all b $\in$ B. Hence, σ^* is an extension of τ, which is an automorphism of A—a contradiction. Thus, $\tau[S^*_{|B}] \neq S^*_{|B}$, and so $A_{|B} \not\mapsto S^*_{|B}$, and so $A \not\Mapsto S^*$, contradicting our assumption that $A \mapsto S$ implies A $\Mapsto$ S.

Hence, no such automorphism τ exists, that is, for all B, B $\subseteq$ A, every automorphism of $A_{|B}$ is extendible to an automorphism of A.

This result has an interesting direct application to the classical geometries. It is well known that Felix Klein in his famous *Erlanger Program* proposed that these geometries be considered as defined by their respective transformation, or automorphism, groups, with geometrical properties and relations now considered to be those properties or relations that were invariant under these transformations. (See Weyl 39, p. 13.) Thus, Klein's proposal, in effect, amounted to the adoption of invariance as the criterion of objective determination, at least in the case of these geometries. And while we have seen that there are good reasons for not regarding invariance as sufficient for objectivity in some cases (e.g., rigid structures), in others, such as Euclidean spaces, invariance "feels right."

The Free Extendibility Theorem (Proposition 5.5) now allows us to replace these intuitions about invariance and Euclidean spaces with an argument. For in Euclidean metric spaces, any metric-preserving bijection between any two of its subsets may be extended to a motion (isometry) of the entire space, that is, to a (global) automorphism (see Blumenthal 6, p. 93). Hence, by the Free Extendibility Theorem, the *free invariants* and *invariants of Euclidean spaces coincide.* Invariance and free invariance, in this case, turn out to coincide as criteria of objectivity.

Not only do (n-dimensional) Euclidean spaces have this extreme mobility, but the same is also true of the (n-dimensional) hyperbolic and spherical geometries (see Blumenthal 6, p. 231). Hence, in all these cases there are grounds for holding to invariance as a sufficient as well as necessary condition for objectivity.[70] Klein's *Erlanger Program* thereby gains at least a partial vindication.

Can we say the same, however, for the invariant approach to Minkowski space-time? Unfortunately, no. In this case, free mobility fails. Consider any light path (null line) L in a Minkowski space-time $M = (E, \omega^2)$, and let ϕ be any permutation of that path that alters the linear order of the events in L. Since the interval between any two points in L is zero, any permutation of L, and, hence, ϕ, will be an automorphism of the Minkowski space-time restricted to L. However, since any (global) automorphism of Minkowski space-time preserves linear order, clearly ϕ cannot be extended to such an automorphism. Owing to the presence of its null lines, Minkowski space-time is not freely mobile.

Fortunately, this does not affect the section 4 results on the conventionality of simultaneity. To see this, consider the general ε-simultaneity relation $(F \rightarrow \mathrm{Sim}_\varepsilon(F))_M$ (see section 4.2), which pairs each inertial frame (F) with its ε-simultaneity relation Sim(F). A typical member of $(F \rightarrow \mathrm{Sim}_\varepsilon(F))_M$ is thus a pair $<F, \mathrm{Sim}_\varepsilon(F)>$. So what happens now if we restrict this relation to a subset K of the universe E of a Minkowski space-time? Clearly, since any inertial frame F is grounded in *all* the events in the universe E, the pair $<F, \mathrm{Sim}_\varepsilon(F)>$ will be in the restriction of $(F \rightarrow \mathrm{Sim}_\varepsilon(F))_M$ to K *just in case* K *is all of* E. In other words, for any K, $K \subseteq E$, the relation $(F \rightarrow \mathrm{Sim}_\varepsilon(F))_M$ will either be unaffected by its restriction to K, or else be equal to the empty relation. As a result, an ε-simultaneity relation $(F \rightarrow \mathrm{Sim}_\varepsilon(F))_M$ will be a free invariant of M if and only if it is an invariant of M. Hence, we have established:

> COROLLARY 5.6. *The* $\varepsilon = \frac{1}{2}$ *simultaneity relation* $(F \rightarrow \mathrm{Sim}_{1/2}(F))_M$ *is the only* ε*-simultaneity relation that is a free invariant of a Minkowski space-time.*[71]

This has been all too brief a treatment of local automorphisms and objectivity. When free mobility fails, currently "equivalent" formulations may look very different locally. For example, formulations of Minkowski space-time in terms of a metric (ω^2), a metric vector space action on a set (see Winnie 43, §6), or in terms of causal structure (as in Robb 28 or Winnie 43) may impose structure on subregions of space-time quite differently, while yielding the same group of (global) automorphisms. As a result, these formulations, while agreeing on matters of global invariants, will disagree on local invariants, and thus on free invariants.

APPENDIX. THEOREMS AND PROOF-SKETCHES (section 2)

LEMMA 2.2. *Where* x *and* y *are structors of* A, *and* μ,ν *are ordinal numbers:*

(*i*) $\mathrm{rnk}_A(x) = 0$ *iff* $x \in A$ or $x = \phi$,
(*ii*) $\mu \leqslant \nu$ *iff* $\mathrm{tier}_A(\mu) \subseteq \mathrm{tier}_A(\nu)$,
(*iii*) *if* $x \in y$ *and* $y \in \mathrm{tier}_A(\mu)$, *then* $x \in \mathrm{tier}_A(\mu)$, *and*
(*iv*) if $x \in y$, *then* $\mathrm{rnk}_A(x) < \mathrm{rnk}_A(y)$.

Proof of (i). At once from Definition 2.1 (i), and the definition of rank.

Proof of (ii). ($\Rightarrow$) By induction on ν.

(i) $\nu = 0$. Since $\mu = \nu$, we have the result at once.
(ii) Assume the result for all $\mu < \nu$.
Suppose $x \in \mathrm{tier}_A(\mu)$, $\mu \leqslant \nu$.
(a) $\nu = \beta + 1$. If $\mu = \nu$, we are done; otherwise, $\mu \leqslant \beta$. In either case, by the induction hypothesis, $x \in \mathrm{tier}_A(\beta)$. By definition, $\mathrm{tier}_A(\nu) = \mathrm{tier}_A(\beta) \cup P(\mathrm{tier}_A(\beta))$, so $x \in \mathrm{tier}_A(\nu)$.
(b) ν a limit ordinal. By definition, $\mathrm{tier}_A(\nu) = \bigcup_{\mu<\nu} \mathrm{tier}_A(\mu)$, and since $\mu \leqslant \nu$, $x \in \mathrm{tier}_A(\nu)$.

($\Leftarrow$) Assume $\mathrm{tier}_A(\mu) \subseteq \mathrm{tier}_A(\nu)$.
Suppose, however, $\nu < \mu$. Then, from ($\Rightarrow$) above, $\mathrm{tier}_A(\nu) \subseteq \mathrm{tier}_A(\mu)$. Hence, $\mathrm{tier}_A(\mu) = \mathrm{tier}_A(\nu)$, $\nu < \mu$. Now for any ν,μ, $\nu < \mu$, $\mathrm{tier}_A(\nu) \in \mathrm{tier}_A(\mu)$. (Exercise.) So $\mathrm{tier}_A(\nu) \in \mathrm{tier}_A(\mu) = \mathrm{tier}_A(\nu)$, that is, $\mathrm{tier}_A(\nu) \in \mathrm{tier}_A(\nu)$, contradicting the axiom of regularity.

Proof of (iii). By induction on μ, using (ii).

Proof of (iv). By induction on $\mathrm{rnk}_A(y)$, using (iii) above.

LEMMA 2.6. *Where* ψ: $A \rightarrow B$ *and* S *is a structor of* A, $\mathrm{rnk}_A(S) = \mathrm{rnk}_B(\psi[S])$.

Proof. By induction on the rank of S, using Definition 2.4.

LEMMA 2.7. *If* $\sigma \in S_A$ *and* B *is a structor of* A, *and we define* $\sigma^*(b) \equiv \sigma[b]$, *all* $b \in B$, *then, for any structor* S *of* B, $\sigma^*[S] = \sigma[S]$.

Proof. As in Lemma 2.6.

PROPOSITION 2.8. *Let structures B and C be structors of A. Then if* $A \mapsto B$ *and* $B \mapsto C$, $A \mapsto C$. *(Also,* $A \mapsto A$.*)*

Proof. Assume $A \mapsto B$ and $B \mapsto C$. Let σ be in Auto(A). Since $A \mapsto$ B, σ^*, as defined in Lemma 2.7, is a permutation of B (exercise). Since $\sigma[B] = B$, by Lemma 2.7, $\sigma^*[B] = B$. Thus, $\sigma^* \in$ Auto(B). Since $B \mapsto C$, $\sigma^*[C] = C$. Hence, again by Lemma 2.7, $\sigma[C] = C$. Thus, $A \mapsto C$.

LEMMA 2.9. *Let* S *be a structor of set* A, *and* $\phi: A \rightarrow B$, $\psi: B \rightarrow C$. *Then:*

(*i*) $(\psi \cdot \phi)[S] = \psi[\phi[S]]$, *and when* ϕ,ψ *are bijections,*

(*ii*) $\phi^{-1}[\phi[S]] = S$; $\phi[S] = T$ *iff* $S = \phi^{-1}[T]$, *and when* B *and* C *are structors of* A,

(*iii*) $\sigma[\psi \cdot \phi] = \sigma[\psi] \cdot \sigma[\phi]$, *where* $\sigma \in S_A$.

Proof of (i). By induction on the rank of S, using Lemma 2.2 (iv).

Proof of (ii). At once from part (i).

Proof of (iii). From Lemmas 2.10 (i) and (2.9) (i), for all $a \in A$, $(\sigma[\psi] \cdot \sigma[\phi])(a) = \sigma[\psi][\sigma[\phi](a)] = \sigma[\psi][\sigma[\phi\sigma^{-1}(a)]] = \sigma[\psi\phi^{-1}[\sigma[\phi\sigma^{-1}(a)]]] = \sigma[\psi[\phi\sigma^{-1}(a)]] = \sigma[\psi\phi\sigma^{-1}(a)] = \sigma[\psi\phi](a)$.

PROPOSITION 2.12. *Let* G *be a subgroup of* S_A. *Then there is a structor* S *of* A *such that:*

(*i*) Auto((A,S)) = G, *and*

(*ii*) $rnk_A(S) = 3$.

Proof. Let A be any structure with universe A. Let A be a well-ordering of A, and $A_* = \{\{a_0\}, \{a_0,a_1\}, \ldots, \{a_0,a_1, \ldots, a_\omega\}, \ldots\}$, that is, $A_* = \{B: B \subseteq A \text{ and } (a_\mu \in B \text{ and } \nu < \mu) \text{ implies } a_\nu \in B\}$. By induction, we can show (exercise) that:

$$\text{If } \sigma_i,\sigma_j \in S_A \text{ and } \sigma_i[A_*] = \sigma_j[A_*], \text{ then } \sigma_i = \sigma_j. \tag{1}$$

Next, let $S_* = \{\sigma[A_*]: \sigma \in \text{Auto}(A)\}$, and $A_* = (A,S_*)$. Since S_* is the closure of a structor of A (A^μ_*) under all automorphisms of A, $A \mapsto S_*$. Hence, $A \mapsto A_*$, and so Auto(A) $\subseteq$ Auto(A_*).

For the converse, suppose $\tau \in$ Auto(A_*). Then $\tau[S_*] = S_*$, that is, $\{\tau\sigma[A_*]: \sigma \in \text{Auto}(A)\} = \{\sigma[A^\mu_*]: \sigma \in \text{Auto}(A)\}$. Thus, taking $\sigma = 1_A$, $\tau[A^\mu_*] = \sigma_i[A^\mu_*]$, where σ_i is some element of

Auto(A). Hence, from (1), $\tau = \sigma_i$, that is, $\tau \in$ Auto A). Thus, Auto (A_*) $\subseteq$ Auto (A).

From the previous paragraph, Auto(A_*) = Auto(A). Inspection now shows that $\text{rnk}_A(S_*) = 3$.

(Note: When we are not concerned with minimizing rank, the closure of A^μ under automorphisms of A will do just as well.)

PROPOSITION 2.16. *Let* $A = (A,R^\beta)$ *and* $B = (B,S^\gamma)$. *Then* ϕ: $A \simeq i\ B$ *iff* $B \mapsto \phi[R^\beta]$ *and* $A \mapsto \phi^{-1}[S^\gamma]$.

Proof. ($\Rightarrow$) Assume $A = (A,R^\beta)$, $B = (B,S^\gamma)$, and ϕ: $A \simeq i\ B$. Then there are invariant extensions A^+, B^+ of A and B, respectively, such that $\phi: A^+ \simeq B^+$. Since R^β is an invariant of A, by Proposition 2.11 (iii). $B \mapsto \phi[R^\beta]$. Since ϕ^{-1}: $B^+ \simeq A^+$, then, for the same reason, $A \mapsto \phi^{-1}[S^\gamma]$.

($\Leftarrow$) Assume that $B \mapsto \phi[R^\beta]$ and $A \mapsto \phi^{-1}[S^\gamma]$, ϕ: A $\simeq$ B a bijection. Let $B^+ = (B,S^\gamma,\phi[R^\beta])$ and $A^+ = (A,\phi^{-1}[S^\gamma],R^\beta)$. (Strictly, we form a single new sequence in each case.) At once, A^+ and B^+ are invariant extensions of A and B, and also at once $\phi[A^+] = B^+$. Hence, ϕ: $A \simeq i\ B$.

COROLLARY 2.17. *Where* $A = (A,R^\beta)$, $B = (B,S^\gamma)$, *and* ϕ: A $\simeq$ B *is a bijection, the following conditions are equivalent:*

(*i*) $A \simeq i\ B$,

(*ii*) ϕ *takes every invariant of A to an invariant of B, and* ϕ^{-1} *takes every invariant of B to an invariant of A,*

(*iii*) $\phi[\text{Auto}(A)] = \text{Auto}(B)$, *and*

(*iv*) $B \mapsto \phi[R^\beta]$ and $A \mapsto \phi^{-1}[S^\gamma]$.

Proof.

(i) $\Rightarrow$ (ii). Suppose $A \mapsto$ T and $\tau \in$ Auto(B). Then $\tau \in$ Auto(B^+), and since ϕ: $A^+ \simeq B^+$, by Proposition 2.11 (ii), $\phi^{-1}[\tau] \in$ Auto(A^+). Hence, $\phi^{-1}[\tau] \in$ Auto(A). Thus, $\phi^{-1}[\tau][T] = T$, and so by Lemma 2.10 (iii), $\phi^{-1}\tau\phi[T] = T$, that is, $\tau[\phi[T]] = \phi[T]$. Hence, $B \mapsto \phi[T]$. The converse is proved similarly.

(ii) $\Rightarrow$ (iii). Since ϕ: $A^+ \simeq B^+$, by Proposition 2.11 (i), (ii), ϕ [Auto(A^+)] = Auto(B^+). But Auto(A^+) = Auto(A) and Auto(B^+) = Auto(B). Hence, ϕ[Auto(A)] = Auto(B).

(iii) $\Rightarrow$ (iv). (a) Suppose $\tau \in$ Auto(B). Then $\phi^{-1}[\tau] \in$ Auto(A), that is, $\phi^{-1}\tau\phi[R^\beta] = R^\beta$, in other words, $\tau[\phi[R^\beta]] = \phi[R^\beta]$. Hence, $B \mapsto \phi[R^\beta]$. Part (b) is proved similarly.

(iv) $\Rightarrow$ (i). Proposition 2.16.

PROPOSITION 2.18. $\simeq$i *is reflexive, symmetrical, and transitive, and hence an equivalence relation.*

Proof. Reflexivity and symmetry are immediate. Suppose $A = (A, R^{\alpha})$, $B = (B, S^{\beta})$, $C = (C, T^{\gamma})$, $A \simeq\text{i}\, B$, and $B \simeq\text{i}\, C$. Let ϕ: $A \simeq\text{i}\, B$ and ψ: $B \simeq\text{i}\, C$. We show that $\psi \cdot \phi$: $A \simeq\text{i}\, C$. Auto(A) is a structor of A; hence, by Lemma 2.9 (i) and Corollary 2.17 (i), (iii), $\psi \cdot \phi[\text{Auto}(A)] = \psi[\phi[\text{Auto}(A)]] = \psi[\text{Auto}(B)] = \text{Auto}(C)$. So, again by Corollary 2.17, $\psi \cdot \phi$: $A \simeq\text{i}\, C$.

PROPOSITION 2.32. *$A \simeq\text{i}\, B$ just in case there is a structure B', such that $A \simeq B'$ and $B' \approx B$.*

Proof. ($\Rightarrow$) *Let* $A = (A, R^{\beta})$, $B = (B, S^{\gamma})$, and ϕ: $A \simeq\text{i}\, B$. Next, let $B' = (B, \phi[R^{\beta}])$. Clearly, ϕ: $A \simeq B'$. By Corollary 2.17 (i)-(iv), $B \mapsto \phi[R^{\beta}]$, and so $B \mapsto B'$. Suppose $\tau \in \text{Auto}(B')$. Then, by Proposition 2.11, $\phi^{-1}[\tau] \in \text{Auto}(A)$, and by Corollary 2.17, since ϕ: $A \simeq \text{i}B$, $\phi[\phi^{-1}[\tau]] \in \text{Auto}(B)$, that is, $\tau \in \text{Auto}(B)$. Hence, $B' \mapsto B$. Since $B \mapsto B'$, $B' \approx \text{B}'$.

($\Leftarrow$) Suppose ϕ: $A \simeq B'$ and $B' \simeq B$. By Proposition 2.31, $\text{Auto}(B) = \text{Auto}(B')$, so by Proposition 2.11, $\phi[\text{Auto}(A)] = \text{Auto}(B)$. Hence, by Corollary 2.17 (i)-(iii), ϕ: $A \simeq\text{i}\, B$.

PROPOSITION 2.37. *Let η_1: $A \simeq\text{n}\, B$. Then η_1 is a unique invariant isomorphism from A to B iff there is no $\tau \in$ Auto(A), $\tau \neq 1_A$, that is an invariant of A.*

Proof. ($\Rightarrow$) Let η_1: $A \simeq\text{n}\, B$. Suppose that $\tau \in \text{Auto}(A)$, $\tau \neq 1_A$, and $A \mapsto \tau$. We now define:

$$\eta_2 \equiv \eta_1 \cdot \tau^{-1}.$$

Clearly $\eta_2 \neq \eta_1$, otherwise $\tau = 1_A$.

(i) First, we show that $A \mapsto \eta_2$. Let $\sigma \in \text{Auto}(A)$. Then $\sigma[\eta_2] = \sigma[\eta_1 \cdot \tau^{-1}] = \sigma[\eta_1] \cdot \sigma[\tau^{-1}] = \eta_1 \cdot \sigma[\tau^{-1}]$. Since $A \mapsto \tau$, $A \mapsto \tau^{-1}$ (exercise). Hence, $\sigma[\eta_2] = \eta_1 \cdot \tau^{-1} = \eta_2$.

(ii) Finally, we show that $\eta_2[A] = B$. $\eta_2[A] = \eta_1 \cdot \tau^{-1}[A] = \eta_1[A] = B$. Hence, η_2: $A \simeq\text{n}\, B$, $\eta_2 \neq \eta$. So η_1 is not unique.

($\Leftarrow$) Let η_1, η_2: $A \simeq\text{n}\, B$, $\eta_1 \neq \eta_2$. We define:

$$\tau = \eta_2^{-1} \cdot \eta_1.$$

Clearly, $\tau \neq 1_A$, else $\eta_1 = \eta_2$.

(i) $\tau(\text{A}) = (\eta_2^{-1} \cdot \eta_1)[A] = \eta_2^{-1}[\eta_1[A]] = \eta_2^{-1}[B] = A$. Thus, $\tau \in \text{Auto}(A)$.

(ii) Let $\sigma \in \text{Auto}(A)$. Then $\sigma[\tau] = \sigma[\eta_2^{-1}\eta_1] = \sigma(\eta_2^{-1}\eta_1)\sigma^{-1} = \sigma\eta_2^{-1}\sigma^{-1}\sigma\eta_1\sigma^{-1} = \sigma[\eta_2^{-1}] \cdot \sigma[\eta_1] = \eta_2^{-1} \cdot \eta_1 = \tau$. Hence, $A \mapsto \tau$.

COROLLARY 2.38. *If $A \simeq n\ B$, then there is exactly one invariant isomorphism from A to B iff $Z(\text{Auto}(A))$ is trivial.*

Proof. At once from Proposition 2.37 and the definition of Z.

PROPOSITION 2.41. *Let $\eta: A \simeq i\ B$ and $\phi: A \simeq C$. Then letting $\phi^*(b) \equiv \phi[b]$,* all $b \in B$, *the following diagram commutes. (Here $D \equiv \phi[\eta][C]$.)*

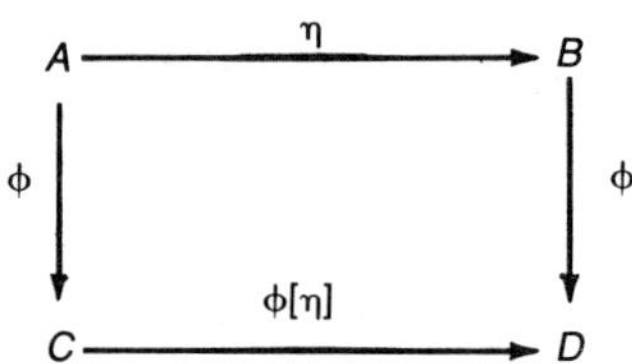

That is, for all $a \in A$, $(\phi^* \cdot \eta)\ (a) = (\phi[\eta] \cdot \phi)\ (a)$.

Furthermore, $\phi[\eta]$, and hence D, is independent of the choice of isomorphism ϕ, that is, for any $\psi: A \simeq C$, $\phi[\eta] = \psi[\eta]$.

Finally, $\phi[\eta]: C \simeq n\ D$.

Proof.

(i) The diagram commutes. Computing as in Lemma 2.10 (iii), $\phi[\eta]\ (x) = \phi[\eta\phi^{-1}(x)]$. Hence, $\phi[\eta] = \phi^*\eta\phi^{-1}$, and so the result at once.

(ii) $\phi[\eta] = \psi[\eta]$ iff $(\psi^{-1}\phi)\ [\eta] = \eta$. Since $\phi, \psi: A \simeq C$, $\psi^{-1}\phi \in$ Auto(A). Hence, since $A \mapsto \eta$, $\psi^{-1}\phi[\eta] = \eta$.

(iii) Finally, we show that $C \mapsto \phi[\eta]$. Let σ' be in Auto(C). Then $\sigma'[\phi[\eta]] = (\sigma'\phi)\ [\eta]$. By Lemma 2.10 (iii) and Proposition 2.11 (ii), $\phi^{-1}\sigma'\phi = \phi^{-1}[\sigma']$ is in Auto(A). Hence $\sigma'[\phi[\eta]] = \phi[(\phi^{-1}\sigma'\phi)[\eta]] = \phi[\eta]$.

COROLLARY 2.43. *Where A and A' have the same universe* A, *if $A \simeq i\ A'$, then $A \approx A'$.*

Proof. Since S is a structor of A iff S is a structor of A', by Proposition 2.42, A and A' have the same invariants. Hence, the result of Proposition 2.31.

PROPOSITION 2.46. *Let A be any structure with* $A = A_0 \dot{\cup} A_1 \dot{\cup} \ldots \dot{\cup} A_\mu \dot{\cup} \ldots$, *each* A_μ *an orbit of* A. *Let* $\bar{a}_0, \bar{a}_1, \ldots, \bar{a}_\mu, \ldots$ *be such that each* $\bar{a}_\mu \in A_\mu$. *Furthermore, let* $S_0, S_1, \ldots, S_\mu, \ldots$ *be distinct structors of* A *such that, for all* $\sigma \in$ Auto(A):

(*i*) $\text{Stab}_A(\bar{a}_\mu) = \text{Stab}_A(S_\mu,)$ *and*

(*ii*) $\sigma(S^\mu) = S_\nu$ *implies* $\mu = \nu$.

If we now define:

$$\eta = \bigcup_\mu \{\sigma[<\bar{a}_\mu, S_\mu>]: \sigma \in \text{Auto}(A)\},$$

then

$$\eta\colon A \simeq n\ \eta[A].$$

Proof. Since η is the closure of a set of structors of A, at once $A \mapsto \eta$. Hence, we must show:

(i) dom (η) = A,

(ii) η is a function, and

(iii) η is one-one (clearly, rng(η) = $\eta[A]$).

Ad (i). Let $a \in A_\mu$. Then for some $\sigma \in \text{Auto}(A)$, $\sigma(\bar{a}_\mu) = a$. Thus, $\sigma[<\bar{a}_\mu,S_\mu>] = <a, \sigma[S_\mu]> \in \eta$. Hence, $a \in \text{dom}(\eta)$. Clearly, $a \in \text{dom}(\eta)$ implies $a \in A$. Hence, dom(η) = A.

Ad (ii). Suppose $<\sigma_i(\bar{a}_\mu),\sigma_i[S_\mu]>$ and $<\sigma_j(\bar{a}_\mu),\sigma_j[S_\mu]>$ are both in η, and $\sigma_i(\bar{a}_\mu) = \sigma_j(\bar{a}_\mu)$, but $\sigma_i[S_\mu] \neq \sigma_j\ [S_\mu]$. Then, $\sigma_j^{-1}\sigma_i(\bar{a}_\mu) = a_\mu$, and $\sigma_j^{-1}\sigma_i \in \text{Stab}_A(a_\mu)$. Thus, $\sigma_j^{-1}\sigma_i[S_\mu] = S_\mu$, that is, $\sigma_i[S_\mu] = \sigma_j[S_\mu]$—a contradiction.

Ad (iii). Suppose $<\sigma_i(\bar{a}_\mu),\sigma_i[S_\mu]>$, $<\sigma_j(\bar{a}_\mu),\sigma_j[S_\nu]> \in \eta$, and (i) $\sigma_i[S_\mu] = \sigma_j[S_\nu]$, but (ii) $\sigma_i(\bar{a}_\mu) \neq \sigma_j(\bar{a}_\mu)$. Since (i), $S_\mu = S_\nu$, and since the S_μ are distinct, $\mu = \nu$. Thus we have: $<\sigma_i(\bar{a}_\mu),\sigma_i[S_\mu]>$, $<\sigma_j(\bar{a}_\mu), \sigma_j[S_\mu]> \in \eta$, (iii) $\sigma_i[S_\mu] = \sigma_j[S_\mu]$, but (iv) $\sigma_i(a_\mu) \neq \sigma_j(\bar{a}_\mu)$. From (iii), $\sigma_j^{-1}\sigma_i[S_\mu] = S_\mu$. Hence, $\sigma_j^{-1}\sigma_i \in \text{Stab}_A(\bar{a}_\mu)$. So $\sigma_1^{-1}\sigma_i(\bar{a}_\mu) = \bar{a}_\mu$, that is, $\sigma_i(\bar{a}_\mu) = \sigma_j(\bar{a}_\mu)$—a contradiction.

LEMMA 2.47. *Let* S *be a structor of* A *which is such that for all* $\sigma \in \text{Auto}(A)$, $\sigma[S] = S$ *implies* $\sigma = 1_A$. *Let* $\bar{a}_0,\bar{a}_1, \ldots, \bar{a}_\mu, \ldots$ *be representatives of each orbit of A, and define:*

$$S\bar{a}_\mu = \{\sigma[<S,\mu>]\colon \sigma \in \text{Stab}_A(\bar{a}_\mu)\}.$$

Then:

(*i*) *the structors* $S\bar{a}_\mu$ *are distinct, and for all* $\sigma \in \text{Auto}(A)$,

(*ii*) $\sigma[S\bar{a}_\mu] = S\bar{a}_\nu$ *implies* $\mu = \nu$,

(*iii*) $\text{Stab}_A(\bar{a}_\mu) = \text{Stab}_A(S\bar{a}_\mu)$.

Proof.

Ad (i), (ii). At once from the indexing of S.

Ad(iii) ($\Rightarrow$) *Assume* $\tau \in \text{Stab}_A(\bar{a}_\mu)$. $\tau[S\bar{a}_\mu] = \{\tau\sigma[<S,\mu>]\colon \sigma \in \text{Stab}_A(\bar{a}_\mu)\}$.
Now, since $\tau \in \text{Stab}_A(\bar{a}_\mu)$, $\{\tau\sigma\colon \sigma \in \text{Stab}_A(\bar{a}_\mu)\} = \text{Stab}_A(\bar{a}_\mu)$. Hence, $\tau[S\bar{a}_\mu] = S\bar{a}_\mu$.
($\Leftarrow$) Assume $\tau[S\bar{a}_\mu] = S\bar{a}_\mu$. Then, if $\sigma_i[<R,\mu>] \in S\bar{a}_\mu$, $\tau\sigma_i[<S,\mu>] = \sigma_j[<S,\mu>]$, $\sigma_i,\sigma_j \in \text{Stab}_A(\bar{a}_\mu)$. Hence, $\sigma_j^{-1}\tau\sigma_i[<S,\mu>] = <S,\mu>$, and so $\sigma_j^{-1}\tau\sigma_i[S] = S$. So $\sigma_j^{-1}\tau\sigma_i = 1_A$, that is, $\tau\sigma_i = \sigma_j$, in other words, $\tau = \sigma_j\sigma_i^{-1}$. Since $\sigma_j,\sigma_i^{-1} \in \text{Stab}_A(\bar{a}_\mu)$, $\tau \in \text{Stab}_A(\bar{a}_\mu)$.

PROPOSITION 2.48. *Let* A^* *be a structure,* η_A: $A^*_c \simeq_n B^*$, *and* A^*, η, *and* G *be defined as in Section 2.5. Then* (I,η,G) *is a natural isomorphism (in the sense of category theory). Furthermore, for all A in the objects of category* A_c, $A \mapsto \eta(A)$.

Proof. That G(A) is a structor of A follows at once from Proposition 2.41. That G is a functor is easily verified by computation. We show that (I,η,G) is natural.

Let B,C be objects in A^*_c and ψ: $B \simeq C$ a morphism. Let ϕ: $A^* \simeq B$, Then, $\psi \cdot \phi$: $A^* \simeq C$. By the definition of η,$\eta_B = \phi[\eta_{A^*}]$, and $\eta_C = (\psi \cdot \phi)[\eta_{A^*}]$. Hence, $\eta_C = \psi[\eta_B]$. As in Lemma 2.10 (iii), and using the fact that ψ is a bijection from B to C, for all x ϵ C, $\eta_c(\psi(x)) = \psi[\eta_B\psi^{-1}(\psi(x))]$, that is, $\eta_C(\psi(x)) = \psi[\eta_B(x)] = G(\psi)(\eta_B(x))$. Hence, $\eta_C\psi = G(\psi)\eta_B$, that is, the diagram in the definition of a natural isomorphism commutes.

Finally, since for any object A in A^*_c, $\eta_A = \phi[\eta_{A^*}]$, where ϕ: $A^* \simeq A$ and η_A: $A^* \simeq_n B^*$, by Proposition 2.41, η_A is a bijection which is an invariant of A.

PROPOSITION 2.51. *If* $A \simeq_n B$, *then* $A \overset{n}{\mapsto} B$.

Proof. By thc discussion in thc tcxt prcccding this proposition wc need only to show that $B \mapsto \eta^*$. Suppose $\tau \in$ Auto(B). Then, by Proposition 2.11 (ii) and Proposition 2.39, there is a $\sigma_i \in$ Auto(A) such that σ_i induces τ on B. Hence, $\tau[\eta^*] = \sigma_i[\eta^*]$. Thus, for all b ϵ B, $\tau[\eta^*](b) = \sigma[\eta^*](b) = \sigma[\eta^*\sigma^{-1}[b]] = \sigma[\eta[\sigma^{-1}[b]]] = \sigma[\eta\sigma^{-1}[b]] = \sigma[\eta][b] = \eta[b] = \eta^*(b)$. Hence, $\tau[\eta^*] = \eta^*$.

PROPOSITION 2.53. *If* $A \overset{n}{\mapsto} B$, *then* $\rho_{A,B}$: Auto(A) $\rightarrow$ Auto(B) *is a bijection, and is thus an isomorphism of* Sym(A) *and* Sym(B).

Proof. From the discussion preceding this proposition in the text, it follows that $\rho_{A,B}$ is a function from Auto(A) to Auto(B). Hence, we must show that $\rho_{A,B}$ is one-one and onto.

First, we show that $\rho_{A,B}$ is one-one. Suppose $\tau \in$ Auto(B), and $\sigma_1,\sigma_2 \in$ Auto(A), with $\rho_{A,B}(\sigma_1) = \rho_{A,B}(\sigma_2) = \tau$. Since $A \overset{n}{\mapsto} B$, let A' be a structure satisfying the conditions defining $\overset{n}{\mapsto}$, and η: $A \simeq_n A'$ be a suitable invariant isomorphism. Since $B \mapsto A'$, τ induces an automorphism $\tau^* \in$ Auto(A'). But then $\rho_{A,A'}(\sigma_1) = \tau^*$ and also $\rho_{A,A'}(\sigma_2) = \tau^*$. Since η is an invariant isomorphism, by Proposition 2.39, $\eta[\sigma_1] = \eta[\sigma_2]$, that is, $\eta^*(\sigma_1) = \eta^*(\sigma_2)$. Hence, since η^* is a bijection (by Proposition 2.11 (iv)), $\sigma_1 = \sigma_2$.

In order to show that $\rho_{A,B}$ is onto, we first show that $\rho_{BA'}$ is one-one. Suppose not. Then there are distinct $\tau_1,\tau_2 \in$ Auto(B) and for all a′ ∈ A′, $\tau_1[a'] = \tau_2[a']$. Hence, $\tau_2^{-1}\tau_1$ induces the identity on A′ and $\tau_2^{-1}\tau_1 \neq 1_B$. Since η: A → A′ and B is a structor of A, by Lemma 2.6, $\eta[B] = B'$is a structor of A′. Furthermore $A' \mapsto B'$.For let σ' be in Auto(A'). Then $\sigma = \eta^{-1}[\sigma'] \in$ Auto(A), and since η is an invariant isomorphism of A and A', then (by Proposition 2.39), $\sigma'[\eta[B]] = \sigma[\eta[B]] = \eta\sigma[B] = \eta[\sigma[B]] = \eta[B]$. Hence, $\sigma'[B'] = B'$, and so $A' \mapsto B'$.

Now let $\tau_* = \tau_2^{-1}\tau_1$ above. Since τ_* induces $1_{A'}$ on A′ and $A' \mapsto B'$, τ_* induces $1_{B'}$ on B′. Thus, $\tau_*[\eta^*] = \tau_*[\{<b,\eta\, b>: b \in B\}] = \{(\tau_*(b),\tau_*[\eta[b]]): b \in B\} = \{(\tau_*(b),\eta[b]): b \in B\}$. Since $\tau_* \neq 1_B$, for some $b_0 \in B$, $\tau_*(b_0) \neq b_0$. Hence, $<b_0,\eta[b_0]>$, $<\tau_*(b_0),\eta[b_0]> \in \eta^*$. But η^* is a bijection (exercise). Hence, a contradiction. Thus, $\rho_{B,A'}$ is one-one.

Finally, suppose $\tau_1 \in$ Auto(B). Since $B \mapsto A'$, τ_1 induces a unique automorphism of A', $\rho_{B,A'}(\tau_1) = \sigma_1'$. Let $\sigma_1 \in$ Auto(A) induces σ_1' (recall, $A \simeq$n A'). Since $A \mapsto B$, σ_1 induces some $\tau_2 \in$ Auto(B). Hence, by Lemma 2.7, if a′ ∈ A′, $\sigma_1[a'] = \tau_2[a']$. Hence, $\tau_2[a'] = \sigma_1'[a']$, that is, τ_2 induces σ_1' on A′. Since the same is true of τ_1 and $\rho_{B,A'}$ is one-one, $\tau_2 = \tau_1$. Thus, σ_1 induces τ_1 on B. Thus, $\rho_{A,B}$ is a bijection. That $\rho_{A,B}$ preserves group composition and the identity is an easy exercise.

Proposition 2.54. *Where A and B are structures:*

(*i*) $A \overset{n}{\mapsto} A$,

(*ii*) *If* $A \overset{n}{\mapsto} B$ *and* S *is a structor of* A *and* B, *then* $A \mapsto$ S *iff* $B \mapsto$ S.

Proof of (i).

Since $A \mapsto A$, the result at once from Proposition 2.51.

Proof of (ii).

(⇒) Assume $A \mapsto$ S. Let τ be in Auto(B). Then, by Proposition 2.53, there is a σ in Auto(A) so that σ induces τ on B, that is $\sigma^* = \tau$. Since S is a structor of B, by Lemma 2.7, since $\sigma[S] = S$, $\tau[S] = S$. Hence, $B \mapsto$ S.

(⇐) Since $A \mapsto B$ and S is an invariant structor of B, $A \mapsto$ S. (See proof of Proposition 2.8.)

Proposition 2.56. $A \overset{n}{\mapsto} B$ *iff* B *is a structor of* A *and* $\rho_{A,B}$: Auto(A) → Auto(B) *is a bijection.*

Proof. (⇒) From Proposition 2.53.

(⇐) Since B is a structor of A, so is B^{ν}—an arbitrary well-ordering of B. Furthermore, if $\tau \in S_B$, clearly $\tau[B^\nu] = B^\nu$ iff $\tau = 1_B$. Now, let σ be in Auto(A). Then $\sigma[B^\nu] = \tau[B^\nu]$, where τ is induced by σ. Hence, $\sigma[B^\nu] = B^\nu$ iff $\tau[B^\nu] = B^\nu$ iff $\tau = 1_B$. Since $\rho_{A,B}$ is a bijection from Auto(A) to Auto(B), $\sigma = 1_A$. Defining $B^\nu{}_{\bar{a}_\mu}$ as in Lemma 2.47 and defining $\eta = \{\sigma[<\bar{a}_\mu, B^\nu_\mu>]: \sigma \in \text{Auto}(A)\}$, we now have, by Proposition 2.46, η: $A \simeq n\, A'$ *where* $A' = \eta[A]$. Furthermore, since η: $A \rightarrow \{\tau[B^\nu_\mu]: \tau \in \text{Auto}(B)\}$, $\eta[A]$, by Lemma 2.6, is a structor of the range of η, and, hence, a structor of B. Hence, we need only to show that: (1) $B \mapsto A'$, and (2) $B \mapsto \eta^*$.

The first is almost immediate. Since η: $A \simeq n\, A'$, $A \mapsto A'$. If $\tau \in$ Auto(B), $\tau[A'] = \sigma[A'] = A'$, where σ induces τ on B. Hence, $B \mapsto A'$.

For the second, we first show that $A \mapsto \eta^*$. Let σ be in Auto(A), and $\eta^* = \{(b,\eta[b])\text{: } b \in B\}$. Then, $\sigma[n^*] = \{(\sigma[b],\sigma[\eta[b]])\text{: } b \in B\} = \{(\sigma[b],\eta\sigma[b])\text{: } b \in B\}$, by Proposition 2.34 (ii) and a simple induction. Since σ induces a bijection on B, let $b' = \sigma[b]$, and so $b = \sigma^{-1}[b']$. Substituting: $\sigma[\eta^*] = \{(b',\eta\sigma[\sigma^{-1}[b']])\text{: } b' \in B\} = \{(b',\eta[b'])\text{: } b' \in B\} = \eta^*$. Hence, $A \mapsto \eta^*$. Thus, by the argument of the preceding paragraph, if $\tau \in$ Auto(B), $\tau[\eta^*] = \eta^*$, that is, $B \mapsto \eta^*$.

PROPOSITION 2.58. *Let* Sym(A) *and* Sym(B) *be isomorphic. Then there is a structure* B' *such that* $A \overset{n}{\mapsto} B'$ *and* $B' \simeq i\, B'$.

Proof. Assume ϕ: Sym(B) $\simeq$ Sym(A). For simplicity's sake, we shall assume that B has a single orbit and indicate how the proof generalizes as we proceed.

Let $B = \{b_0, \ldots, b_\mu, \ldots\}$, $\mu \in I$. (*In the general case,* index each orbit.) Let B_{0,b_μ} be the set of all automorphisms of B that take b_0 to b_μ. Thus, B_{0,b_0} is just the stability group of b_0, and, it is easy to show, B_{0,b_μ} is a (left) coset of B_{0,b_0}. Thus, the sets B_{0,b_μ} are disjoint, exhaust Auto(B), and the mapping Γ defined by $\Gamma(b_\mu) = B_{0,b_\mu}$ is one-one. (Exercise.) (*In the general case,* this holds for *each* orbit of B.) Now we construct a structure B' so that $A \overset{n}{\mapsto} B'$ and $B' \simeq i\, B$.

First, we define:

(i) $b'_\mu = \{\sigma[A^\gamma]: \sigma \in \phi[B_{0,b_\mu}]\}$, where ϕ is the above isomorphism of Sym(B) and Sym(A), and A^γ is any well-ordering of the universe A of A. (*In the general case,* index A^γ by the orbits of B, obtaining A^γ_0, A^γ_1, etc., and suitably index the b's obtaining b^0_μ, b^1_μ, etc.) Now let:

(ii) $B' = \{b'_\mu: \mu \in I\}$.

Finally, let:

(iii) $B' = (B';G')$,

where G′ is the set of all permutations of B′ induced by Auto(A). (Notice that B′ is a structor of A.)

Next, we show that $A \overset{n}{\mapsto} B'$. By Proposition 2.56, it suffices to show that $\rho_{A,B'}$ is a bijection. First, we show that B′ is an invariant of A (and so $\rho_{A,B'}$: Auto(A) → $S_{B'}$).

Assume $\tau \in$ Auto(A). From (ii), $\tau[B'] = \{\tau[b_\mu']: \mu \in I\} = \{\tau[\{\sigma[A^\gamma]: \sigma \in \phi[B_{0,b\mu}]\}]: \mu \in I\} = \{\{\tau\sigma[A^\gamma]: \sigma \in (\phi[B_{0,b\mu}]\}: \mu \in I\}, = \{\{\sigma[A^\upsilon]: \sigma \in \tau\ (\phi[B_{0,b\mu}])\}: \mu \in I\}$, where $\tau(\phi[B_{0,b\mu}])$ is left multiplication by τ. Since ϕ is a group isomorphism, ϕ preserves subgroups and their cosets; hence, the $\phi[B_{0,b\mu}]$ are left cosets of $\phi[B_{0,b0}]$ in Auto (A), and so $\{\tau(\phi[B_{0,b\mu}]): \mu \in I\}$ is just $\{\phi[B_{0,b\mu}], \mu \in I\}$. Hence, $\tau[B'] = B'$. and so $A \mapsto B'$ and $\rho_{A,B'}$: Auto (A) → S_B.

Next, we show that $\rho_{A,B'}$ is one-one. Suppose not. Then the kernel of $\rho_{A,B'}$ is not trivial, and thus there is a $\sigma_* \in$ Auto(A) such that $\sigma_* \neq 1_A$ but σ_* induces $1_{B'}$. Suppose $\sigma_*[b_\mu'] = b_\mu'$, all $\mu \in I$. Then, from (i),

$$\sigma_*[b'_\mu] = \{\sigma_*\sigma[A^\gamma]: \sigma \in \phi[B_{0,b\mu}]\}$$
$$= \{\sigma[A^\gamma]: \sigma \in \sigma_*(\phi[B_{0,b\mu}])\},$$

all $\mu \in I$. And so,

$$\{\sigma[A^\gamma]: \sigma \in \phi[B_{0,b\mu}]\} = \{\sigma[A^\gamma]: \sigma \in \sigma_*(\phi[B_{0,b\mu}])\},$$

all $\mu \in I$. Since $\sigma_1[A^\gamma] = \sigma_2[A^\gamma]$ iff $\sigma_1 = \sigma_2$, we can have the above only if $\phi[B_{0,b\mu}] = \sigma_*(\phi[B_{0,b\mu}])$, all $\mu \in I$; that is, only if $\sigma^* \in [B_{0,b\mu}]$, all $\mu \in I$ (since the $\phi[B_{0,b\mu}]$ are cosets). Since cosets are disjoint, we must have $B_{0,b0} =$ Auto(B). Hence, Auto(B) = $\{1_B\}$. Since, Sym(A) ≃ Sym(B), Auto(A) = $\{1_A\}$. Thus, $\sigma_* = 1_A$—a contradiction. Thus, $\rho_{A,B'}$ is one-one, and so $A \overset{n}{\mapsto}$ (B′;G′).

Finally, we show that $B' \simeq$i B. By Corollary 2.17, it suffices to find a bijection ψ: B′ → B such that ψ[Auto(B')] = Auto(B). We define such a ψ by:

(iv) $\psi(b') = b_\mu,$

for all $b_\mu' \in B'$. First, we show that ψ^{-1}: B → B′ is a bijection.

First we show that ψ^{-1}: B → B′. Since $\psi^{-1}(B_\mu) = B_\mu'$, this follows at once from the definition (i) above of b'_μ.

Now suppose $\psi^{-1}(b_\mu) = \psi^{-1}(b_\nu)$. Then $\{\sigma[A^\gamma]: \sigma \in \phi[B_{0,b\mu}]\} = \{\sigma[A^\gamma]: \sigma \in \phi[B_{0,b\nu}]\}$. But this requires $B_{0,b\mu} = B_{0,b\nu}$, which is impossible unless $b_\mu = b_\nu$. Hence, ψ^{-1} is a bijection.

That ψ^{-1} is onto B′ follows at once from (i) above. Hence, ψ: B′ → B is a bijection.

The last step is to show that ψ[Auto(B')] = Auto(B). Since

ϕ: Sym(B) $\simeq$ Sym(A) and Auto(B') contains just the A-induced permutations on B′, it suffices to show that, for all $\tau \in$ Auto(B), $\psi[\phi(\tau)] = \tau$.

By Lemma 2.10 (iii), $\psi[\phi(\tau)](b_\mu) = \psi\phi(\tau)\psi^{-1}(b_\mu) =$

$$\psi\phi(\tau)(b'_\mu) = \psi\phi(\tau)\ (\{\sigma[A^\gamma]: \sigma \in \phi[B_{0,b\mu}]\}) =$$
$$\psi(\{\phi(\tau)\sigma[A^\gamma]: \sigma \in \phi[B_{0,b\mu}]\}) =$$
$$\psi(\{\sigma[A^\gamma]: \sigma \in \phi(\tau)\ (\phi[B_{0,b\mu}])\}) =$$
$$\psi(\{\sigma[A^\gamma]: \sigma \in \phi[\tau(B_{0,b\mu})]\}) = \text{(exercise!)} =$$
$$\psi(\{\sigma[A^\gamma]: \sigma \in \phi[B_{0,\tau(b\mu)}]\}) = \psi(\tau(b\mu)') = \tau(b\mu).$$

Hence $\psi[\phi(\tau)] = \tau$, and so ψ[Auto(B')] $= B$, yielding $B' \simeq$i B.

COROLLARY 2.59. $A \approx$s B *iff* Sym(A) $\simeq$ Sym(B).

Proof. At once from Proposition 2.58 and the discussion following Proposition 2.57.

NOTES

This work has been partially supported by the National Science Foundation (NSF Soc 78-07281).

1. For a clear account of theories as definitions, see Suppes (33) and (34).
2. As emphasized by Suppes in (33).
3. Looking ahead a bit: The vector space V is n-dimensional if and only if the set of vectors $<v_1, \ldots, v_n>$ spanned by the linearly independent vectors, $v_1, \ldots v_n$ is an invariant of V.
4. Cf. especially, the inability to define the set N of natural numbers in standard languages and the situation in $L_{\infty,\omega}$. Definability is also treated more fully in section 1.2.
5. Thus $\sigma(1) = 3$, $\sigma(2) = 4$, etc. Permutations will often be represented in this way throughout this essay.
6. It is by no means clear, however, that Weyl held that invariance *sufficed* for objectivity, in all structures. Hence, the following developments are probably more accurately seen as an extension of Weyl's views, rather than a description. Of course, the emphasis on invariance, especially in geometry, was around long before Weyl (see sec. 5.2).
7. For the details, see sec. 1.2.
8. For the precise account, see sec. 2, Coroll. 2.17 and Prop. 2.31.
9. For details, see also the discussion of sec. 5.1.
10. The proofs of this section may be omitted on first reading without serious loss of continuity. They draw freely on the concepts and results of the model theory of elementary languages. See Bell and Slomson (2) or Chang and Keisler (7).
11. Some of the following results are scattered throughout in literature disguised by variations in terminology. Others (like Prop. 1.10, 1.11) I have not found in print, but they may well be "known" to those who specialize in these matters.
12. In the more general case, $\underline{R}$ would be replaced by a sequence of relation predicates,

$\underline{R}_1, \ldots, \underline{R}_\mu$ of various degrees; the following results are not affected by the simplification adopted here.

13. More commonly, "Beth's theorem" is used to refer to the statement, equivalent to Prop. 1.1, that implicit and explicit definability coincide.

14. See the discussion of Scott's isomorphism theorem and its consequences in Barwise (1).

15. Precisely, ϕ_A would be such as to satisfy the condition: For any possible model *B*, *B* is a model of ϕ_A iff *B* is isomorphic to *A*. See Barwise (1) for more on Scott sentences.

16. A weak version of this proposition is obtained by replacing $\equiv$ by $\simeq$ throughout, and follows at once from the definition of theoretical invariance.

17. Assuming that such contrary-to-fact conditionals make some sense.

18. In Sec. 5, these matters are discussed further, and an alternative to invariance (and definability) as criteria of objectivity is developed.

19. Throughout, we use a metalanguage containing Zermelo-Fraenkel set theory with individuals and the axiom of regularity (cf. Suppes 35), but any system containing the regularity axiom will do, and some others would be more convenient at some points in the following.

20. Throughout this chapter, proofs that are brief *and* instructive will be found in the main text. Otherwise, proofs or proof sketches will be found in the appendix.

21. Defining $<a,b> = \{\{a\},\{a,b\}\}$, so that $<a,b> \in \text{tier}_A(2)$ when $a,b \in A$ (and similarly, for ordered n-tuples).

22. Two such structors need not be invariants of *A;* however, it is easy to show (exercise) that the class of all automorphic images of *any structor* of A will always be an invariant of *A*.

23. As we shall see later (sec. 2.6), even this account turns out to be overly restrictive.

24. Often, when dealing with a set A structured by a single structor S, we shall write $A = (A,S)$ rather than the more awkward—but accurate—version: $A = (A,\{<0,S>\})$.

25. Cf. Blumenthal (5), (6), for the appropriate definitions, and Sec. 3 below.

26. Cf. Yale (44), chap. 2.

27. However, consider Suppes' remark: "A satisfactory general definition of isomorphism for two set-theoretical entities of any kind is difficult if not impossible to formulate" (34, p. 262).

28. Notice that the proof of the nonexistence of a set Inv(*A*) does not require the axiom of regularity, since this axiom is not needed to show the nonexistence of the set of all ordinals.

29. But also see the discussion and results of sec. 5 on this point.

30. It might seem, at this point, that clarity would benefit from defining the invariant structure of *A* as the set of all structures invariantly equivalent to *A*. But this proposal suffers from the same defect had by the set of all invariants of *A:* No such set exists. Perhaps this provides good reason for using a less restrictive set theory in our metalanguage, say Gödel-Bernays-von Neumann set theory.

31. Nevertheless, after all this talk of generality, the following account only applies to global coordinate systems for structures. Hence, for structures coordinatized by overlapping patches—like Lorentz space-times—these results have only a local interpretation.

32. Strictly, $\{(\phi,B)\}_A$.

33. Strictly, we should write Z(Sym(*A*)), where, recall, $\text{Sym}(A) = <\text{Auto}(A), \circ>$ is the symmetry group of *A*.

34. This section may be omitted without loss to any following developments.

35. This is not to say that category theory fails to offer valuable insights into this and other problems.

36. For the details on this and the following definitions, see any standard account of category theory; Herlich and Strecker (18) is probably closest in notation and approach to what follows.

37. These can be defined as those morphisms e that are such that, for any morphism ϕ, $\phi \circ e = \phi$, whenever $\phi \circ e$ is in M.

38. As required by the axioms defining a category (see 18, p. 22).

39. Two categories are held to be *equivalent* just in case, after all isomorphic objects have been replaced by arbitrarily chosen single representatives, the resulting categories are isomorphic in the usual sense. As we have already seen, the invariant approach yields an equivalence relation between structures more general than isomorphisms, and in the following section this generality will be increased. It remains to be seen just what the categories will turn out to be equivalent using this approach.

40. See Eilenberg and MacLane (11), introductory section.

41. This is a special case of the standard definition of a natural isomorphism of any functors F and G from category K_1 to category K_2 (cf. 18, 13.1).

42. It appears that we cannot go further and generate a homorphism category from A^* since, in general, a structure A^* will not sufficiently restrict the admissible morphisms so as to yield invariance for each η_A.

43. And hence not preserved by equivalences between categories. For the relevant definitions, see (18), §14.

44. The reader may have noted that we have not shown that $A' \mapsto B'$ holds; this will be established later in this section. For now, computing B' in the example should be convincing enough.

45. Hilary Putnam, in effect, puts this fact to use to argue against scientific realism in Putnam (25). For a (correct, I believe) reply to Putnam, see Merrill (22).

46. A choice of τ which was not a homeomorphism, would, of course, also yield *a* Euclidean topology for E_2, but not the *same* Euclidean topology as that on E_1.

47. Cf. the discussion preceeding Prop. 2.25.

48. Grünbaum (17, p. 99) speaks of the "standard congruence given by $ds^2 = dx^2 + dy^2$" as if the form of the metric in Cartesian coordinates thereby distinguished it from "nonstandard" congruence. But, for any "nonstandard" Euclidean congruence, there exists a Cartesian coordinate system in which that metric takes the Pythagorean form.

49. The choice of a metric d that yields a geometry of a different type will be discussed later.

50. See especially the earlier discussions of Def. 2.30, 2.3, and Prop. 2.57.

51. Cf. Weyl (39, §14).

52. See Modenov and Parkhomenko (23, app.). Thus, any automorphism of E', since it preserves all unit intervals, is an isometry of E. And any automorphism of E, since it preserves all unit intervals, must take all other k_o-intervals of k_o-intervals, and so be an automorphism of E'.

53. In the definiens, "<" is the usual less than relation for integers.

54. However, it may have been so (initially) to Riemann and many earlier geometers. Without precise foundations, the belief that Euclidean geometry is an invariant of a point-set of suitable cardinality and "topology" would not be readily refuted by anything but counterexamples, which is just what Riemann provided. See Torretti (37, §2.25).

55. In the notation of sec. 2.4, $\mathrm{Stab}_A(S) \subseteq \mathrm{Stab}_A(T)$.

56. See Reichenbach (26) for the notion of "descriptive simplicity," Wesley Salmon's (31, chap. 4) provides an excellent survey of the conventionality thesis, and examines a number of (unsuccessful) attempts to refute it.

57. In the four-dimensional case, there are three independent choices of ϵ—one for each independent spatial direction.

58. For the automorphisms of Minowski space-time, see Robertson and Noonan (29, §13.5). The proof of Proposition 4.6, while using a temporal inversion, could just as well have used a spatial reflection or—in the four-dimensional case—a spatial rotation. The proof of Proposition 4.7 carries over to the four-dimensional case as is.

59. Although this has not been shown here, it is possible to prove—using the same methods—the direct analogue of Malament's result; that is, it can be shown that the $\epsilon = 1/2$ simultaneity relation is the only transitive simultaniety relation that is an invariant of a Minkowski space-time.

60. I include here Grünbaum (17), Salmon (31), and Winnie (42).

61. Certainly by Reichenbach, Grünbaum, Salmon, Winnie, and others.

62. There are many examples in the literature. One of the best is Burke Townsend's incisive account (38).

63. For simplicity's sake, a two-dimensional space-time is considered. The points made here carry over to the four-dimensional case routinely.

64. The required transformation is given by $x^{\epsilon 0} = x^0 + x^1(2\epsilon - 1)$.

65. A coordinate system is said to be *adapted* to a frame just in case the set $\{e: x^i(e) = a_i, i = 1,2,3\}$ is a trajectory of *F*, for all $a_i \in$ Reals.

66. See Fraïssé (12, Vol 1, §4.1) for the same idea restricted to relational systems.

67. Cf. Fraïssé's notion of "free interpretability," in (12, Vol. 1, §4.2).

68. Here, only for convenience's sake, the structors of B are regarded as a set.

69. Quantification is the problem here. Free invariance *is* preserved by the sentential connectives.

70. The elliptic space case is unclear (to me). There are examples of congruence between distinct subsets of an elliptic space that do not extend to isometries (Blumenthal 6, §89), but I know of no such example involving a congruence between a set and itself.

71. Actually, we can do even better by not requiring that *F* be a global inertial frame and, instead, consider mere intersections of these with open, connected subsets of E. Now rerestrictions of this localized simultaniety relation will either again be empty, or else the automorphisms of the restricted structure $M_{|K}$ will be extendible to automorphisms of *M*. Once again, invariance and free invariance of local ϵ-simultaneity will coincide.

REFERENCES

1. Barwise, K. J. "Back and Forth Through Infinitary Logic." In *Studies in Model Theory*, ed. M. D. Morley, pp. 5–34. MAA Studies 8. Mathematical Association of America, 1973.
2. Bell, J. L., and Slomson, A. *Models and Ultraproducts: An Introduction*. Amsterdam: North-Holland, 1969.
3. Bernays, P., and Fraenkel, A. A. *Axiomatic Set Theory*. Amsterdam: North Holland, 1968.
4. Beth, E. W. "On Padoa's Method in the Theory of Definition." *Indag. Math.* 15 (1953): 330–39.
5. Blumenthal, L. *A Modern View of Geometry*. San Francisco: W. H. Freeman, 1961.
6. ———. *Distance Geometry*. 2nd ed. New York: Chelsea, 1970.
7. Chang, C. C., and Keisler, H. J. *Model Theory*. Amsterdam: North-Holland, 1973.
8. De Bouvère, K. L. "A Mathematical Characterization of Explicit Definability." *Nederl. Akad. Wetensch. Proc. Ser. A*, 66 *(Indag. Math.* 25) (1963): 264–74.
9. Dickman, M. A. *Large Infinitary Languages*. Amsterdam: North-Holland, 1975.
10. Earman, J. "How to Talk About the Topology of Time." *Nous* 11 (1977): 211–26.
11. Eilenberg, S., and McLane, S. "General Theory of Natural Equivalences." *Trans. Amer. Math. Soc.* 58 (1945): 231–94.
12. Fraissé, R. *Course of Mathematical Logic*. Vols. 1 and 2. Dordrecht: D. Reidel, 1974.
13. Geroch, R. *General Relativity from A to B*. Chicago: University of Chicago Press, 1978.
14. Glymour, C. *Theories* Ph.D. diss. Indiana University, 1969.

15. Goodman, N. *Ways of Worldmaking.* Indianapolis: Hackett, 1978.
16. Gregory, J. "Beth Definability in Infinitary Languages." *J. Symb. Logic* 39 (1974): 22–27.
17. Grünbaum, A. *Philosophical Problems of Space and Time.* 2nd ed. Dordrecht: Reidel, 1973.
18. Herrlich, H., and Strecker, G. *Category Theory.* Boston: Allyn and Bacon, 1973.
19. Keisler, H. J. "Ultraproducts and Elementary Classes." *Indag. Math.* 23 (1961): 477–95.
20. Latzer, R. "Non-directed Light Signals and the Structure of Time." *Synthèse* 24 (1972): 263–80.
21. Malament, D. "Causal Theories of Time and the Conventionality of Simultaneity." *Nois* 11, (1977): 293–300.
22. Merrill, G. H. "The Model-Theoretic Argument Against Realism." *Phil. Sci.* 47 (1980): 69–81.
23. Modenov, P. S., and Parkhomenko, A. S. *Geometric Transformations.* Vol. 1. New York: Academic Press, 1965.
24. Montague, R., and Vaught, R. L., "Natural Models of Set Theory." *Fund. Math.* 47 (1957): 219–42.
25. Putnam, H. "Realism and Reason." *Proceedings and Addresses of the American Philosophical Association* 50 (1977): 493–98.
26. Reichenbach, H. *The Philosophy of Space and Time.* New York: Dover, 1958.
27. ———. *The Axiomatization of the Theory of Relativity.* Berkeley and Los Angeles: University of California Press, 1969.
28. Robb, A. A. *A Theory of Time and Space.* Cambridge: Cambridge University Press, 1914.
29. Robertson, H. P., and Noonan, T. W. *Relativity and Cosmology.* Philadelphia: W. H. Saunders, 1968.
30. Rose, J. S. *A Course on Group Theory.* Cambridge: Cambridge University Press, 1978.
31. Salmon, W. *Space, Time, and Motion: A Philosophical Introduction.* 2nd rev. Minneapolis: University of Minnesota Press, 1980.
32. Snapper, E., and Troyer, R. *Metric Affine Geometry.* New York: Academic Press, 1971.
33. Suppes, P. "Set-theoretical Structures in Science." Ms.
34. ———. *Introduction to Logic.* Princeton, N.J.: Van Nostrand, 1957.
35. ———. *Axiomatic Set Theory.* Princeton, N.J.: Van Nostrand, 1960.
36. Svenonius, L. "A Theorem on Permutations in Models." *Theoria* 25 (1959): 173–78.
37. Torretti, R. *Philosophy of Geometry from Riemann to Poincaré.* Dordrecht: Reidel, 1978.
38. Townsend, B. "Jackson and Pargetter on Distant Simultaneity." *Phil. of Sci.* 47 (1980): 646–53.
39. Weyl, H. *Philosophy of Mathematics and Natural Science.* Princeton, N.J.: Princeton University Press, 1949.
40. ———. *Symmetry.* Princeton, N.J.: Princeton University Press, 1952.
41. Whitehead, A. N. *The Concept of Nature.* Cambridge: Cambridge University Press, 1926.
42. Winnie, J. A. "Special Relativity Without One-Way Velocity Assumptions: Parts I and II." *Phil. Sci.* 37, nos. 1 and 2 (March and June 1970): 81–99, 223–38.
43. ———. "The Causal Theory of Space-Time." In *Foundations of Space-Time Theories,* Minnesota Studies in the Philosophy of Science, vol. 8, ed. J. Earman, C. Glymour, and J. Stachel. Minneapolis: University of Minnesota Press, 1977.
44. Yale, P. B. *Geometry and Symmetry.* San Francisco: Holden-Day, 1968.

DAVID B. MALAMENT
University of Chicago

Newtonian Gravity, Limits, and the Geometry of Space

> Gravitation simply represents a continual effort of the universe to straighten itself out.
>
> —Edmund Whittaker

1. INTRODUCTION

With hindsight it is possible to reformulate Newtonian gravitational theory so that it strikingly resembles general relativity. The reformulation was first presented in the 1920s by Cartan (1) and Friedrichs (4), and since then further developed by Havas (5), Trautman (10), Künzle (7), and others. Their work is important for several reasons.

First, it shows that various features of general relativity, once thought to be uniquely characteristic of it, do not distinguish it from Newtonian theory. On reformulation the latter, too, is a "generally covariant" theory of four-dimensional space-time structure. As in general relativity, gravity is interpreted geometrically. Rather than thinking of particles as being deflected from straight trajectories by the presence of a gravitational potential, one thinks of them simply as traversing geodesic trajectories in curved space-time. What seemed a mysterious "force" is now nothing but a manifestation of space-time curvature. Also as in general relativity, space-time becomes "dynamical" under the reformulation. Rather than being a fixed, invariant backdrop against which physics unfolds, space-time affine structure participates in the unfolding. Curvature is dynamically correlated with the presence of matter according to Poisson's equation.

Second, the work of Cartan, Friedrichs, and others clarifies the gauge status of the Newtonian gravitational potential. In the geometric formulation of Newtonian theory one works with a single (curved) affine structure. It can be decomposed into two pieces—a flat affine structure and a gravitational potential—to recover the standard formulation of the theory. But in the absence of special boundary conditions, the decomposition will not be unique. Physically, there is no unique way to divide into inertial and gravitational components the forces experienced by particles. Neither has any direct physical significance. Only their "sum" does. (This is one version of the "equivalence principle.") It is an attractive feature of the geometric reformulation that it trades two gauge quantities for this sum.

Third, the work under discussion provides the means with which to make clear geometric sence of the standard claim that Newtonian gravitational theory is the "classical limit" of general relativity. One considers an appropriate one-parameter family of relativistic models $(M, g_{ab}(\lambda), T_{ab}(\lambda))$ satisfying Einstein's equation, defined for $\lambda > 0$, and then proves that in the limit as $\lambda \to 0$ a classical model $(M, t_a, h^{ab}, \nabla_a, \rho)$ satisfying (the recast version of) Poisson's equation is defined. Intuitively, as $\lambda \to 0$, the null cones of the $g_{ab}(\lambda)$ "flatten" until they become degenerate.

One of my goals in what follows is simply to describe this geometric limiting process in some detail. (My account is a variant of that given by Künzle 8.) I also want to bring the discussion of "limits" to bear on a special topic.

Much has been written about the geometry of space, but almost exclusively in connection with general claims for or against conventionality. There has been little consideration of reasons "internal to physics" why space should have one geometric structure rather than another. My remarks concern one such. It seems to me that there is an interesting sense in which Newtonian physics *must* posit that space (but not space-time) is Euclidean.

The real world (presumably) is governed by general relativity, at least in the large. Newtonian physics is successful experimentally (and distinguished theoretically) because it is the classical limit of that theory. But precisely *to the extent to which it is the classical limit of general relativity, Newtonian gravitational theory is "forced" to posit that space is Euclidean.* The very limiting process that produces Newtonian theory "squeezes out" all spatial curvature. (This is perhaps surprising. One might have thought that any Newtonian assumptions about spatial geometry would appear arbitrary from the vantage point of general relativity.)

It is not difficult to make the italicized statement precise. As noted, there is a clear sense in which the geometric formulation of Poisson's equation $R_{ab} = K\rho t_a t_b$ is the limiting form of Einstein's equation. But against the background of classical space-time structure, this restriction of the Ricci tensor *entails* that space be flat. (This follows from the proposition in section 3.)

In addition to sections on spatial flatness (3) and "limits" (5), I have included background expository sections on classical space-time structure (2) and Newtonian gravitational theory (4). I hope these will be of some interest in their own right.

2. CLASSICAL SPACE-TIME STRUCTURE

We shall take a classical space-time model to be a structure $(M, t_a, h^{ab}, \nabla_a)$ where

(*i*) M is a smooth, connected, four-dimensional differentiable manifold that represents the totality of all "point event locations";
(*ii*) t_a is a smooth, nonvanishing covariant vector field on M which serves as a *temporal metric;*
(*iii*) h^{ab} is a smooth, nonvanishing, symmetric contravariant tensor field on M of signature (0,1,1,1) which serves as a *spatial metric;*
(*iv*) ∇_a is a smooth derivative operator on M which represents the *affine structure* of space-time;

and where the following two conditions are met:

Orthogonality $h^{ab}t_b = 0$
Compatibility $\nabla_a t_b = \nabla_a h^{bc} = 0.$

This characterization is quite general. No global restrictions on M are included. No assumption is made that ∇_a (or even just its "spatial part") is flat.[1]

The objects t_a, h^{ab},∇_a collectively represent the space-time structure which is presupposed by classical (i.e., Galilean relativistic) dynamics. According to that theory, it makes sense to say of a point particle that (A) it is accelerating at 32 ft/sec^2, but not to say that (B) it is at rest (i.e., not moving at all). This fact alone commits classical dynamics to the above listed combination of metrical and affine structures. This should become clear as the separate roles of t_a, h^{ab}, and ∇_a are discussed.

Let us first consider t_a. It endows M with an observer invariant temporal structure. It does so by assigning to any vector ξ^a at a point in M a "temporal length" $t_a\xi^a$. Hence, t_a induces a degenerate cone structure on the tangent space at each point. We can think of vectors as *spacelike* or *timelike* according to whether their temporal lengths vanish or not. It follows from the signature of h^{ab} and the orthogonality condition that the subspace of spacelike vectors at any point is three dimensional.

From the compatibility condition it follows that t_a is closed, that is, $\nabla_{[a}t_{b]} = 0$. So at least locally it must be exact. There must exist a *time function* t satisfying $t_a = \nabla_a t$. If M is topologically $\mathbf{R}^4$ or satisfies a suitable global condition (e.g., if it is simply connected), then a globally defined time function $t\colon M \rightarrow \mathbf{R}$ must exist. In this case space-time can be decomposed into a one-parameter family of global (t = constant) time slices. One can speak of "space" at a given "time." A different choice of time function would result in a different zero-point for the time scale but would induce the same time slices, and the same elapsed intervals between them. (We could have taken a time function as primitive rather then the field t_a, but this way we avoid global restrictions on temporal structure.)

With t_a we can characterize "timelike curves" and use them to represent the "worldlines" of point particles. Suppose I is an open interval in **R** and γ: I→M is a smooth[2] curve with tangent field ξ^a. We say that γ is *timelike* (respectively *spacelike)* if ξ^a is everywhere timelike (respectively spacelike). Timelike curves cannot be tangent to time slices, but they can be arbitrarily "flat." This reflects geometrically the fact that in classical physics there is no upper bound to the speed with which point particles can travel (as measured by any observer).

Given any timelike curve, it can always be reparametrized so that its tangent vectors ξ^a satisfy the normalization condition $t_a\xi^a = 1$. In this case we say that the curve is "parametrized by time" and call the ξ^a "four-velocity" vectors. It is convenient to work with this parametrization, and we shall do so in what follows.

Next, let us consider the role of the derivative operator ∇_a. It imparts an affine structure to M; without it we could not attribute acceleration to timelike curves (and so represent assertion A).

Quite generally, to have an affine structure on a manifold is just to have a notion of constancy of vector fields along curves (or a notion of equality between vectors residing in distinct, but "infinitesimally close," tangent spaces). In our notation the definitions come out as follows. If γ is a curve with tangent field ξ^a, and if λ^a is a vector field along (the image of) γ, then $\xi^n\nabla_n\lambda^a$ is a vector field along γ which gives "the rate of change of λ^a in the direction ξ^a." If $\xi^n\nabla_n\lambda^a = 0$, we say that the field λ^a is *(covariantly) constant* along γ. If $\xi^n\nabla_n\xi^a = 0$, we say the curve is a *geodesic.*

In the special case where γ is a (normalized) timelike curve in a classical space-time model, the vectors $\xi^n\nabla_n\xi^a$ give the *(four)-acceleration* of the curve. (Hence, timelike geodesics have vanishing acceleration, i.e., constant velocity.) It is important that *acceleration vectors are spacelike.* This follows since

$$\begin{aligned} 0 &= \xi^n\nabla_n(1) = \xi^n\nabla_n(t_a\xi^a) = t_a(\xi^n\nabla_n\xi^a) + \xi^a(\xi^n\nabla_n t_a) \\ &= t_a(\xi^n\nabla_n\xi^a). \end{aligned}$$

(For the final equality we need the compatibility condition.)

In terms of ∇_a alone one can assert that a timelike curve has vanishing acceleration, constant acceleration, uniformly changing acceleration, and so forth. But h^{ab} is needed if acceleration *magnitudes* are to be ascribed. Hence, it is needed if A above is to be representable.

h^{ab} serves as a spatial metric, but just how it does so is a bit tricky. Since it carries contravariant indices, it does not directly assign lengths to (contravariant) vectors at all. The first thing to consider is why one does not simply work with a covariant spatial metric h_{ab}. It would straightforwardly assign a spatial length $(h_{ab}\xi^a\xi^b)^{1/2}$ to any vector ξ^a.

Actually, this is just the problem. Suppose, again, we have a timelike curve γ with four-velocity field ξ^a representing the worldline of some point particle. In terms of h_{ab}, we could assert that the particle had 0 velocity (i.e., that it was at rest): $h_{ab}\xi^a\xi^b = 0$. So clearly h_{ab} would introduce a level of space-time structure that has no physical significance within classical dynamics. It would allow us to represent not only A above, but also B.

The contravariant "metric" h^{ab} gives us just what we want—A without B. It does so by *indirectly* assigning lengths to spacelike vectors but no lengths whatsoever to timelike vectors. This works out perfectly since (as noted) acceleration vectors are always spacelike (and velocity vectors never are). To see how h^{ab} does this it suffices to take note of several simple algebraic facts.

Basic Facts

1. Given any vector ρ^a, ρ^a is spacelike iff there exists a vector λ_b satisfying $h^{ab}\lambda_b = \rho^a$.
2. For all vectors $\overset{1}{\lambda}_b$ and $\overset{2}{\lambda}_b$, $h^{ab}\overset{1}{\lambda}_b = h^{ab}\overset{2}{\lambda}_b$ iff $(\overset{2}{\lambda}_b - \overset{1}{\lambda}_b)$ is a multiple of t_b.
3. For all vectors $\overset{1}{\lambda}_b$ and $\overset{2}{\lambda}_b$, if $h^{ab}\overset{1}{\lambda}_b = h^{ab}\overset{2}{\lambda}_b$ then $h^{ab}\overset{1}{\lambda}_a\overset{1}{\lambda}_b = h^{ab}\overset{2}{\lambda}_a\overset{2}{\lambda}_b$.

These are verified using the signature of h^{ab} and the orthogonality condition.[3] The first two assertions are captured in figure 1. Notice that *all* vectors in the cotangent space V_b (at some point in M) are mapped by h^{ab} into *spacelike vectors* in the tangent space V^a (at that point).

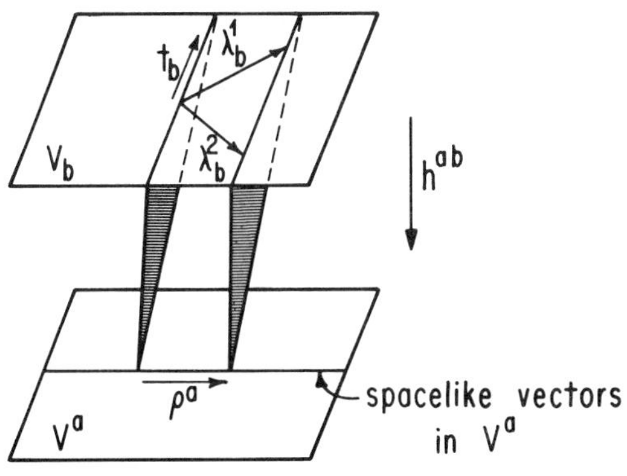

Figure 1

The mapping is not injective. The preimage of any spacelike vector is a set of covariant vectors, any two of which differ by a multiple of t_b.

Now suppose that ρ^a is spacelike. Here is how h^{ab} indirectly assigns a length to ρ^a. By (1) there must exist a vector λ_b satisfying $h^{ab}\lambda_b = \rho^a$. We take the *length* of ρ^a to be just $(h^{ab}\lambda_a\lambda_b)^{1/2}$. By (3) this assignment is independent of our choice λ_b. So the length is well defined. Notice that the procedure does not work for timelike vectors, since they have no preimage under h^{ab} at all.

Finally, let us consider the significance of the compatibility condition. Quite generally, whenever metrical and affine structures coexist on a manifold, each induces a notion of "constancy." Compatibility conditions ensure that they agree.

The condition $\nabla_a h^{bc} = 0$ is equivalent to the requirement that the h^{ab}-length of all ∇_a-constant *spacelike* vector fields along arbitrary timelike curves is constant. It has immediate physical significance. Imagine that we have an "infinitesimal" measuring rod before us in some state of motion. h^{ab} determines whether the rod is changing in length; but ∇_a determines whether one end of the rod is moving relative to the other. Spatial compatibility is just the condition that *if* there is no relative motion, *then* there is no change in length. It is difficult to see how the condition can be violated without our notions of length and motion breaking down altogether.

The temporal compatibility condition $\nabla_a t_b = 0$ can be treated analogously. It is equivalent to the requirement that the t_a-length of all ∇_a-constant vector fields along arbitrary timelike curves be constant.

3. SPATIAL FLATNESS

Suppose $(M, t_a, h^{ab}, \nabla_a)$ is a classical space-time model. In this section we shall discuss different ways of formulating the condition that "space is flat" at some point of M.

We begin by recording for reference several facts about the curvature tensor $R^a{}_{bcd}$. In the case of a classical space-time model, the situation is just a bit different from that in the more familiar case where the derivative operator is derived from a (nondegenerate) Riemannian metric.

The curvature tensor is defined by the conditions $R^a{}_{bcd}\xi^b = -2\nabla_{[c}\nabla_{d]}\xi^a$ or $R^a{}_{bcd}\xi_a = 2\nabla_{[c}\nabla_{d]}\xi_b$ (for arbitrary ξ^a and ξ_a), and satisfies the constraints

$$R^a{}_{b(cd)} = 0 \tag{1}$$

$$R^a{}_{[bcd]} = 0 \tag{2}$$

$$\nabla_{[m}R^a{}_{|b|cd]} = 0. \qquad \text{(Bianchi's identity)} \tag{3}$$

(In the third statement the vertical lines around b indicate that the index is excluded from the antisymmetrization.) All this is quite general and independent of the presence of a metric. The compatibility condition imposes two further constraints:

$$t_a R^a{}_{bcd} = 0 \tag{4}$$

$$h^{b(m} R^{a)}{}_{bcd} = 0. \tag{5}$$

(To verify these note that $0 = 2\nabla_{[c}\nabla_{d]}t_b = R^a{}_{bcd}t_a$, and $0 = 2\nabla_{[c}\nabla_{d]}h^{am} = -R^a{}_{bcd}h^{bm} - R^m{}_{bcd}h^{ab}$.)

Since h^{ab} is not invertible we cannot use it to raise *and* lower indices as one does standardly with a Riemannian metric. But we can *raise* indices with h^{ab}, and it is convenient to do so. For example, we can write $R^a{}_{bcd}h^{mb}$ as $R^{am}{}_{cd}$. Then (5) can be expressed

$$R^{(am)}{}_{cd} = 0. \tag{5a}$$

If we raise all three covariant indices, we form a tensor field R^{abcd}. It has the symmetries

$$R^{ab(cd)} = 0 \tag{6}$$

$$R^{a[bcd]} = 0 \tag{7}$$

$$R^{(ab)cd} = 0. \tag{8}$$

These imply the further condition

$$R^{abcd} = R^{cdab}. \tag{9}$$

(The argument is exactly as in the Riemannian case.)

Now consider the Ricci tensor $R_{bc} = R^a{}_{bca}$ and the scalar curvature $R = R_{bd}h^{bd}$. To investigate their properties we introduce an arbitrary unit timelike vector field ξ^a ($t_a\xi^a = 1$). We can lower indices *relative to* ξ^a. Let $\hat{h}_{ab}$ be the symmetric field defined[4] by

$$\hat{h}_{ab}\xi^b = 0 \tag{10a}$$

$$\hat{h}_{ab}h^{bc} = \delta_a{}^c - t_a\xi^c. \tag{10b}$$

(It is called the *spatial projection field relative to* ξ^a.)
Using (4) and (10) we can easily check that

$$R^a{}_{abd} = \hat{h}_{ar}R^{ar}{}_{bd} = 0 \tag{11}$$

$$R^{bc} = \hat{h}_{rs}R^{rbcs} \tag{12}$$

$$R = \hat{h}_{rs}R^{rs}. \tag{13}$$

(E.g., $\hat{h}_{ar}R^{ar}{}_{bd} = \hat{h}_{ar}h^{rs}R^a{}_{sbd} = (\delta_a{}^s - t_a\xi^s)R^a{}_{sbd} = R^s{}_{sbd}$. This, with 5a, gives 11.) Furthermore, it follows from (1) and (2) that

$$R_{bc} - R_{cb} = R^a{}_{bca} - R^a{}_{cba} = R^a{}_{bca} + R^a{}_{cab} = -R^a{}_{abc}.$$

So R_{bc} must be symmetric,

$$R_{[bc]} = 0. \tag{14}$$

A final fact about the curvature tensor we shall need is

$$R^{abcd} = (h^{bc}R^{ad} + h^{ad}R^{bc} - h^{ac}R^{bd} - h^{bd}R^{ac}) + \tfrac{1}{2}\,(h^{ac}h^{bd} - h^{ad}h^{bc})R. \tag{15}$$

This relation is familiar from the case of a three-dimensional Riemannian metric. It follows from the symmetries (6)–(9), (14), and the crucial fact that all the indices in R^{abcd} are spacelike (i.e., contracting with t_a on any index gives 0).[5]

Now we turn to the consideration of spatial flatness. Let p be a point in our classical space-time model $(M, t_a, h^{ab}, \nabla_a)$. We say that *space-time is flat* at *p* if $R^a{}_{bcd} = 0$ there. In parallel, we say that *space is flat* if the "spatial part" of the curvature tensor vanishes there, that is, if $R^{abcd} = 0$ at p. The next proposition serves to motivate the definition. A *spacelike hypersurface* is an imbedded, three-dimensional submanifold of M with the property that all curves (with images) confined to the submanifold are spacelike.

Spatial Flatness Proposition.[6] Let $(M, t_a, h^{ab}, \nabla_a)$ be a classical space-time model. The following conditions are equivalent at any point in M:
(i) Space is flat (i.e., $R^{abcd} = 0$)
(ii) $R^{bd} = 0$
(iii) $R_{bd} = t_{(b}\phi_{d)}$ for some vector ϕ_d.
Furthermore, given any simply connected spacelike hypersurface S, these conditions hold throughout S iff parallel transport of spacelike vectors within S is path independent.

Proof. Let ξ^a be an arbitrary unit timelike vector at p with corresponding projection tensor $\hat{h}_{ab}$. The equivalence of (i) and (ii) follows from (12), (13), and (15). To verify the equivalence of (ii) and (iii) consider the vector

$$\phi_d = 2R_{dn}\xi^n - t_d(R_{mn}\xi^m\xi^n).$$

We have $\phi_d\xi^d = R_{mn}\xi^m\xi^n$ and $\phi_d h^{dr} = 2R_{dn}h^{dr}\xi^n$. Therefore, at any point where $R^{bd} = 0$ it must be the case that $R_{bd} = t_{(b}\phi_{d)}$. (Both sides agree in their action on $\xi^b\xi^d$, $\xi^b h^{dr}$, and $h^{bs}h^{dr}$. Hence, they agree on $\xi^b\xi^d$, all pairs $\xi^b\sigma^d$, and all pairs $\sigma^b\lambda^d$ where σ^b and λ^d are spacelike. So they must agree on *all* pairs of vectors.) The converse (iii) $\Rightarrow$ (ii) is immediate.

Now, let ξ^a be an arbitrary unit timelike field on S with corresponding field $\hat{h}_{ab}$. Given an arbitrary tensor field on S we say that it is *spacelike* (relative to ξ^a) if all its contractions with t_a and ξ^a vanish (e.g., $\alpha^a{}_{bc}$ is spacelike if $t_a\alpha^a{}_{bc} = \alpha^a{}_{bc}\xi^b = \alpha^a{}_{bc}\xi^c = 0$). Clearly h^{ab} and $\hat{h}_{ab}$ are spacelike. We also define $\hat{\delta}_a{}^b = \delta_a{}^b - t_a\xi^b$. We can think of it as the delta (identity) function on S. It is spacelike:

$$\hat{\delta}_a{}^b\xi^a = 0 = \hat{\delta}_a{}^b t_b.$$

And it preserves all *spacelike* vectors:

$$\hat{\delta}_a{}^b\sigma^a = \sigma^b \qquad \text{for all spacelike } \sigma^a$$

$$\hat{\delta}_a{}^b\sigma_b = \sigma_a \qquad \text{for all spacelike } \sigma_b.$$

Furthermore, we can think of $\hat{h}_{ab}$ as a (nondegenerate) Riemannian metric on S. Note that $\hat{h}_{ab}$ does not annihilate any nonzero, *spacelike* vector, and it has the (spatial) inverse h^{ab}, that is,

$$\hat{h}_{ab}h^{bc} = \hat{\delta}_a{}^c.$$

(This is just 10b.) It follows that there exists a unique *three-dimensional* derivative operator D_a on S compatible with $\hat{h}_{ab}$ ($D_m\hat{h}_{ab} = 0$). We can characterize it in terms of ∇_a.

D_a acts on spacelike fields on S by first applying ∇_a and then projecting all covariant indices with $\hat{\delta}_a{}^b$, for example,

$$D_m\alpha^a_{bc} = \hat{\delta}_m{}^p\hat{\delta}_b{}^r\hat{\delta}_c{}^s\nabla_p\alpha^a_{rs}.$$

The projection ensures that the resultant field is spacelike. (There is no need to project the contravariant indices. The condition $\nabla_a t_b = 0$ ensures that these remain spacelike even after ∇_a is applied.) It is not difficult to check that D_a satisfies all the defining conditions for a derivative operator and that $D_m(\hat{\delta}_a{}^b) = D_m(\hat{h}_{ab}) = D_m(h^{ab}) = 0$.[7]

Notice, now, that for all spacelike fields σ^a, λ^a on S, $\sigma^n D_n\lambda^a = \sigma^n\hat{\delta}_n{}^r\nabla_r\lambda^a = \sigma^r\nabla_r\lambda^a$. Hence, D_a and ∇_a induce the same conditions of parallel transport (for spacelike vectors) on S. Thus, for the second half of the proposition it suffices for us to prove: $R^{abcd} = 0$ *on S iff all parallel transport* (of spacelike vectors) *with respect to* D_a *is path independent.* But this formulation is quite manageable. Suppose $\mathsf{R}^a{}_{bcd}$ is the curvature tensor on S associated with D_a. The second clause is equivalent to the vanishing of $\mathsf{R}^a{}_{bcd}$ on S. So it suffices, finally, for us to prove that at all points of S

$$\mathsf{R}^a{}_{bcd} = 0 \text{ iff } R^{abcd} = 0. \tag{16}$$

This just involves a bit of computation. The left-hand condition of (16) is equivalent to the assertion that for all spacelike σ^b

$$0 = \mathsf{R}^a{}_{bcd}\sigma^b = -2D_{[c}D_{d]}\sigma^a = 2\hat{\delta}_c{}^m\hat{\delta}_d{}^n\nabla_{[m}\nabla_{n]}\sigma^a = -\hat{\delta}_c{}^m\hat{\delta}_d{}^n R^a{}_{rmn}\sigma^r.$$

Hence, it is equivalent to the condition

$$0 = \hat{\delta}_c{}^m \hat{\delta}_d{}^n R^a{}_{rmn} h^{rb} = \hat{\delta}_c{}^m \hat{\delta}_d{}^n R^{ab}{}_{mn}.$$

Contracting this equation with $h^{cr}h^{ds}$ yields $R^{abrs} = 0$. Conversely, contracting $R^{abrs} = 0$ with $\hat{h}_{cr}\hat{h}_{ds}$ yields $\hat{\delta}_c{}^m \hat{\delta}_d{}^n R^{ab}{}_{mn} = 0$. So we have (16).

4. NEWTONIAN GRAVITATIONAL THEORY

Let us now consider Newtonian gravitational theory. The *standard*, nongeometrized version is formulated in terms of a gravitational potential V as well as the background space-time affine structure $\overset{S}{\nabla}_a$. The theory asserts that any (point) particle subject only to gravitational force satisfies

$$\xi^n \overset{S}{\nabla}_n \xi^a = -h^{ar} \overset{S}{\nabla}_r V. \text{ (equation of motion)} \quad (17)$$

(Here, ξ^a is the four velocity of the particle.) It also asserts that V satisfies

$$h^{ab} \overset{S}{\nabla}_a \overset{S}{\nabla}_b V = K\rho.^8 \text{ (Poisson's equation)} \quad (18)$$

(Here, ρ is the scalar mass density function, and K is some constant.) It is usually assumed, in addition, that $\overset{S}{\nabla}_a$ is flat. But it is worth noting that both (17) and (18) make perfectly good sense even without this assumption (or any others).[9]

For the geometric reformulation we need to recall a few facts about derivative operators on a manifold M. Given any two such ∇_a and $\overline{\nabla}_a$, there exists a symmetric field C^a_{bc} on M such that for any field $\alpha^{a_1 \ldots a_r}_{b_1 \ldots b_s}$ on M,

$$\overline{\nabla}_m \alpha^{a_1 \ldots a_r}_{b_1 \ldots b_s} = \nabla_m \alpha^{a_1 \ldots a_r}_{b_1 \ldots b_s} + \alpha^{a_1 \ldots a_r}_{nb_2 \ldots b_s} C^n_{b_1 m} + \ldots + \alpha^{a_1 \ldots a_r}_{b_1 \ldots b_{s-1} n} C^n_{b_s m} - \alpha^{n a_2 \ldots a_r}_{b_1 \ldots b_s} C^{a_1}_{nm} - \ldots - \alpha^{a_1 \ldots a_{r-1} n}_{b_1 \ldots b_r} C^{a_r}_{nm}. \quad (19)$$

(The components of C^a_{bc} in a particular coordinate system are given by subtracting the Christoffel symbols of $\overline{\nabla}_a$ from those of ∇_a.) Conversely, given any derivative operator ∇_a on M and a symmetric field C^a_{bc}, (19) defines a new derivative operator on M. If operators ∇_a and $\overline{\nabla}_a$ are related by (19) we shall write $\overline{\nabla}_a = (\nabla_a, C^a_{bc})$. Suppose this relation obtains and that $R^a{}_{bcd}$ and $\bar{R}^a{}_{bcd}$ are the curvature tensors associated with ∇_a and $\overline{\nabla}_a$. Then we have[10]

$$\bar{R}^a{}_{bcd} = R^a{}_{bcd} + 2\nabla_{[c}C^a_{d]b} + 2C^n_{b[c}C^a_{d]n}. \qquad (20)$$

$$\bar{R}_{bc} = R_{bc} + 2\nabla_{[c}C^a_{a]b} + 2C^n_{b[c}C^a_{a]n}. \qquad (21)$$

To "geometrize" Newtonian gravity, we look for a new derivative operator ∇_a whose timelike geodesics are precisely those trajectories that satisfy the equation of motion (17). It will allow us to construe as "free particles" those that were previously described as subject to (no forces except) gravity. The basic existence and uniqueness fact is the following.

Geometrization Lemma. Suppose $(M, t_a, h^{ab}, \overset{S}{\nabla}_a)$ is a classical space-time model, and V is a function on M. Further suppose $\nabla_a = (\overset{S}{\nabla}_a, -t_b t_c h^{ar}\overset{S}{\nabla}_r V)$. Then

(i) $(M, t_a, h^{ab}, \nabla_a)$ is a classical space-time model.

(ii) ∇_a is the unique derivative operator on M such that for all timelike curves with four velocity ξ^a,

$$\xi^n\nabla_n\xi^a = 0 \text{ iff } \xi^n\overset{S}{\nabla}_n\xi^a = -h^{ar}\overset{S}{\nabla}_r V.$$

(iii) $R_{ab} = \overset{S}{R}_{ab} + (h^{mn}\overset{S}{\nabla}_m\overset{S}{\nabla}_n V)\, t_a t_b$.

Proof. The only thing to be checked for (i) is that ∇_a satisfies the compatibility condition. But this follows by orthogonality:

$$\nabla_a t_b = \overset{S}{\nabla}_a t_b + t_n(-t_a t_b h^{nr}\overset{S}{\nabla}_r V) = 0$$

$$\nabla_a h^{bc} = \overset{S}{\nabla}_a h^{bc} - h^{nc}(-t_n t_a h^{br}\overset{S}{\nabla}_r V) - h^{bn}(-t_a t_n h^{cr}\overset{S}{\nabla}_r V) = 0.$$

For (ii) suppose that $\overline{\nabla}_a = (\overset{S}{\nabla}_a, C^a_{bc})$ is an *arbitrary* derivative operator on M. Then $\xi^n\overline{\nabla}_n\xi^a = \xi^n(\overset{S}{\nabla}_n\xi^a - \xi^m C^a_{mn})$. So $\overline{\nabla}_a$ will satisfy the condition in (ii) iff for all timelike curves with four velocity ξ^a, $C^a_{mn}\xi^m\xi^n = -h^{ar}\overset{S}{\nabla}_r V$. One can show that this latter equation holds iff $C^a_{mn} = (-h^{ar}\overset{S}{\nabla}_r V)t_m t_n$.

Finally, for (iii) note that if $C^a_{bc} = (-h^{ar}\overset{S}{\nabla}_r V)t_b t_c$ then $C^n_{b[c}C^a_{a]n} = 0$ and $\overset{S}{\nabla}_{[c}C^a_{a]b} = -t_b t_{[a}\overset{S}{\nabla}_{c]}(h^{ar}\overset{S}{\nabla}_r V) = \tfrac{1}{2}t_b t_c(h^{ar}\overset{S}{\nabla}_a\overset{S}{\nabla}_r V)$. Therefore (iii) follows from (21).

No special assumptions are needed for the geometrization lemma. In particular, it need not be assumed that the original derivative operator $\overset{S}{\nabla}_a$ is flat. But if this is the case, or just if $\overset{S}{R}_{ab}$ vanishes everywhere, then by (iii) Poisson's equation can be recast in terms of ∇_a: $R_{ab} = (K\rho)t_a t_b$. The geometric formulation of Newtonian gravitational theory *begins* with this equation, shown in figure 2. (When additional constraints (24) and (25) are added one can actually work backward to recover the standard formulation.[11])

Geometric Formulation of Newtonian Gravitational Theory

Künzle version (22)–(24)	$\xi^n \nabla_n \xi^a = 0$	(equation of motion)	*Trautman*[12] *version* (22)–(25)	(22)
	$R_{ab} = K\rho t_a t_b$	(Poisson's equation)		(23)
	$R^{[a}{}_{(b}{}^{c]}{}_{d)} = 0$			(24)
	$R^{ab}{}_{cd} = 0$			(25)

Figure 2

Two points are to be stressed, both of which are immediate consequences of our proposition in the previous section. First, the standard formulation of Newtonian theory can be recast geometrically if, *but only if*, space is initially assumed to be flat. (The original derivative operator $\overset{S}{\nabla}_a$ must be Ricci flat if Poisson's equation $h^{ab}\overset{S}{\nabla}_a\overset{S}{\nabla}_b V = K\rho$ is to be expressed in terms of ∇_a alone.)

Second, however we arrived at it, *Poisson's equation (23) implies spatial flatness.* No classical gravitational theory that includes it can allow for spatial curvature.[13] In the next section we show that, in a precise sense, (23) is the classical limit of Einstein's equation. This will complete our account of why space "must" be Euclidean within classical physics.

5. LIMITS

Let us first consider limits of tensor fields quite generally.

Suppose I is some interval in **R**, and $\alpha^a_{bc}(\lambda)$ is a one-parameter family of tensor fields on a manifold M, defined for $\lambda \in$ I. (Fields with other index structures are treated analogously.) Suppose λ_0 is in the closure of I. We say that β^a_{bc} is the *(pointwise) limit* of $\alpha^a_{bc}(\lambda)$ as $\lambda \to \lambda_0$ if for all fields δ^{bc}_a on M,

$$\lim_{\lambda \to \lambda_0} \left[\alpha^a_{bc}(\lambda)\, \delta^{bc}_a \right] = \beta^a_{bc}\, \delta^{bc}_a .$$

This makes sense since $\alpha^a_{bc}(\lambda)\delta^{bc}_a$ is just a real-valued function on M for each λ.

In what follows we shall always work with families $\sigma^{ab}_c(\lambda)$ that are sufficiently well behaved that their associated (covariant) derivative fields also have limits. If we write '$\alpha^{ab}_c(\lambda) \to \beta^a_{bc}$' it should be understood that

$$\beta^a_{bc} \text{ is the limit of } \alpha^a_{bc}(\lambda),$$

$$\nabla_m \beta^a_{bc} \text{ is the limit of } \nabla_m \alpha^a_{bc}(\lambda),$$

$$\nabla_m \nabla_n \beta^a_{bc} \text{ is the limit of } \nabla_m \nabla_n \alpha^a_{bc}(\lambda),$$

and so forth to all orders. (Notice that these derivative conditions do not presuppose any additional structure on our manifold. Derivative operators always exist, at least locally; and the conditions will hold with respect to one of them iff they hold with respect to all.[14])

Suppose that $\alpha^a_{bc}(\lambda) \rightarrow \beta^a_{bc}$ as $\lambda \rightarrow \lambda_0$. We say that $\alpha^a_{bc}(\lambda)$ is *differentiable* (*in* λ) at λ_0 if the difference quotient

$$\frac{1}{\lambda - \lambda_0}\left[\alpha^a_{bc}(\lambda) - \beta^a_{bc}\right]$$

has a limit (in the sense above) as $\lambda \rightarrow \lambda_0$. Clearly, the definition can be iterated. We can say what it means for $\alpha^a_{bc}(\lambda)$ to be *n^{th}-order differentiable* (in λ) at λ_0 for arbitrary n. Just for convenience, we shall work with limits that are differentiable (in λ) to *all orders*. This, too, should be understood when we write '$\sigma^a_{bc}(\lambda) \rightarrow \beta^a_{bc}$'. (We have now imposed two different differentiability restrictions on our limits. Neither is really needed in full. In each case the second-order version of the restriction would probably suffice.)

Now let us consider our special case. A *relativistic space-time model* will be taken to be a structure (M, g_{ab}) where M is a connected, four-dimensional differentiable manifold, and g_{ab} is a (nondegenerate) Riemannian metric on M of Lorentz signature $(+1, -1, -1, -1)$. Our goal is to describe a ("covariant") geometric limiting process in which relativistic models give birth to their classical counterparts. Intuitively, it will involve the opening of null cones, more and more, until they finally become degenerate, that is, tangent to a spacelike hypersurface.

For purposes of motivation, let's first consider a very special case. Suppose t, x, y, z are standard coordinates in Minkowski space-time, and one takes the metric to have components $g_{ab} = \text{diag}(1, -\frac{1}{c^2}, -\frac{1}{c^2}, -\frac{1}{c^2})$ with respect to those coordinates. Then g_{ab} has a well defined limit as $c \rightarrow \infty$. Its inverse $g^{ab} = \text{diag}(1, -c^2, -c^2, -c^2)$ does not have a limit but can be rescaled $\frac{1}{c^2}\, g^{ab}$ so that it does.

The limiting process is special in several respects. The null cones are taken to open symmetrically around the t axis at each point. The opening occurs uniformly across the underlying manifold. And background affine structure is kept fixed (and flat) throughout the process. These features cannot be retained when one considers limits of arbitrary (curved) relativistic space-time models. One could require that the null cones open symmetrically around *some* unit timelike vector field. It seems preferable, however, not to restrict the exact manner in which the cones open. What is essential is that eventually they become degenerate, and that they do so in such a way as to allow both g_{ab} and a rescaled version of g^{ab} to have limits. We shall simply characterize the "$c \rightarrow \infty$" limiting process in terms of these requirements.

Let $g_{ab}(\lambda)$ be a one-parameter family of (nondegenerate) Lorentz metrics on a given manifold M. Assume that λ ranges over some interval $(0, k)$. We are interested in the case where $g_{ab}(\lambda)$ satisfies two conditions:

(C_1) $g_{ab}(\lambda) \rightarrow t_a t_b$ as $\lambda \rightarrow 0$ for some closed field t_a;

(C_2) $\lambda g^{ab}(\lambda) \rightarrow -h^{ab}$ as $\lambda \rightarrow 0$ for some field h^{ab} of signature (0,1,1,1).

(λ corresponds to $\frac{1}{c^2}$.) Consider the geometric significance of (C_1). Since t_a is closed (i.e., $\nabla_{[a}t_{b]} = 0$ for *any* derivative operator ∇_a on M), it is locally exact. So locally there exists a function t satisfying $t_a = \nabla_a t$. The t = constant hypersurfaces are precisely the ones to which the cones of $g_{ab}(\lambda)$ become tangent as $\lambda \rightarrow 0$. [To see this suppose that η^a is any vector at some point tangent to a t = constant hypersurface. So $t_a\eta^a = \eta^a\nabla_a t = 0$. The (squared) length of η^a with respect to $g_{ab}(\lambda)$ is $g_{ab}(\lambda)\eta^a\eta^b$. As $\lambda \rightarrow 0$, this number approaches $t_a t_b \eta^a\eta^b = 0$. So in the limit η^a becomes a null vector!]

Starting with M and $g_{ab}(\lambda)$ as above, we have the beginnings of a classical space-time model $(M, t_a, h^{ab}, —)$. It remains to be *proved* that in the limit described, the derivative operators $\overset{\lambda}{\nabla}_a$ [corresponding to the $g_{ab}(\lambda)$] themselves converge to a derivative operator ∇_a which is compatible with t_a and h^{ab}. [This makes sense. Let $\tilde{\nabla}_a$ be any (auxiliary) derivative operator on M and suppose $\overset{\lambda}{\nabla}_a = (\tilde{\nabla}_a, \overset{\lambda}{C}{}^a_{bc})$, $\nabla_a = (\tilde{\nabla}_a, C^a_{bc})$. Then we can take $\overset{\lambda}{\nabla}_a \rightarrow \nabla_a$ to mean $\overset{\lambda}{C}{}^a_{bc} \rightarrow C^a_{bc}$ as $\lambda \rightarrow 0$. The condition is independent of our choice of $\tilde{\nabla}_a$.]

Proposition on Limits (1). Suppose $g_{ab}(\lambda)$ is a one-parameter family of Lorentz metrics on a manifold M, and t_a, h^{ab} are as in (C_1) and (C_2) above. Then

(i) There is a derivative operator ∇_a on M satisfying $\overset{\lambda}{\nabla}_a \rightarrow \nabla_a$ as $\lambda \rightarrow 0$.

(ii) $(M, t_a, h^{ab}, \nabla_a)$ is a classical space-time model that satisfies condition (24), $R^{[a}{}_{(b}{}^{c]}{}_{d)} = 0$.

Proof. Since $g_{ab}(\lambda) \rightarrow t_a t_b$ (smoothly), there must exist fields h_{ab}, $p_{ab}(\lambda)$, p_{ab} satisfying

$$g_{ab}(\lambda) = t_a t_b - \lambda h_{ab} + \lambda^2 p_{ab}(\lambda) \tag{26}$$

$$\rho_{ab}(\lambda) \rightarrow p_{ab} \text{ as } \lambda \rightarrow 0.$$

[This is just one way of asserting that the limit $g_{ab}(\lambda) \rightarrow t_a t_b$ is twice differentiable.] Similarly, there must exist fields r^{ab}, $s^{ab}(\lambda)$, s^{ab} satisfying

$$\lambda g^{ab}(\lambda) = -h^{ab} + \lambda r^{ab} + \lambda^2 s^{ab}(\lambda) \tag{27}$$

$$s^{ab}(\lambda) \to s^{ab} \text{ as } \lambda \to 0.$$

Since $\lambda\delta_a{}^c = \lambda g_{ab}(\lambda) g^{bc}(\lambda)$ it follows from (26) and (27) that

$$\lambda\delta_a{}^c = -t_a t_b h^{bc} + \lambda(t_a t_b r^{bc} - h_{ab}h^{bc}) + \lambda^2(\ .\ .\ .\ ).$$

Hence (taking the limit $\lambda \to 0$),

$$t_b h^{bc} = 0. \tag{28}$$

Now we prove (i).

Let $\tilde{\nabla}_a$ be any (auxiliary) derivative operator on M, and suppose $\overset{\lambda}{\nabla}_a = (\tilde{\nabla}_a, C^a_{bc}(\lambda))$ for all λ. It is a basic result of Riemannian geometry that

$$C^a_{bc}(\lambda) = \frac{1}{2} g^{ad}(\lambda) \left[\tilde{\nabla}_d\, g_{bc}(\lambda) - \tilde{\nabla}_b\, g_{dc}(\lambda) - \tilde{\nabla}_c\, g_{db}(\lambda) \right].$$

On substituting (26) and (27), and using the fact that $\nabla_{[a} t_{b]} = 0$, we obtain

$$C^a_{bc}(\lambda) = \frac{1}{2}\left(\frac{-h^{ad}}{\lambda} + r^{ad} + \lambda\, s^{ad}(\lambda) \right) \left[-2t_d \tilde{\nabla}_b t_c - \lambda M_{dbc} + \lambda^2 N_{dbc}(\lambda) \right],$$

where

$$M_{dbc} = \tilde{\nabla}_d h_{bc} - \tilde{\nabla}_b h_{dc} - \tilde{\nabla}_c h_{db}$$

$$N_{dbc}(\lambda) = \tilde{\nabla}_d p_{bc}(\lambda) - \tilde{\nabla}_b p_{dc}(\lambda) - \tilde{\nabla}_c p_{db}(\lambda).$$

If we define

$$C^a_{bc} = -r^{ad} t_d \tilde{\nabla}_b t_c + \frac{1}{2} h^{ad} M_{dbc},$$

then by (28) it follows that $C^a_{bc}(\lambda) \to C^a_{bc}$ as $\lambda \to 0$. Hence, if $\nabla_a = (\tilde{\nabla}_a, C^a_{bc})$, then $\overset{\lambda}{\nabla}_a \to \nabla_a$ as $\lambda \to 0$.

Notice next that if

$$\alpha^{a\,.\,.\,.\,b}_{c\,.\,.\,.\,d}(\lambda) \to \alpha^{a\,.\,.\,.\,b}_{c\,.\,.\,.\,d} \text{ as } \lambda \to 0,$$

then it follows that

$$\overset{\lambda}{\nabla}_n\, \alpha^{a\,.\,.\,.\,b}_{c\,.\,.\,.\,d}(\lambda) \to \nabla_n\, \alpha^{a\,.\,.\,.\,b}_{c\,.\,.\,.\,d} \text{ as } \lambda \to 0.$$

[Suppose, for example, $\alpha^a_{cd}(\lambda) \to \alpha^a_{cd}$. Then

$$\overset{\lambda}{\nabla}_n \, \alpha^a_{cd}(\lambda) = \overset{\sim}{\nabla}_n \, \alpha^a_{cd}(\lambda) + \alpha^a_{pd}(\lambda)\, C^p_{nc}(\lambda) + \alpha^a_{cp}(\lambda)\, C^p_{nd}(\lambda) - \alpha^p_{cd}(\lambda)\, C^a_{pn}(\lambda).$$

The limit of the right hand side as $\lambda \to 0$ is

$$\nabla_n \, \alpha^a_{cd} = \overset{\sim}{\nabla}_n \, \alpha^a_{cd} + \alpha^a_{pd}\, C^p_{nc} + \alpha^a_{cp}\, C^p_{nd} - \alpha^p_{cd}\, C^a_{pn}.$$

So, as claimed, $\overset{\lambda}{\nabla}_n \, \alpha^a_{cd}(\lambda) \to \nabla_n \, \alpha^a_{cd}$.] Therefore

$$\overset{\lambda}{\nabla}_n \, g_{ab}(\lambda) \to \nabla_n \, (t_a t_b)$$

and

$$\overset{\lambda}{\nabla}_n \, \lambda \, g^{ab}(\lambda) \to -\nabla_n \, h^{ab} \qquad \text{as } \lambda \to 0.$$

But for all λ, $\overset{\lambda}{\nabla}_n \, g^{ab}(\lambda) = 0 = \overset{\lambda}{\nabla}_n \, g^{ab}(\lambda)$. So $\nabla_n \, h^{ab} = 0$ and $\nabla_n (t_a t_b) = 0$. The latter condition implies $\nabla_a t_b = 0$. Hence, we may conclude that $(M, t_a, h^{ab}, \nabla_a)$ is a classical space-time model.

It remains to verify that the model satisfies condition (24). When working with $g_{ab}(\lambda)$ we can raise *and* lower indices. It is not only true that $\overset{\lambda}{R}{}^{abcd} = \overset{\lambda}{R}{}^{cdab}$, but also that $\overset{\lambda}{R}{}^{a}{}_{b}{}^{c}{}_{d} = \overset{\lambda}{R}{}^{c}{}_{d}{}^{a}{}_{b}$. (Only the former relation (9) holds automatically for classical space-time models.) Thus, for all λ we have

$$\overset{\lambda}{R}{}^{a}{}_{bnd}\, g^{nc}(\lambda) = \overset{\lambda}{R}{}^{c}{}_{dnb}\, g^{na}(\lambda). \tag{29}$$

But now $\overset{\lambda}{R}{}^{a}{}_{bcd} \to R^a{}_{bcd}$ as $\lambda \to 0$, where $R^a{}_{bcd}$ corresponds to ∇_a. [This follows from (20)]. For all λ,

$$\overset{\lambda}{R}{}^{a}{}_{bcd} = \tilde{R}{}^{a}{}_{bcd} + 2\overset{\sim}{\nabla}_{[c} C^a_{d]b}(\lambda) + 2C^n_{b[c}(\lambda) C^a_{d]n}(\lambda).$$

The limit of the right-hand side is just

$$R^a{}_{bcd} = \tilde{R}{}^{a}{}_{bcd} + 2\overset{\sim}{\nabla}_{[c} C^a_{d]b} + 2C^n_{b[c} C^a_{d]n}.$$

Thus, $\overset{\lambda}{R}{}^{a}{}_{bcd} \to R^a{}_{bcd}$ as claimed.] So, if we multiply both sides of (29) by λ and take the limit as $\lambda \to 0$, we derive $R^a{}_{bnd} h^{nc} = R^c{}_{dnb} h^{na}$, or $R^a{}_b{}^c{}_d = R^c{}_d{}^a{}_b$. Condition (24) is an immediate consequence.

Let us now consider the material content of space-time and determine the role played by Einstein's equation in the limiting process.

Suppose that for each λ we have not only a Lorentz metric $g_{ab}(\lambda)$, but also a symmetric (energy-momentum) tensor $T_{ab}(\lambda)$. Further suppose that two new conditions are met:

(C_3) Einstein's equation $\overset{\lambda}{R}_{ab} = 8\pi(T_{ab}(\lambda) - \frac{1}{2} g_{ab}(\lambda)\overset{\lambda}{T})$ holds for all λ.

(C_4) $T^{ab}(\lambda) \to T^{ab}$ as $\lambda \to 0$ for some field T^{ab}.

[Here $\overset{\lambda}{T} = T_{ab}(\lambda)\, g^{ab}(\lambda)$ and $T^{ab}(\lambda) = T_{mn}(\lambda)\, g^{ma}(\lambda)\, g^{nb}(\lambda)$.] As we shall soon prove, the conditions guarantee that Poisson's equation $R_{ab} = 4\pi\rho t_a t_b$ will hold in the limit for suitable ρ. (If 8π is taken as the proportionality factor in Einstein's equation, then K in 23 comes out to 4π.)

Before giving the proof, let us recall the physical interpretation of the energy-momentum tensor T_{ab} and use it to motivate (C_4). Suppose ξ^a is the four velocity of an observer 0 at a point in space-time where the metric is g_{ab}. 0 will decompose T_{ab} into its "temporal and spatial parts" by contracting each index with $\xi^a\xi^m$ or $(\xi^a\xi^m - g^{am})$. (The latter is the "spatial metric" as determined by 0.) Each of the components has direct physical significance.

$$T_{ab}\xi^a\xi^b = \text{mass-energy density (relative to 0)}$$

$$T_{ab}\xi^a(\xi^b\xi^n - g^{bn}) = \text{three-momentum density (relative to 0)}$$

$$T_{ab}\,(\xi^a\xi^m - g^{am})\,(\xi^b\xi^n - g^{bn}) = \text{three-dimensional stress tensor (relative to 0.)}$$

(See for example, Misner, Thorne, and Wheeler 9, p. 131.) It seems reasonable to require of our limiting process that it assign limiting values to these quantities (as determined by some 0). But if this is to be the case, then (C_4) must hold. [Consider a family of coalligned vectors $\xi^a(\lambda)$, each of unit length with respect to $g_{ab}(\lambda)$. (It has a limit.) For each λ, perform the decomposition. If $T_{ab}(\lambda)\ \xi^a(\lambda)\ \xi^b(\lambda)$ and the other two components have limits, then so does $T_{ab}(\lambda)\, g^{am}(\lambda)\, g^{bn}(\lambda)$.]

Proposition on Limits (2). Suppose $g_{ab}(\lambda)$ is a one-parameter family of Lorentz metrics on a manifold M which, together with the symmetric family $T_{ab}(\lambda)$, satisfies conditions (C_1)–(C_4). Further suppose $(M, t_a, h^{ab}, \nabla_a)$ is the limit classical space-time model described in the previous proposition. Then there is a function ρ on M satisfying

(i) $T_{ab}(\lambda) \to \rho t_a t_b$ as $\lambda \to 0$.

(ii) $R_{ab} = 4\pi\rho t_a t_b$. (Here, R_{ab} is associated with ∇_a.)

Proof. Since $T^{ab}(\lambda) \to T^{ab}$ and $g_{ab}(\lambda) \to t_a t_b$ as $\lambda \to 0$, we have $T_{ab}(\lambda) = T^{mn}(\lambda)\, g_{am}(\lambda)\, g_{bn}(\lambda) \to (T^{mn}t_m t_n)\, t_a t_b$ as $\lambda \to 0$. So for (i) it suffices to take $\rho = T^{mn}t_m t_n$.

For each λ we have $\overset{\lambda}{R}_{ab} = 8\pi(T_{ab}(\lambda) - \frac{1}{2} g_{ab}(\lambda)\, \overset{\lambda}{T})$. The limit of the left-hand side as $\lambda \to 0$ is R_{ab}. (In the proof of the previous proposition we checked that $\overset{\lambda}{R}{}^a{}_{bcd} \to R^a{}_{bcd}$.) But $\overset{\lambda}{T} = T^{ab}(\lambda) g_{ab}(\lambda) \to T^{ab}t_a t_b = \rho$ as $\lambda \to 0$. So the limit of the right-hand side is $8\pi(\rho t_a t_b - \frac{1}{2} t_a t_b\, \rho) = 4\pi\rho t_a t_b$. This gives us (ii).

As far as the statement of the proposition goes, ρ is just some function on M. But it "inherits" an interpretation from the energy-momentum tensors $T_{ab}(\lambda)$. Suppose $\xi^a(\lambda)$ is a one-parameter family of coalligned vectors, each of unit length relative to $g_{ab}(\lambda)$, with limit ξ^a. We have

$$T_{ab}(\lambda)\ \xi^a(\lambda)\ \xi^b(\lambda) \to \rho t_a t_b\ \xi^a \xi^b = \rho.$$

So we can think of ρ as the limit mass (-energy) as determined by *any* observer.

Notice that nothing much is changed if a "cosmological constant" is introduced into Einstein's equation. The limiting form of

$$\overset{\lambda}{R}_{ab} = 8\pi(T_{ab}(\lambda) - \frac{1}{2} g_{ab}(\lambda)\ \overset{\lambda}{T}) + C\ g_{ab}(\lambda)$$

is

$$R_{ab} = (4\pi\rho + C)\ t_a t_b.$$

In effect one transfers the constant C to Poisson's equation, but spacelike hypersurfaces still have to be flat.

Now all the pieces are in place. The propositions in this section indicate the sense in which the geometric formulations of Newtonian gravitational theory is the classical limit of general relativity. The proposition in Section 3 shows that spatial flatness is a necessary consequence of that formulation.

NOTES

I wish to thank Robert Geroch for several helpful suggestions. My research was supported by a grant from the National Science Foundation (SOC–7825046).

1. Readers familiar with this characterization of classical space-time structure can skip to sec. 3. With one exception our notation and sign conventions follow Trautman (10). We write h^{ab} where he writes g^{ab}. We reserve the latter for Lorentz metrics in relativity theory.

2. It is cumbersome to repeat the assumption of smoothness every time reference is made to some geometrical object or other. In what follows it should simply be taken for granted wherever meaningful.

3. (Proof) The "if" clauses of (1) and (2) are immediate consequences of orthogonality. The "only if" clause of (1) follows by dimensionality considerations. Construe h^{ab} as a linear map h^{ab}: $V_b \to V^a$ from the cotangent space V_b to the tangent space V^a at some point in M. By the signature of h^{ab} there must exist a three-dimensional subspace $\overline{V}_b \subseteq V_b$ over which h^{ab} is definite. It follows that the image $h^{ab}\,[\overline{V}_b]$ must be a three-dimensional subspace of V^a. But by the "if" clause of (1), it contains only spacelike vectors; and the entire subspace of spacelike vectors in V^a is three dimensional. Hence, $h^{ab}\,[\overline{V}_b]$ must include *all* spacelike vectors.

For the "only if" clause of (2), suppose that $h^{ab}\overset{1}{\lambda}_b = h^{ab}\overset{2}{\lambda}_b$. Then h^{ab} annihilates both t_b and $(\overset{1}{\lambda}_b - \overset{2}{\lambda}_b)$. But the space of all vectors annihilated by h^{ab} is only one dimensional. So $(\overset{1}{\lambda}_b - \overset{2}{\lambda}_b)$ must be a multiple of t_b.

Finally, (3) is a direct consequence of (2). Suppose $h^{ab}\overset{1}{\lambda}_b = h^{ab}\overset{2}{\lambda}_b$. Then $\overset{2}{\lambda}_b = \overset{1}{\lambda}_b + kt_b$ for some k, and, hence, $h^{ab}\overset{2}{\lambda}_a\overset{2}{\lambda}_b = h^{ab}(\overset{1}{\lambda}_a + kt_a)(\overset{1}{\lambda}_b + kt_b) = h^{ab}\overset{1}{\lambda}_a\overset{1}{\lambda}_b$ by orthogonality.

4. Equivalently, one can define $\hat{h}_{ab}$ as follows. Let $\overset{i}{\sigma}{}^a$ $i = 1,2,3$ be any three linearly independent spacelike vectors at a point. Then $\xi^a, \overset{1}{\sigma}{}^a, \overset{2}{\sigma}{}^a, \overset{3}{\sigma}{}^a$ is a basis for the tangent space at that point. To define $\hat{h}_{ab}$ it suffices to specify its action on each of these vectors. For each i, let $\overset{i}{\lambda}_b$ be *any* vector satisfying $\overset{i}{\sigma}{}^a = h^{ab}\overset{i}{\sigma}_b$. Then we stipulate

$$\hat{h}_{ab}\xi^b = 0$$

$$\hat{h}_{ab}\overset{i}{\sigma}{}^b = \overset{i}{\lambda}_a - t_a(\overset{i}{\lambda}_n\xi^n) \qquad \text{for } i = 1,2,3.$$

5. (Proof sketch.) We prove (15) at an arbitrary point p by introducing an appropriate basis there and considering the resulting component relations. Let ξ^a be any unit timelike vector at p with corresponding projection tensor $\hat{h}_{ab}$. Since $\hat{h}_{ab}$ is symmetric we can find linearly independent vectors $\overset{i}{\sigma}_a$ $i = 1,2,3$ at p satisfying $\hat{h}_{ab} = \sum_{i=1}^{3} \overset{i}{\sigma}_a\overset{i}{\sigma}_b$. Since all the indices in R^{abcd} and R^{ab} are spacelike, both tensors are determined by their actions on the basis vectors $\overset{i}{\sigma}_a$. Consider the components

$$\overset{ij}{R} = R^{ab}\overset{i}{\sigma}_a\overset{j}{\sigma}_b \qquad i,j = 1,2,3$$

$$\overset{ijkl}{R} = R^{abcd}\overset{i}{\sigma}_a\overset{j}{\sigma}_b\overset{k}{\sigma}_c\overset{l}{\sigma}_d. \qquad i,j,k = 1,2,3.$$

Because of the symmetries of R^{ab} and R^{abcd}, each has only six independent components:

$$\overset{11}{R}, \overset{12}{R}, \overset{13}{R}, \overset{22}{R}, \overset{23}{R}, \overset{33}{R}$$

$$\overset{1212}{R}, \overset{1313}{R}, \overset{2323}{R}, \overset{1213}{R}, \overset{1223}{R}, \overset{1323}{R}.$$

Now by (12) we have $R^{ab} = R^{rabs}\hat{h}_{rs} = \sum_{i=1}^{3} R^{rabs}\overset{i}{\sigma}_r\overset{i}{\sigma}_s$. Hence, for all $j,k = 1,2,3$ we have the component relation $\overset{jk}{R} = \sum_{i=1}^{3}\overset{ijkl}{R}$. Also, by (13) $R = R^{ab}\hat{h}_{ab} = \sum_{i=1}^{3} R^{ab}\overset{i}{\sigma}_a\overset{i}{\sigma}_b = \sum_{i=1}^{3}\overset{ii}{R}$.

So

$$\overset{11}{R} = -\overset{1212}{R} - \overset{1313}{R} \qquad \overset{12}{R} = -\overset{1323}{R}$$

$$\overset{22}{R} = -\overset{1212}{R} - \overset{2323}{R} \qquad \overset{13}{R} = \overset{1223}{R}$$

$$\overset{33}{R} = -\overset{1313}{R} - \overset{2323}{R} \qquad \overset{23}{R} = -\overset{1213}{R}$$

$$R = \overset{11}{R} + \overset{22}{R} + \overset{33}{R}.$$

Using these we can check that the two sides of (15) agree in their action on any quadruple $\overset{i}{\sigma}_a\overset{j}{\sigma}_b\overset{k}{\sigma}_c\overset{l}{\sigma}_d$. As an example, consider $\overset{1}{\sigma}_a\overset{2}{\sigma}_b\overset{1}{\sigma}_c\overset{2}{\sigma}_d$. Since $\hat{h}_{ab} = \sum_{i=1}^{3}\overset{i}{\sigma}_a\overset{i}{\sigma}_b$ we have $h^{ab}\overset{1}{\sigma}_a\overset{1}{\sigma}_b = h^{ab}\overset{2}{\sigma}_a\overset{2}{\sigma}_b = 1$ and $h^{ab}\overset{1}{\sigma}_a\overset{2}{\sigma}_b = 0$. So it suffices for us to show

$$\overset{1212}{R} = (-\overset{22}{R} - \overset{11}{R}) + \frac{1}{2}R.$$

But this follows from the entries in our table.

6. A somewhat different version is given in Künzle (7).

7. Note that $D_m(\hat{\delta}_a{}^b) = \hat{\delta}_m{}^r\hat{\delta}_a{}^s\nabla_r\hat{\delta}_s{}^b = \hat{\delta}_m{}^r\hat{\delta}_a{}^s\nabla_r(\delta_s{}^b - t_s\xi^b) = -\hat{\delta}_m{}^r\,\hat{\sigma}_a{}^s\,t_s\nabla_r\xi^b = 0$. Hence, $0 = D_m(\hat{h}_{ab}h^{bc}) = h^{bc}D_m\hat{h}_{ab}$.
But we also have $\xi^b D_m\hat{h}_{ab} = \xi^b\hat{\delta}_m{}^r\hat{\delta}_a{}^s\hat{\delta}_b{}^u\nabla_r\,\hat{h}_{su} = 0$. So $D_m\,\hat{h}_{ab} = 0$.

8. Poisson's equation will look more familiar if we introduce an arbitrary unit timelike field and recast the equation in terms of the associated induced spatial derivative operator D_a. (See the proof of the proposition in sec. 3.) We have $D^2V = D^nD_nV = h^{mn}D_mD_nV = h^{mn}\,\hat{\delta}_m{}^r\,\nabla_r\,(\hat{\delta}_n{}^s\,\nabla_s\,V) = h^{mn}\,\nabla_m\,\nabla_n\,V = K\rho$.

9. Thus, the standard (but not geometric) formulation of Newtonian gravitational theory *does* lend itself to a generalization in which space is non-Euclidean. In some accounts of "Newtonian cosmology" the generalization is pursued. (See, for example, Heckmann 6, p. 8.) This does not contradict our claims in sec. 1. The generalizations are ruled out *if* one demands that one's classical gravitational theory qualify as a limiting version of general relativity.

10. To prove (20) it suffices to compute $\overline{\nabla}_{[c}\overline{\nabla}_{d]}\lambda^a$ using (19) for an arbitrary λ^a.

11. The recovery theorems of Künzle and Trautman are presented in (7) and (10) respectively. As will be seen in the next section, (23) and (24) are "forced" when one considers the classical limit of general relativity. This is not the case with Trautman's (25). It is simply an added condition that assures the possibility of recovering a flat derivative operator. It can be thought of as a requirement that spacelike hypersurfaces are flat and, in a sense, "parallel to one another." More precisely, *given any open, simple connected region 0 in a classical space-time model, (25) holds in 0 iff parallel transport of spacelike vectors in 0 is path independent.* This is proved much as was the proposition in sec. 3. (There one considered parallel transport of spacelike vectors within a particular spacelike hypersurface; here one considers their transport throughout an open region.)

Note added (September 1983): Jürgen Ehlers has pointed out to me that Künzle's recovery theorem, as formulated in (7), is defective. One can recover standard formulations of Newtonian gravitational theory only if one first supplements (23) and (24) with a global condition of asymptotic flatness. Thus, Künzle's theory should be seen as a kind of generalization of Newtonian gravitational theory. It reduces to the latter in the special case of greatest physical and historical interest (i.e., the case of a gravitational field surrounding an "isolated system" such as our solar system.) But it does not do so in general. In contrast, Trautman's theory *does* properly deserve to be thought of as a version of Newtonian gravitational theory. These points are discussed with great care in J. Ehlers, "Über den Newtonschen Grenzwert der Einsteinschen Gravitationstheorie," in *Grundlagen probleme der modernen Physik,* J. Nitsch, J. Pfarr, and E. W. Stachow, eds. Bibliographisches Institut, 1981. Ehler's work goes beyond my own in several respects. Unfortunately, I did not know about it when this paper was written in 1979.

12. In (10), Trautman's axiom 2 is $t_{[n}R^a{}_{b]cd} = 0$. This is equivalent to (25). His axiom 3 is $R^a{}_b{}^c{}_d = R^c{}_d{}^a{}_b$, which is a slightly stronger version of (24). (Actually the two are equivalent in the presence of 25.) His axiom 1 is $R^a{}_{abd} = 0$. This relation is automatic on our account of classical space-time structure. (Recall 11.)

13. This is a point about which Earman and Friedman (3, p. 335) are mistaken. They claim that once (24) and (25) are dropped, one can no longer prove that spacelike hypersurfaces are Euclidean. (The oversight does not affect the principal claims of their paper.)

14. This follows because any two derivative operators ∇_a and $\overline{\nabla}_a$ on M will satisfy (19) for some field $C^a_{b\,c}$.

REFERENCES

1. Cartan, E. "Sur les Variétés a Connexion Affine et la Théorie de la Relativité Généralisée." *Annales Scientifiques de l'Ecole Normale Supérieure* 40 (1923): 325–412, and 41 (1924): 1–25.
2. Dautcourt, G. "Die Newtonsche Gravitationstheorie als Strenger Grenzfall der Allgemeinen Relativitätstheorie." *Acta Physics Polonica* 65 (1964): 637–46.
3. Earman, J. and Friedman, M. "The Meaning and Status of Newton's Law of Inertia and the Nature of Gravitational Forces." *Philosophy of Science* 40 (1973): 329–59.
4. Friedrichs, K. "Eine invariante Formulierung des Newtonschen Gravitationsgesetzes und des Grenzüberganges vom Einsteinschen zum Newtonschen Gesetz." *Mathematische Annalen* 98 (1927): 566–75.
5. Havas, P. "Four-Dimensional Formulations of Newtonian Mechanics and Their Relation to the Special and General Theory of Relativity." *Reviews of Modern Physics* 36 (1964): 938–65.
6. Heckmann, O. *Theorien der Kosmologie.* 2nd ed. Vienna: Springer-Verlag, 1968.
7. Künzle, H. "Galilei and Lorentz Structures on Space-time: Comparison of the Corresponding Geometry and Physics." *Annales Institute Henri Poincaré* 17 (1972): 337–62.
8. ———. "Covariant Newtonian Limit of Lorentz Space-Times." *General Relativity and Gravitation* 7 (1976): 445–57.
9. Misner, C., Thorne, K., and Wheeler, J. *Gravitation.* San Francisco: W. H. Freeman, 1973.
10. Trautman, A. "Foundations and Current Problems of General Relativity." In *Lectures in General Relativity,* ed. S. Deser and K. Ford. Englewood Cliffs, N.J.: Prentice-Hall, 1965.

MICHAEL R. GARDNER
Intellimac, Inc. Rockville, Md.

Quantum Mechanics and the Received View of Theories

> The new line in philosophy of science is the necessary result of the new trends in physics, particularly of the theory of relativity and the quantum theory.
>
> —Phillip Frank

> The quantum-mechanical formalism . . . represents a purely symbolic scheme permitting only predictions . . . as to results obtainable under conditions specified by means of classical concepts.
>
> —Niels Bohr

1. INTRODUCTION

In the last few years, we have heard a great deal of critical discussion of something known as the Received (or Logical Positivist) View of scientific theories. Lately, various criticisms have become so widely accepted that it seems fair to say that the Received View is no longer received in anything but name. The Received View asserts that the nonlogical vocabulary of a scientific theory can be divided into two parts, consisting of observation terms and theoretical terms; that the theory admits of a canonical formulation which contains semantical rules only for the observation terms; and that the theoretical terms are partially interpreted through the theory's postulates and the semantical rules for the observation terms. The observation terms are said to be those variables and constants that range over or denote observable objects and those predicates that ascribe observable attributes to such objects. The theoretical terms comprise the remaining nonlogical vocabulary.[1]

It is not entirely clear whether an instrumentalist conception of the purpose of scientific theories is to be regarded as a corollary of the

Received View's conception of their structure. I define instrumentalism as the doctrine that the purpose of a theory containing theoretical terms is not to describe or explain unobservable phenomena, but only to permit the prediction of observable phenomena. If we suppose a theory contains no terms that designate or are true of unobservable objects, then obviously it can be no part of the purpose of a theory to describe, or explain the properties of, such objects. But this supposition is not the only reason a proponent of the Received View might have for asserting that the reconstruction of a scientific theory contains no semantical rules for the theoretical terms. He might instead think that theoretical terms have denotations, ranges, or extensions, but that they are only partially specified by the postulates of the theory and the semantical rules for the observation terms. Thus, the Received View has an instrumentalist and also a realist interpretation, depending on whether it is associated with the first or the second line of thought.

As it happens, many proponents of the Received View explicitly espoused instrumentalism and thus made clear which interpretation they held. Phillip Frank, for example, writes that according to the logical positivist ("Received") view, which he endorses, "the system of science . . . is an instrument to be invented and constructed in order to find one's way among experiences." "The symbols used in the principles [have] in themselves no meaning beyond their value for the derivation of facts."[2] Similarly, Hempel's well-known discussion of "The Theoretician's Dilemma"[3] is based entirely on the presupposition that "scientific systematization is ultimately aimed at establishing explanatory and predictive order among . . . the phenomena that can be 'directly observed' by us." The Kemeny-Oppenheim model of intertheoretic reduction also presupposes that the entire factual content of a theory is contained in its observational theorems.[4] In what follows I always have in mind this instrumentalist version of the Received View.

A great deal of the criticism aimed at the Received View has centered on the observational/theoretical distinction, in terms of which it is stated. Critics have held that the distinction cannot be made tolerably precise in a way compatible with the intentions of its users. Achinstein, for example, has argued that one of Carnap's explications—a predicate is theoretical if its applicability is not quickly decidable with a few observations unaided by instruments—leads to the consequence that the predicate "is chopped sirloin" is theoretical.[5] It follows that "This is chopped sirloin" is not a statement of fact but a device useful for predicting how a substance will look and taste after being cooked, for example. Hilary Putnam points out in lectures that if we happened to have a singular term in our theory that would naively be said to denote a particle that continually oscillates in size around the threshold of observability, then on the Received View we would have to be continually deleting and replacing the semantical rule giving the term's denotation. Again, Putnam is able to score heavily against the Received View of the function of scientific theories by making the simple point that "some scientists are, indeed, primarily interested in prediction and control of human experience; but most scientists are interested in such objects as viruses and radio stars in their own right . . . [They want] to learn more about them, to explain their behavior and properties better."[6]

When a philosophical view can be made to seem ludicrous with such apparent ease, we may well begin to wonder how our distinguished forebearers could possibly have held it. And in view of their manifest intelligence, we may even begin to wonder whether we have properly understood their views and their reasons for holding them. I want to argue that a full appreciation of the strengths and limitations of the Received View requires that we eschew the abstraction from actual scientific theories in which the positivists themselves and most historians and critics of the movement have tended to discuss it. (See for example the histories cited by Suppe, p. 15.) If we, unlike the positivists, think of the philosophy of science as at least in part an empirical discipline whose subject matter is science, then it would be surprising if its historical development were not heavily influenced by the historical progress of science itself. Thus it seems likely that if we wish to understand why the Received View was held, we shall have to discover which scientific theories were most on its proponents' minds and what they thought about them. Aside from the historical interest of such an inquiry, it will also at least provide a starting point for determining what, if anything, those particular theories can teach us philosophically about scientific theories in general. I say "starting point," since it may, of course, turn out that the positivists misinterpreted the science of their time, or else that they overgeneralized from the peculiarities of particular theories that impressed them most strongly.

The evidence suggests that the theories that the Vienna Circle found most philosophically interesting were quantum mechanics and the theory of relativity (special and general). Reichenbach, of course, wrote a book on each. Carnap also remarks upon the great philosophical significance of the theory of relativity and upon his interest in quantum mechanics.[7] And Phillip Frank, a physicist who belonged to the Circle, remarks in his own history of its doctrines that "the new line in philosophy is the necessary result of the new trends in physics, particularly the theory of relativity and the quantum theory."[8] A historical understanding and philosophical evaluation of the Received View requires keeping in mind that it developed largely in response to these particular physical theories.

One of the consequences of the Received View is that if there are two or more incompatible theories with the same observational consequences, it is a matter of convention rather than empirical fact which theory we accept, since the theory's empirical content is contained in the observational consequence alone. Clearly, in order for this conventionalist doctrine to get off the ground, a firm observational/theoretical distinction must be presupposed. Lawrence Sklar has asked why conventionalism has been espoused most often in the case of the physical theories that involves a *geometry*.[9] His answer is that for geometric theories, of the modern sort anyway, it is relatively clear what counts as the observational language: it consists of the terms needed to describe the behavior of point masses influenced only by gravitation, of light rays, and of idealized rods and clocks. This remark, in addition to answering Sklar's own question, may also provide part of the reason why the other aspects of the Received View that depend on the observational/theoretical distinction seemed plausible to philosophers who were powerfully impressed by a particular geometrical theory, general relativity, in connection with which this distinction is relatively easy to make. (Of course, I am not suggesting that relativity's "observation vocabulary" precisely fits any of the positivists' explications of this phrase, but only that these explications may have been rather crude attempts to generalize a distinction that is admittedly clear in this one case.)

Elsewhere in his book Sklar makes a different point especially relevant to my main topic, the logical and historical relations between quantum mechanics and the Received View. Sklar seems not to feel that reflection upon the theory of relativity is fully adequate to establish the intelligibility of the observational/theoretical distinction in general or even in the special case of that theory, despite its unusual plausibility there. Accordingly, he considers the possibility that the controversies about conventionalism and related doctrines can be undercut simply by denying the intelligibility of the distinction. Eventually, however, he concludes that this move is unacceptable, on the ground that the scientific community "unquestioningly" accepts this distinction, at least in the case of quantum mechanics, since it is a presupposition of the universally accepted thesis of the observational equivalence of the Schrödinger and Heisenberg representations of the theory. "And," Sklar says, "if it makes sense here, why should it be denied intelligibility for physical theories in general?"[10]

I will discuss Sklar's argument in detail later. I mention it now only to illustrate my claim that examining the Received View in the light of quantum mechanics may be expected to throw light not only on the historical origins of the Received View, but also on the question of the extent to which quantum mechanics can be used to defend it. In what follows I will argue that quantum mechanics provided the positivists with a large share of their reasons for adopting the partial-interpretation aspects of Received View, and that these reasons were, without exception, poor ones. I shall also adduce evidence that makes it likely that the instrumentalist aspect of the Received View also received support from quantum theory via Bohr's doctrine of complementarity. I shall then try to show that one of the lines of thought in Bohr's doctrine can, in fact, be used to defend a highly qualified version of instrumentalism. These interactions of science and philosophy do not all go in one direction, though; and in the interests of symmetry I shall also make a few remarks about the influence of twentieth-century philosophy of science (specifically, operationalism) upon the development of quantum mechanics. Finally, I shall try to show how these issues relate to questions about the foundations of quantum theory which are still very controversial.

2. FROM SPECIAL RELATIVITY TO PARTIAL INTERPRETATION

The story begins with Einstein's epoch-making 1905 paper on special relativity.[11] Einstein made a number of statements in that paper that suggested that he thought certain physical-magnitude terms require definitions in terms of physical operations: that time is, by definition, a number obtained as a result of certain operations involving clocks, and that distance is a number obtained using a rigid rod and perhaps also some clocks. These remarks by Einstein, and the crucial role they seemed to play in his theory, suggested to Bridgman that in a sound physical theory every concept is defined in terms of a set of operations.[12]

It is by now fairly uncontroversial to say that this was the wrong conclusion to draw. As Putnam and others have argued, a clock is often defined as something that keeps time and a rigid rod as one in which distances between neighboring points remain constant. If so, of course, Einstein's "operational definitions" are circular. On the other hand, if these terms are defined in some other way—for example, by describing their referents' structure—then the objects thus indicated can (and usually will) give distorted readings for time and length, which can only be corrected by the circular procedure of using laws involving the very notions allegedly being defined. Whatever Einstein himself may have thought when he wrote the 1905 paper, the only tenable view of the procedures he describes there is that they serve not to define anything, but to illustrate in a heuristically useful manner the consequences of the theory in certain idealized experimental situations.[13] The proper view of an "operational definition" or "coordinative definition"—such as "Voltage is the number yielded by a meter of kind K under circumstances C"—is not that it holds by definition or convention, but that it holds (at least approximately) in virtue of a theory involving the term "voltage" which has been applied to a description of meters of kind K in circumstances C.[14]

Though operationalism was not a correct philosophical moral of the special theory of relativity, it nonetheless played a crucial role in the development of quantum mechanics, as other chroniclers of the latter theory's history have noted.[15] By 1925 Heisenberg had become convinced of the inadequacy of the early Bohr theory of the atom, according to which its orbital electrons follow trajectories corresponding to angular momenta and energies in certain discrete sets. Instead, Heisenberg decided that "one ought to ignore the problem of electron orbits inside the atom, and treat the frequencies and amplitudes associated with the line intensities as perfectly good substitutes. In any case, these magnitudes could be observed directly, and as my friend Otto [Laporte] had pointed out when expounding on Einstein's theory [of relativity] during our bicycle tour round Lake Walchensee, physicists must consider none but observable magnitudes when trying to solve the atomic puzzle."[16] Though he apparently acquired the idea from his friend Otto rather than from Bridgman, Heisenberg drew essentially the same inference as Bridgman from special relativity: that every physical-magnitude term in a theory must be definable in a language suited to describing observables (such as operations). Inspired partly by this philosophy, Heisenberg wrote his paper in 1925, in which all magnitudes are measurable ("observable") and are represented by certain matrices.[17] His philosophical error, therefore, led to an important step toward modern quantum theory, in which so-called observables are represented by operators in a Hilbert space.

The part of this story that is not well known is that Einstein himself told Heisenberg in 1926 that his philosophical interpretation of special relativity was untenable. As Heisenberg reconstructed the conversation, he remarked to Einstein that he had been following Einstein's reasoning in the special theory in assuming that all physical magnitudes must be observable. Einstein's reply was:

> Possibly I did use this kind of reasoning, . . . but it is nonsense all the same. Perhaps I could put it more diplomatically by saying that it may be heuristically useful to keep in mind what one has observed. But on principle it is wrong to try founding a theory on observable magnitudes alone. In reality the very opposite happens. It is the theory which decides what we can observe. You must appreciate that observation is a very complicated process. The phenomenon under observation produces certain events in our measuring apparatus, . . . which eventually and by complicated paths produce sense impressions and help fix the effects in our consciousness. Along this whole path . . . we must be able to tell how nature functions, must know the natural laws at least in practical terms, before we can claim to have observed anything at all. Only theory, that is, knowledge of natural laws, enables us to deduce the underlying phenomena from our sense impressions.[18]

Clearly, Einstein here (as Heisenberg reconstructs his remarks) takes the same position I endorsed earlier: what it generally requires a body of intervening theory, including a theory of the measuring instruments, to connect a value of a magnitude with an observable reading on an instrument, and that the standard physical-magnitude terms therefore cannot be defined in observational terms: "It is the theory which decides what we can observe."

An additional historical irony is that after his fortunate misinterpretation of special relativity, Heisenberg also misinterpreted Einstein's efforts to set him straight and was led by this second misinterpretation to yet another great advance in quantum theory. His immediate response to Einstein was one of confusion since, he said, he had long thought that relativity was inspired by the Machian notion that a theory is merely a "condensation of observations." Nearly a year later, Heisenberg was wrestling with the problem of how the wave function could possibly represent the path of an electron, which seemed to be observable in a cloud chamber. With Einstein's dictum, "It is the theory which decides what we can observe," echoing in his mind, it occurred to Heisenberg that perhaps an electron's trajectory could not be observed after all, that its velocity and position could be measured only within certain limits, and that quantum theory—as he thought Einstein's remark suggested—could determine those limits. He then proceeded to derive his celebrated uncertainty principle.[19]

It is clear that Einstein's dictum did not mean what Heisenberg evidently thought it meant: that a theory sets limits to the accuracy of measurements. Einstein's meaning was instead that a so-called "operational definition" is not a definition at all but rather an intratheoretic statement—that is, a conclusion drawn from several other general and singular assumptions. But once again we can be glad that Heisenberg missed this philosophical point. We may, however, regret that Heisenberg was led to the uncertainty principle via his misunderstanding of Einstein's dictum rather than some other route, since it was no doubt at least partly for that reason that he interpreted the principle as a restriction on simultaneous measurability rather than on statistical scatter.[20]

During this same period, the twenties and thirties, the members of the Vienna Circle were, like Heisenberg, struggling to determine what modifications would need to be made in their initially Machian point of view in order to accommodate the physics of their own time. When the Circle first formed, many of its most influential members followed Mach in holding that the concepts of science could be translated into a phenomenalistic language, which referred exclusively to experiences. By about 1930, however, the group came to espouse physicalism, the belief that meaningful scientific propositions could be generated by definitional transformations beginning with sentences ascribing observable properties to observable physical objects.[21]

Physicalism eventually came to be associated in the positivists' minds with the doctrines of those notorious English physicists whom Duhem criticized[22] for insisting that one could understand a physical theory only by visualizing a mechanical model or analogy for it. Frank (pp. 141–42) remarked that the advent of electromagnetic theory had already rendered this thesis dubious, since efforts to construct a satisfactory mechanical model for it had been unsuccessful. But, he said,

> Heinrich Hertz finally cut the Gordian Knot, so to speak, when he said: "Maxwell's theory is nothing else than Maxwell's equations. That is, the question is not whether these equations are pictural [*anschaulich*], that is, can be interpreted mechanistically, but only whether pictural conclusions can be derived from them which can be tested by means of gross mechanical experiments."
>
> These words gave birth to what we call today the "positivistic conception" of physics.

Frank's claim, then, is that abandonment of the requirement for a "pictural" interpretation of scientific statements amounted to abandonment of the physicalistic requirement of explicit interpretation of *each* scientific concept in familiar physical terms, and thus led to the Received View. Evidently, he tacitly conflated the demand for a model with the demand for a pysicalistic translation; otherwise, this inference would have no plausibility at all.

If Frank thinks that Hertz's reflections on electromagnetic theory gave rise to the positivistic conception of theories, however, why does he say several times that "the rise of twentieth-century physics, of relativity and the quantum theory, was closely connected with a new view of the basic principles" (p. 246)? The answer must be that what had earlier been clear to Hertz forced itself upon the positivists primarily as a result of their reflections on the newer theories of general relativity and quantum mechanics, where the difficulties of visualizing models seemed especially clear. This suggestion is confirmed by Frank's immediately succeeding remark that as a result of these later physical theories, "it was no longer taken for granted that the principles from which the facts had to be derived should contain a specific analogy, either to an organism or a mechanism. Nothing was required except that the observed phenomena could be derived from the principles in a consistent way and as simply as possible. . . . the symbols used in the principles had in themselves no meaning beyond their value for the derivation of facts."

We can see the same line of thought in Carnap's 1939 monograph,[23] his earliest full-scale exposition and defense of the partial-interpretation thesis. In a passage (§24) I shall discuss later Carnap says that his decision not to include semantical rules for the theoretical terms is controlled by a desire to have a theory's formal reconstruction reflect the process by which we come to understand the theory. And in his final section he makes some important remarks on understanding in physics. He says that the increasingly formal character of physical theories, especially since Maxwell's, has made it more and more possible to forego an "intuitive understanding" of the abstract terms and axioms and theorems formulated with their help. The possibility and even necessity of abandoning the search for an understanding of that kind was not realized for a long time. When abstract, nonintuitive formulas, as, e.g., Maxwell's equations of electromagnetism, were proposed as new axioms, physicists endeavored to make them "intuitive" by constructing a "model", i.e., a way of representing electromagnetic micro-processes by an analogy to known macro-processes, e.g., movements of visible things. Many attempts have been made in this direction, but without satisfactory results. It is important to realize that the discovery of a model has no more than an aesthetic or didactic or at best a heuristic value, but is not at all essential for a successful application of the physical theory. The demand for an intuitive understanding of the axioms was less and less fulfilled when the development led to the general theory of relativity and then to quantum mechanics, involving the wave function.

Like Frank, then, Carnap here suggests that the development of Maxwell's theory had rendered the requirement of a visualizable model dubious, but that general relativity and quantum mechanics finally clinched the point. And like Frank, again, he infers that since visualizable models are inessential for understanding (and since the semantical rules should reflect the process of understanding), we need not interpret the theoretical vocabulary in familiar physical terms.

Though the partial-interpretation thesis was thus suggested to the positivists primarily by general relativity and quantum mechanics, they viewed it as a thesis about scientific theories in general, which some had earlier seen to be true of electromagnetic theory, and which they in retrospect thought to have been true all along even of mechanical theories. Carnap remarked in the same monograph that his conclusions about quantum theory also held for Maxwell's equations; and Hempel, some years later, applied them to Newton's gravitational theory.[24]

Leaving general relativity aside again, it is worth asking why the difficulties of visualization should have seemed especially acute in the case of quantum mechanics. An explanation leaps immediately to mind: acceptance by some of the positivists of Bohr's doctrine of complementarity. This is a complex doctrine, but the heart of it is that there are certain pairs of concepts that can be used to express information obtainable under incompatible experimental conditions but that cannot be "combined into a self-contained picture of the object" under investigation, as Bohr put it in a lecture in 1938.[25] One of these complementary pairs consisted of *position* and *momentum*, which Bohr—following Heisenberg—thought to be applicable only under incompatible experimental conditions. Another pair consisted of *wave* and *particle*. Bohr thought that the wavelike aspects of "atomic objects" manifest themselves under certain conditions, such as the two-slit interference experiment, and the particlelike aspects under others, such as those allowing determination of a trajectory. But the two kinds of experiment are incompatible, Bohr held, so that the "wave picture" and "particle picture" cannot be combined into a single "picture."[26]

Wave-particle complementarity is evidently quite harmonious with the line of thought leading to the partial-interpretation thesis; for it implies that these two familiar physical concepts are inadequate to provide a visualizable interpretation of a term like "electron," and thus may seem to suggest that such a term must be uninterpreted. This seems to be the argument Carnap has in mind when he says (§25): "Some [persons], especially philosophers, go so far as even to contend that those modern theories, since they are not intuitively understandable, are not at all theories about nature but 'mere formalistic constructions', mere calculi." This argument, to which Carnap raises no objection except insofar as it implies that the *entire* calculus is uninterpreted, rather obviously reflects—if somewhat inaccurately—the influence of Bohr. For Bohr not only held that *wave* and *particle* are inadequate as complete pictures of atomic objects, but also drew the instrumentalist conclusion that some quantum concepts are merely "artifices" or "expedients" of a merely "symbolical character," rather than descriptions of "the real essence of the phenomena."[27] I say the reflection of Bohr is inaccurate, since he did not accept the extreme view Carnap is discussing. The likely historical sequence, then, is that Bohr asserted that a term like "electron" is a predictive rather than referential device; the unnamed philosophers Carnap mentions concluded that the entire quantum theory is an uninterpreted calculus; and Carnap reminded them that only part of the theory is left uninterpreted. Here, consciously or not, Carnap agreed with Bohr.

Additional evidence for Bohr's influence is that at least some of the positivists knew of complementarity and were partly sympathetic to it.[28] Indeed, Frank himself remarks that it is "fully compatible with . . . logical empiricism."[29] However, he nearly always refers to *position* and *momentum* rather than to *wave* and *particle*. Once he remarks that "quantum mechanics, like relativity theory, is said to introduce a system of purely mathematical formulas, which cannot be interpreted either in terms of waves, particles, or any other visualizable things."[30] This looks like the argument Carnap is discussing, but the context leaves unclear to what extent Frank accepts it.

Turning from a historical question to a logical one, does complementarity after all really provide any grounds for the partial-interpretation thesis? In brief, no. First of all, the issues of models and of (full or partial) interpretation are confused with each other here. As Duhem made quite clear, those English physicists did not propose their models in order to give the meanings of terms in their theories. Rather, in giving a model for a theory they were describing phenomena in some respects analogous to those treated by the theory, with the object of thereby making the theory easier to understand. Therefore, abandoning the notion that a theory requires a model does not, as the positivists supposed, entail abandoning the notion that theoretical terms have meanings and denotations specifiable by semantical rules. Connecting these two notions requires conflating "interpret" in the sense of "give a model (analogy)" with "interpret" in the sense of "give the semantical rules."

Moreover, the doctrine of wave-particle complementarity, which seems very likely to have been the positivists' reason for thinking "pictural" models for quantum theory did not exist, was itself the product of a confusion. It is one aspect of what Popper has called "the great quantum muddle," which is the error of interpreting a probability amplitude as yielding in certain confused ways the physical properties of individual elements of the population.[31] In his papers of 1926,[32] Schrödinger did, indeed, view the wave-function as yielding the charge-density of an electron thought of as spatially "smeared-out" in the manner of a wave, and the fact that the wavelike phenomenon of interference occurs in some experiments with electrons tended to confirm this picture. Unfortunately, there was also evidence that an electron is, at least sometimes, a particle. Bohr sought to resolve this dilemma by claiming that wavelike and particlelike aspects appeared under incompatible experimental conditions. But this is not the right solution. First, in the two-slit experiment, where the wavelike phenomenon of interference occurs, the electron still shows up at a particular location on the screen and thus displays particlelike properties, as well. And second, in a one-slit experiment, where the particlelike property of a trajectory can be determined, diffraction occurs, which is a wavelike phenomenon. Since Born's key paper of 1926,[33] it has been clear that the only sense in which an electron has wavelike aspects is that the function that determines the probability of finding it in a given region obeys an equation somewhat like the classical wave equation. As far as we know, an electron is never, under any experimental conditions, like a wave in the sense of being spread out in space. Of course, its probability amplitude may well have nonzero values over a large spatial region, and even be subject to interference, but that is not to say that it must then be visualized as like a wave and unlike a particle.[34] Wave-particle complementarity was therefore an unfortunate relic of Schrödinger's interpretation of the wave-function and could not provide a good reason for the Received View.

Of course, it is also possible that it was ψ itself which the positivists meant to deny is visualizable. But since all functions, those of classical and quantum physics alike, are abstract and therefore nonvisualizable, this supposition would not explain the special role of quantum mechanics in the positivists' thinking.

I do not wish to deny that the doctrine that visualizable models are inessential for understanding theories is a significant contribution to the philosophy of science. I am only arguing that it is merely a historical accident that this thesis was suggested to the positivists partly by quantum mechanics, since only acceptance of the false doctrine of wave-particle complementarity could make visualizability seem especially problematic for that particular theory.

The positivists had another line of argument leading to the partial-interpretation thesis which had nothing to do with visualizability. A 1947 paper by Frank[35] makes it quite clear that this thesis was sometimes seen as naturally arrived at by weakening operationalism in a way demanded by such theories as general relativity and quantum mechanics, which contain magnitudes like the metric tensor and wave function, which cannot themselves be measured or *a fortiori* operationally defined.

If we understand a so-called "operational definition" in the way Putnam suggests—as an intratheoretic statement specifying a method of measurement—then the positivists' doctrine that not every physical-magnitude term requires an "operational definition" is not only true, but constitutes an important development in the philosophy of science. But, like their earlier observations about models, it provides no support for the partial-interpretation thesis. Since "operational definitions" for the fundamental magnitudes are not semantical rules, their nonexistence has no tendency to show that physical-magnitude terms lack meaning and denotation. It is entirely consistent to deny that "ψ" is defined by some measuring procedure and yet to hold that it has a denotation—namely, the wave function.

In one of his earliest defenses of the partial-interpretation thesis, Carnap agreed with this point—that semantical rules can exist even when measuring procedures and visualizable models do not. His reason for holding that the theory's rational reconstruction does not contain semantical rules for the theoretical terms, then, was not that the theoretical terms lack meaning, but that a semantical rule—like "'ψ'denotes the wave-function"—would not by itself explain the meaning of the term to a layman previously unfamiliar with it. A layman comes to understand "ψ" not through such a semantical rule—much less via a mechanical model or a definition in antecedently familiar terms—but by learning how to use quantum theory to deduce predictions of experimental results he already knows how to describe. Thus, Carnap held, while we *can* give semantical rules for the theoretical terms—which seems to imply they do have meaning—such rules would not accurately reflect the process of understanding.[36]

Now this thesis about understanding may very well be true of at least some theoretical terms. The trouble is that it does not at all support the other doctrines based on the observational/theoretical distinction which later became associated with the Received View in the minds of many of its prominent advocates. For example, it does not support the Kemeny-

Oppenheim view (sec. 1, above) that the entire factual content of a theory is expressible in the observation language or Nagel's claim that theoretical statements cannot be either true or false because of "the purely formal difficulty that . . . [the theoretical] terms are in effect variables"—that is, uninterpreted schematic letters.[37] Such conclusions plainly require the premise, stronger than Carnap's original one, that theoretical terms lack semantical rules *because* they lack meaning and denotation. This is surely a case where the Vienna Circle's insistence on casting its philosophical doctrines in formal terms proved to be seriously misleading.

I conclude that all of the arguments for the partial-interpretation thesis which the positivists based upon quantum theory were defective, or at least highly misleading.

3. QUANTUM THEORY AND THE OBSERVATIONAL/THEORETICAL DISTINCTION

I should like now to consider an argument by a postpositivist philosopher who thinks that at least some aspects of the positivist conception of theories can be salvaged through reflections on quantum theory. Lawrence Sklar has argued that something like the Received View is a presupposition of a thesis universally accepted by the scientific community—the equivalence of the Heisenberg and Schrödinger representations of quantum theory. His argument is as follows:

> In the Schrödinger representation, the state of a system . . . changes according to a dynamical law . . . The mathematical representatives of the observables are operators, constant in time. . . . In the Heisenberg picture the state of a system . . . is represented by . . . the unchanging state-vector for the prepared system. The operators for each observable, on the other hand, are time-dependent mathematical structures. . . .
>
> Since the evolution of the probabilities predicted by both the Heisenberg and Schrödinger representations are identical, . . . it is universally accepted that the Schrödinger and Heisenberg theories constitute merely alternative representations of the same theory. The moral is clear. "States" and "observables" are not themselves directly inspectable quantities. They are, rather, constituents of the theoretical superstructure of the theory. So the fact that states are time-dependent in the one picture and time-independent in the other is irrelevant.

That this apparent contradiction is regarded as compatible with the two representations' equivalence, shows (Sklar claims) that "the intelligibility of the observation basis/theoretical superstructure distinction is assumed by all," since only if it exists does it make sense to hold, as everyone does, that the equivalence of the two representations is guaranteed by their observational equivalence, despite apparent incompatibilities in their theoretical superstructures. Sklar concludes, if the distinction "makes sense here, why should it be denied in general?"[38]

This example, then, may seem to support two aspects of the Received

View: not only the observational/theoretical distinction, but also the thesis that all of a theory's content is contained in its observational consequences—that sentences containing theoretical terms lack truth-values, except perhaps by convention and, thus, cannot genuinely conflict with each other.

We should not, however, be too hasty about arranging a disinterment for the Received View. For one thing, as Sklar himself is the first to admit, the distinction allegedly presupposed by the equivalence in question is not really that between observable and unobservable entities and properties. It is rather the distinction between certain functions (wave functions and operators) and, on the other hand, the probabilities of certain properties. Whether we consider probabilities of observable properties to be observable or not, it seems clear that the properties in question are not, in general, observable. An electron's spin or kinetic energy—and *a fortiori* associated probabilities—is a paradigm of properties the positivists would have considered unobservable. The fact that Heisenberg called them "observables" may lend a spurious plausibility to Sklar's argument, but does not make them observable in the sense required by the Received View. Moreover, even if this distinction were essential to quantum theory, it is unclear how it is supposed to follow that it is intelligible for scientific theories in general. After all, Sklar himself (pp. 116–17) says that the distinction has a clarity in geometric theories which it lacks in other cases. Why does quantum theory provide a better basis for generalization?

In fact, however, no dichotomy in quantum theory's vocabulary is required to resolve the apparent contradiction and establish the equivalence. When Heisenberg says his wave functions ψ_H are constant in time, he seems to be contradicting Schrödinger's view that *his* wave functions ψ_S (t) vary with time only if one ignores the fact that ψ_H is defined as ψ_S (t_0). Plainly, there is no contradiction in saying that ψ_S (t) changes with time, whereas $\psi(t_0)$, its value at some initial instant t_0, does not. Similarly, Heisenberg's operators $A_H(t)$ are defined as U^t (t, t_0) A_S U (t, t_0), where U is the evolution operator for ψ_S (t); and, hence, there is no contradiction in saying that A_S, but not A_H (t), is constant in time. Moreover, the principal equations of the Heisenberg representation are obtained by deriving them mathematically from those of the Schrödinger representation after supplementing the latter with the above definitions. And if one theory is obtained from another simply by adding some additional defined terms, we do not need the observational/theoretical distinction to show they are really the same theory. So Sklar's argument from quantum theory to certain aspects of the Received View is no more effective than those of the positivists.

As a letter from Sklar suggests, he might wish to reply to my last argument as follows: despite the possibility of establishing the two representations' equivalence through the use of definitions and mathematics alone, many writers do, in fact, appeal to the identity of their measurement-predictions as an *additional and sufficient* ground for the equivalence, and thus reveal their agreement with the positivists. For example, E. Merzbacher in *Quantum Mechanics* (New York: John Wiley and Sons, 1970) argues: "The possibility of formulating quantum dynamics in different pictures [representations] arises because the mathematical entities such as state vectors and operators are not the quantities which are directly accessible to physical measurement. Rather, the comparison with observation is made in terms of eigenvalues and expansion coefficients whose equality therefore guarantees equivalence" (p. 351).

My response to this defense of Sklar is to reiterate that the distinction concededly used in such arguments as Merzbacher's is between "measurable" and "unmeasurable," not between "observable" and "unobservable." I do not see that the distinctions are enough alike to justify inferring that there must be something to the positivists' account of theories after all. Secondly, I see no reason to suppose that when Merzbacher says the two representations' identical sets of eigenvalues and expansion coefficients guarantee that they are "equally acceptable" or "equivalent," he means anything more than that they give the same solutions to the problems quantum theory was designed to solve—namely, the probabilities of the possible values of various quantities. He is not obviously committed to the inference from "observational" equivalence to logical equivalence. (This case lends support to Dudley Shapere's view that the feature of actual scientific practice which looks most like the observational/theoretical distinction is the distinction between (1) the information for which a theory is expected to account and (2) the theory itself. (See his "Scientific Theories and Their Domains," in Suppe, *Scientific Theories*, pp. 518–89.)

4. COMPLEMENTARITY AND INSTRUMENTALISM

So far, I have considered the relevance of quantum theory to the Received View's partial-interpretation thesis and observational/theoretical distinction. I now want to turn to the question of quantum theory's influence upon the acceptance by many of the positivists of instrumentalism. I shall not be arguing that quantum theory provided many of the positivists with the reasons for their *original* adoption of instrumentalism. The fact that in the movement's earliest days they were much impressed by Mach, who was an instrumentalist, is probably the best explanation for their initial acceptance. Rather, what I want to suggest is that a large part of the reason they *continued* to hold this unnatural and implausible view, even into the 1960s (Hempel, *Scientific Explanation*), is that modern physics—and complementarity in particular—seemed to provide powerful support for it.

I am unable to cite specific passages in which evidence of complementarity's influence in this regard leaps to the eye. Instead, I simply rely on the evidence already cited that many of the positivists knew of Bohr's doctrine and were sympathetic to it and, on the evidence to be presented, that complementarity has instrumentalist aspects. I want to suggest, somewhat speculatively, that there were two main reasons why instrumentalism seemed plausible to many of the positivists, both reasons deriving from the character of the science of their own time. One—which I shall simply mention—is that they had reason to suppose (see sec. 2) that the spatial and temporal concepts of relativity theory could be defined in terms of certain measuring procedures, and therefore to suppose that the purpose of relativity theory is to enable one to predict the results of those procedures, rather than, say, to describe the actual structure of space-time. The second, I suggest, is that Bohr's doctrine provided them with reason to think that the purpose of quantum theory is also to enable one to predict (statistically) the results of measurements, rather than to describe the microcosm as it is in itself.

In the remainder of this chapter I shall discuss the degree to which Bohr's views resemble the positivists'—a question that also bears on the likelihood of historical influence. But I shall also discuss the broader and nonhistorical question of the extent to which quantum mechanics does *in fact* support instrumentalism.

In his explanations of complementarity, Bohr frequently makes statements that sound almost exactly like the formulations of instrumentalism used by advocates of the Received View, such as Frank. For example, Bohr writes: "When speaking of a conceptual framework, we refer merely to the unambiguous logical representation of relations between experiences" or, as he says instead one paragraph later—and as if it made no difference—"between measurements." He suggests that we restrict the application of the term "phenomenon" to "observations obtained under specified circumstances," and then claims that "the appropriate physical interpretation of the symbolic quantum-mechanical formalism amounts only to predictions, of determinate or statistical character, pertaining to individual phenomena." The interpretative rule Bohr has in mind, of course, is the Born rule for the expectation value of observable A in state ψ: $\langle A \rangle = \langle \psi, A\psi \rangle$. Again, "the quantum-mechanical formalism . . . represents a purely symbolic scheme permitting only predictions . . . as to results obtainable under conditions specified by means of classical concepts."

At times, Bohr seems to endorse not only the Received View's instrumentalism but also its partial-interpretation thesis. If we identify, at least tentatively, the observation language with the part of ordinary language used to describe phenomena (in Bohr's sense), then his position is that the formalism cannot be interpreted in any but observational terms, that is, it cannot be interpreted as stating anything about properties or events undetected by measurement. Sentences purporting to describe undetected events or properties, he holds, are meaningless: quantum-mechanical experiments cannot be "interpreted as giving information about independent properties of the objects," for we cannot thus "distinguish sharply between the behavior of the objects and the means of observation."[39]

Someone who doubted my claim that there is a close parallel between complementarity and instrumentalism might question whether "phenomena" (or "observations" or "measurements") in Bohr's sense are the sort of thing one could describe in anything like the positivists' observation language, in terms of which instrumentalism is stated. The objector might say that a measurement could be a microscopic event, like the scattering of a photon, which must be amplified to be noticed by an experimenter. I will concede later what the objector would be right about measurement and that quantum theory supports an instrumentalism of a modified sort. But what is relevant just now is that the objector would be wrong about Bohr. One of the latter's fundamental tenets is that measuring apparatus must be sufficiently large to make Planck's constant negligible and to make experimental outcomes describable using "everyday concepts, perhaps refined by the terminology of classical physics."[40] Though the positivists never said precisely what the observation terms are, the terms Bohr refers to are enough like them to make his view at least very similar to theirs.

One of the many difficulties in interpreting Bohr is that there is a contrasting strand in his thinking that is at least *prima facie* inconsistent with the instrumentalist assertions just discussed. He remarks that despite "the difficulties in talking about properties of atomic objects independent of the conditions of observation . . . an electron may be called a . . . material particle [presumably even when unobserved], since measurements of its inertial mass always give the same result."[41] Apparently, then, Bohr holds that if the probability of a given result of a measurement of observable A is 1, then—even before the measurement takes place—A already has the value that would result; that is,

$$\text{If } A\psi = a\psi, \text{ then } A = a.$$

That Bohr also accepts the converse is evident from his reply[42] to Einstein, Podolsky, and Rosen,[43] in which he states that the "conditions which define the possible types of predictions regarding the future behavior of the system . . . constitute an inherent element of the description of any phenomenon to which the term physical reality can be properly attached." This presumably means that if a value for an observable exists (has "physical reality"), then it is possible to predict with certainty via the wave function that that value will be obtained upon measurement—that is:

$$\text{If } A = a, \text{ then } A\psi = a\psi.$$

Conjoining these last two statements, we have

$$\text{“}A = a\text{” is true} \equiv A\psi = a\psi.$$

Since this last thesis says, in effect, that a system has a property if and only if its quantum-mechanical state determines (with certainty) that it does, it can also be expressed by saying "Quantum mechanics is a complete theory," which is what Bohr was asserting and Einstein, Podolsky, and Rosen (EPR) were denying in their famous debate.

The difficulty in formulating a consistent interpretation of complementarity, however, is that Bohr's completeness claim is inconsistent with his instrumentalism. For it is, in effect, a rule giving a physical interpretation to certain formulas by connecting them to physical properties which are not necessarily being measured, despite Bohr's repeated assertions that the only consistent physical interpretations for quantum-mechanical formulas connect them to probabilities or mean values of measurement results. One way that Bohr could render his position consistent—and his lately quoted remark about electronic mass suggests this is the way he would have taken if pressed—is by saying that quantum mechanics indeed tells us nothing about unmeasured magnitudes *except* where the system is in an eigenstate, though it does tell us something about our language; namely, that statements about unmeasured magnitudes are meaningless, again except where eigenstates are involved. In other words, "$A = a$" is true just when $A\psi = a\psi$, and is meaningless otherwise. Though this position is consistent, it would require him to abandon his unqualified assertions of instrumentalism.

The other way out is to follow EPR's suggestion and abandon the completeness claim. Doing so would leave him with the quantum theory formalism—as axiomatized, for example, by G. Mackey[44]—supplemented by the semantical rule that the theory's probability measures give the probabilities of various measurement-outcomes. The completeness of the theory's descriptions of systems via wave functions would *not* be asserted. Such a point of view is known as the "minimal statistical interpretation" of quantum theory, which arises from insisting more strictly than Bohr himself does on his instrumentalism—that is, on his doctrine that the "appropriate physical interpretation of the symbolic quantum-mechanical formalism amounts only to predictions, of determinate or statistical character, pertaining to individual phenomena [measurement-results]." On the minimal statistical interpretation (MSI) the quantum theory involves no attempt to specify the general conditions under which an observable has a value.

It is this view of quantum theory, which was one of the contradictory strands in Bohr's thought, on which I want to focus the remainder of this chapter. For this quasi-instrumentalist view of quantum theory, unlike the completeness assumption, was what I am claiming influenced the positivists. (On "quasi," see sec. 5.) Moreover, it can, I think, be used to defend part of what the positivists were saying. But first I need to say a few things in defense of the MSI itself.

The MSI is certainly adequate to yield the sort of empirical predictions whose success have been the reasons for quantum theory's continuing acceptance—predictions, that is, of various probabilities and expectations, and of the confinement of each observable to a certain subset of the real line known as its spectrum. The completeness assumption is in no way required for the derivation of such predictions. Since that additional assumption serves no useful purpose, why, one might well ask, isn't everyone an advocate of the MSI?

One reason is surely the great authority of Bohr, deriving from a feeling that his views have been experimentally confirmed. But I have already noted that completeness is not required in the characteristic quantum-theoretical predictions. A second reason is that many people make the same thought transition Einstein made from the denial of completeness to the existence, or at least the necessity of a search for, hidden variables. Einstein wrote, thirteen years after the EPR paper, that "if the statistical quantum theory does not pretend to describe the individual system (and its development in time) completely, it appears unavoidable to look elsewhere for a complete description of the individual system. . . . [For] the conceptual system of the statistical quantum theory . . . could not [then] serve as the basis of theoretical physics."[45] Since many people believe that hidden variables have been shown to be nonexistent, however, they feel they must assert that quantum theory is complete as it stands. (E. P. Wigner, for example, expressed this feeling in answer to a question at the conference on quantum theory at the University of Western Ontario in 1971.)

It is not necessary, though, to make the move Wigner is concerned about from the denial of completeness to the belief in hidden variables. EPR recognized in their cited article that the two questions are quite separate and explicitly said that they left open the question on whether quantum theory can be supplemented by hidden variables. More recently, L. Ballentine[46] has advocated the minimal statistical interpretation despite sharing Wigner's skepticism about hidden variables, and there is no reason to consider his position inconsistent.

The final criticism of the MSI I shall consider relates to its instrumentalist aspect, which I have not thus far emphasized, but which has an important bearing on our main problem—the extent to which quantum mechanics supports the Received View of theories. The MSI is similar to instrumentalism in that it insists that the empirical interpretation of quantum theory relates the formalism only to the results of measurements—confining these results to certain spectra and assigning them certain probabilities—and that the theory therefore provides no information about any other phenomena.[47] It may be tempting to say that the references to measurements in the interpretation of the theory are superfluous and that $\langle A \rangle$ can simply be identified as the expectation value of A, rather than of results of measuring A. Such a view would presumably be based on the assumption that any observable has a value at all times and that a measurement simply reveals the value that that observable would have possessed if it had not been subjected to a measurement process. However, such values would be hidden variables in Kochen and Specker's sense—that is, quantities that determine a measurement-result for each observable—and are incompatible with quantum theory, as those authors have conclusively shown.[48] It is clear, then, that the advocate of the MSI must hold that quantum theory yields no information about physical systems except the probabilities of measured values of observables, and that the reference to measurements here is essential.

The MSI is therefore obviously reminiscent of instrumentalism as I defined it earlier: the view that scientific theories are designed to predict observable phenomena rather than to describe unobservable objects "as they are in themselves." And some writers find the MSI objectionable for this reason. Michael Friedman, for example, has suggested in conversation that the MSI is unacceptable for the same reasons that instrumentalism is unacceptable as a general theory of theories. Though he did not say, I suppose he had in mind such familiar objections as that instrumentalism relies on the observational/theoretical distinction and is empirically false as an account of the motivations of scientists, who are in fact interested in unobservable things in themselves. But neither of those objections tells against the MSI. To make such particularized claims as that "$\langle A \rangle$" is to be interpreted as the expected value of A-measurements, whereas "$A\psi = a\psi$" is not to be interpreted as a necessary and sufficient condition for "$A = a$," it is not necessary to rely on an across-the-board distinction between observational and theoretical terms of all of science. Nor is it necessary to make any general claim about the purposes of scientific theories other than quantum mechanics. The MSI is therefore entirely compatible with the fact that scientists employing other theories are "interested in such objects as radio stars and viruses in their own right," as Putnam put it (Putnam, "*Craig's Theorem*).

Another relevant objection to the MSI which is directed at its quasi-instrumentalist character has been stated by Einstein and Putnam. Einstein remarks that one of his objections to complementarity is that, in his own view, statements concerning measurements "can occur only as special instances viz., parts, of physical description, to which I cannot ascribe any exceptional position above the rest."[49] Though he does not say here exactly how he interprets complementarity—indeed, he remarks that he has been unable to give it a sharp formulation—his objection seems to be aimed at the instrumentalist aspect of complementarity, which I suggested earlier is equivalent to the MSI. For this view certainly assigns to measurements "an exceptional position" among physical processes. The basic law of the theory are formulated explicitly in terms of measurements rather than in more general physical terms applicable to measurements as special cases.

Putnam also objects to the MSI on the same ground. He writes: "The term 'measurement' plays *no fundamental role in physical theory as such.* Measurements are a subclass of physical interactions— no more or less than that. They are an important subclass, to be sure, and it is important to study them, to prove theorems about them, etc.; but 'measurement' can never be an *undefined* term in a satisfactory physical theory, and measurements can never obey any 'ultimate' laws other than the laws 'ultimately' obeyed by *all* physical interactions."[50]

In addition to his general antioperationalist point of view, which he feels supports the above picture of the relations of measurements to theories, Putnam has also remarked in conversation that he thinks instrumentalism, in general or even just in quantum theory, embodies an objectionable element of anthropocentrism. Heisenberg, for example, has explicitly asserted that quantum theory must be formulated partly in terms of classical concepts, whose nature is determined by the human cognitive apparatus. He concludes that the theory thus involves "a reference to ourselves" and so "is not completely objective.[51] Putnam's objection to this view is that it is incompatible with a minimal version of the thesis of the unity of science. This thesis asserts at least that human beings, like measuring devices, have no distinguished place in relation to the laws of nature—except of course, that humans are the laws' propounders, testers, and so forth. The particles composing human bodies obey the very same laws obeyed by the rest of the particles in the universe. But if, as seems plausible, we define a measurement roughly as a process that enables *us* to determine, at least approximately, the value of a physical magnitude, and if the fundamental laws of nature involve references to measurements, then it appears that we humans do after all have a special place in relation to the laws of nature. That "measurement" should be defined in terms of human discriminatory capacities is entirely natural in the context of a theory of confirmation, where measurements are viewed simply as processes used in testing and applying scientific theories; for we are the only species to do these things. But it is another matter if fundamental scientific laws contain the term "measurement." Can we really believe, Putnam asks, that the most basic physical processes obey laws that just happen to contain essential references to the discriminatory capacities of human beings? Why didn't they instead turn out to refer to the discriminations that some species of beetles can make?

Most of this problem is unnecessary. In the context of the interpretation of quantum theory, it is irrelevant that for the purposes of a theory of experimental testing it may be appropriate to define "measurement" by references to human capabilities. The relevant question is, rather, how it must be defined to achieve the goals of quantum theory, particularly of the quantum theory of measurement. It is sufficient for these purposes to define a measurement of observable A at time t in system S by means of the device M as follows: an interaction between S and M such that there is a one-to-one correlation between the values of A at t and of some observable G of M at the end of the interaction. Whether the values of G are distinguishable by humans is inessential, and hence M might just be an electron scattered from an atom. In the case of repeatable measurements—which yield almost the same result if repeated quickly enough—this definition entails that if the initial state of S is eigenstate α_k of A, the final state of $S+M$ is $\alpha_k \otimes \aleph_k$, where $\aleph_k$ is the eigenstate of G correlated to α_k. Park [52] has shown that this assumption is sufficient to achieve what I take to be the principal desideratum of the quantum theory of (repeatable) measurements: the derivation of the transition of S to a mixture of eigenstates of A.

I conclude that the minimal statistical interpretation, which—in harmony with one strain in Bohr's thought—holds that quantum theory merely gives information concerning measurement-results, is not after all committed to saying that quantum theory is anthropocentric. Neither is it subject to Putnam's objection that "measurement" must not be taken as a primitive term. But part of the Einstein-Putnam objection must simply be conceded. For the MSI does imply that measurements, as I defined them, have an exceptional position in relation to the laws of quantum theory, in that those laws serve simply to predict the statistics of interactions of this sort. This peculiar fact must simply be accepted, however difficult it may be to swallow, let alone explain.

5. CONCLUSION

Having defended the MSI against several objections, in conclusion, I now return to my main topic and assess its relevance to the Received View of theories. I have already said that the MSI is in harmony with instrumentalism in that it implies that the purpose of quantum theory is to enable one to derive certain facts about the results of measurements—that they are confined to certain spectra and have certain probabilities—and that it gives no information about unmeasured entities. Does quantum mechanics on this interpretation therefore support the positivists' instrumentalist view of scientific theories in general? It does not. First, the MSI implies that the purpose of quantum theory is to predict the statistics of measurements—which need not be observable, as an instrumentalist, strictly speaking, would insist. So the MSI is instrumentalist in a somewhat broadened sense, very similar in spirit, though, to the positivists'. Second, the considerations favoring an instrumentalist view of quantum mechanics are quite peculiar to that particular theory. Despite these two points, however, one can easily see how philosophers who were powerfully impressed by quantum mechanics may well have felt it supported a generalized instrumentalism.

While the MSI supports an instrumentalist view, at least of quantum theory, it does not support the partial-interpretation aspect of the Received View even in this special case. For if, as the MSI asserts, ψ is not to be interpreted as determining what properties some system has, it does not follow, of course, that it has no interpretation at all—no meaning or denotation. In fact, it does have a denotation: the wave function. Thus the MSI does not support, even in the case of quantum theory, the version of instrumentalism discussed by Nagel: the claim that theories are *merely* instruments of predictions and not as wholes true or false, since some of their formulas contain uninterpreted schematic letters.[53] A theory can be, and quantum mechanics is, both an instrument of prediction *and* a system of true or false statements.

In short, then, the Received View is both an oversimplification of, and overgeneralization from, certain quite peculiar features of quantum mechanics and, to recall Sklar's point, relativity theory. Despite its limitations, however, the Received View is still worth our attention as a case study in the interaction of science and philosophy, as well as because so much of current philosophy of science is a reaction against it. I hope that the foregoing examination of its relations to quantum mechanics helps to place it in an accurate historical and philosophical perspective, and that it has made along the way some points about quantum theory that are of interest in their own right.

NOTES

I wish to thank Michael Friedman, Lawrence Sklar, Lindley Darden, Dudley Shapere, and Frederick Suppe for helpful criticisms of earlier drafts of this essay.

1. See introduction to F. Suppe, ed., *The Structure of Scientific Theories* (Urbana: University of Illinois Press, 1974), for references and a survey of the rise and fall of the Received View. I am referring throughout this paper only to what Suppe calls its final version, and

not earlier versions requiring that theoretical terms be explicitly defined or introduced by reduction sentences.

2. P. Frank, *Modern Science and Its Philosophy* (New York: Collier, 1961), pp. 110, 247.

3. In C. G. Hempel, *Aspects of Scientific Explanation* (New York: Free Press, 1965), p. 177.

4. J. Kemeny and P. Oppenheim, "On Reduction," *Phil. Studies* 7 (1956): 6–19.

5. P. Achinstein, *Concepts of Science* (Baltimore: Johns Hopkins University Press, 1968), p. 176.

6. H. Putnam, "Craig's Theorem," *Jour. of Phil.* 62 (1965): 251–60.

7. R. Carnap, "Intellectual Autobiography," in P. A. Schilpp, ed., *The Philosophy of Rudolf Carnap* (La Salle, Ill.: Open Court, 1963), pp. 14–15, 21. His doctoral dissertation *Der Raum* (Jena: Universitët Jena, 1922) dealt with general relativity.

8. Frank, *Modern Science*, pp. 49–50.

9. L. Sklar, *Space, Time, and Spacetime* (Berkeley and Los Angeles: University of California Press, 1974), pp. 115–17.

10. Ibid., pp. 144–45.

11. Reprinted in A. Einstein et al., *The Principle of Relativity* (New York: Dover Press, 1952).

12. P. W. Bridgman, *The Logic of Modern Physics* (New York: Macmillan, 1927), chap. 1.

13. H. Putnam, "Is Logic Empirical?" in R. Cohen and M. Wartofsky, eds., *Boston Studies in the Philosophy of Science* (Dordrecht: Reidel, 1970), vol. 5, sec. 8.

14. H. Putnam, "A Philosopher Looks at Quantum Mechanics," in R. Colodny, ed., *Beyond the Edge of Certainty* (Englewood Cliffs, N.J.: Prentice-Hall, 1965), p. 76.

15. E.g., Frank, *Modern Science*, pp. 282–83.

16. W. Heisenberg, *Physics and Beyond* (New York: Harper & Row, 1971), p. 60.

17. W. Heisenberg, "The Interpretation of Kinematic and Mechanical Relationships According to the Quantum Theory," in G. Ludwig, ed., *Wave Mechanics* (Oxford: Pergamon Press, 1968), pp. 168–82.

18. Heisenberg, *Physics and Beyond*, p. 63.

19. Ibid., pp. 64, 77–79.

20. In my view K. R. Popper showed the latter to be the only correct interpretation in his *Logik der Forschung* (Vienna: Springer, 1935), §75.

21. Carnap, "Intellectual Autobiography," p. 50.

22. P. Duhem, *The Aim and Structure of Physical Theory* (New York: Atheneum, 1962), pp. 69–104.

23. R. Carnap, *Foundations of Logic and Mathematics* (Chicago: University of Chicago Press, 1939).

24. Hempel, *Scientific Explanation*, p. 113.

25. Reprinted in N. Bohr, *Atomic Physics and Human Knowledge* (New York: Science Editions, 1961), p. 26.

26. N. Bohr, *Atomic Theory and the Description of Nature* (London: Cambridge University Press, 1934), pp. 56, 59. See also E. Scheibe, *The Logical Analysis of Quantum Mechanics* (Oxford: Pergamon Press, 1973), on Bohr.

27. Bohr, *Atomic Theory*, pp. 17–18.

28. E.g., H. Reichenbach, *Philosophic Foundations of Quantum Mechanics* (Berkeley and Los Angeles: University of California Press, 1944), p. 2.

29. Frank, *Modern Science*, pp. 117–78.

30. Ibid., p. 153.

31. K. R. Popper, "Quantum Mechanics Without 'The Observer'," in M. Bunge, ed., *Quantum Theory and Reality* (New York: Springer, 1967), p. 19.

32. E. Schrödinger, "Quantization as an Eigenvalue Problem" and "On the Relationship of the Heisenberg-Born-Jordan Quantum Mechanics to Mine," reprinted in Ludwig, *Wave Mechanics*, pp. 94–167.

33. E. Born, "Quantum Mechanics of Collision Processes," in Ludwig, *Wave Mechanics,* pp. 206–25.
34. I have discussed the application of Born's view to the two-slit experiment in more detail in "Is Quantum Logic Really Logic?" *Philosophy of Science* 38 (1971): 508–29.
35. In Frank, *Modern Science,* pp. 277–92. See also Hempel, *Scientific Explanation,* pp. 109–13.
36. Carnap, *Foundations,* pp. 62–68.
37. E. Nagel, *The Structure of Science* (New York: Harcourt, Brace and World, 1963), p. 141.
38. Sklar, *Space, Time,* pp. 144–45.
39. Bohr, *Atomic Physics,* pp. 68, 64, 40, 25–26. Essays quoted date from 1938–54.
40. Ibid., p. 26.
41. Ibid., p. 98.
42. N. Bohr, "Can Quantum-Mechanical Description of Physical Reality Be Considered Complete?" *Physical Review* 48 (1935): 696–702.
43. A. Einstein, B. Podolsky, N. Rosen, "Can Quantum-Mechanical Description of Physical Reality Be Considered Complete?" *Physical Review* 47 (1935): 777–80.
44. G. W. Mackey, *Mathematical Foundations of Quantum Mechanics* (New York: Benjamin, 1963), pp. 56–114.
45. In P. A. Schilpp, ed., *Albert Einstein: Philosopher-Scientist* (La Salle, Ill.: Open Court, 1949), p. 672.
46. L. E. Ballentine, "The Statistical Interpretation of Quantum Mechanics," *Reviews of Modern Physics* 42 (1970): 358–81.
47. I say "similar" because for the moment I am leaving open the question of whether the MSI should include Bohr's view that measurements are observable.
48. I discuss their proof in "Quantum-Theoretic Realism: Popper and Einstein vs. Kochen and Specker," *British Journal for the Philosophy of Science* 23 (1972): 13–23.
49. In Schilpp, *Albert Einstein,* p. 674.
50. Putnam, "Quantum Mechanics," p. 77.
51. W. Heisenberg, *Physics and Philosophy* (New York: Harger, 1958), pp. 55–56.
52. J. L. Park, "Quantum Theoretical Concepts of Measurement," *Philosophy of Science,* 35 (1968): 205–31, 389–411.
53. Nagel, *Structure of Science,* p. 141.

JOHN STACHEL
Boston University

Do Quanta Need a New Logic?

> The question of the logical forms which are best adapted to quantum theory is in fact a practical problem, concerned with the choice of the most convenient manner in which to express the new situation that arises in this domain.
>
> —Niels Bohr
> *New Theories in Physics*

1. INTRODUCTION

We live in a time of revolution. Political revolutions have toppled countless regimes; social revolutions have dispossessed entire classes; the sexual revolution promises to deprive man (in the narrow sense of the word) of his age-old prerogatives. The world of culture has not been slow to respond to the challenge of the new: "Art Nouveau" is already ancient, cubism has become classic, the "new novel" is no novelty. Even staid old science has not been exempt from the powerful currents of change in our century. We'll soon celebrate the centennial of the first talk of a "Revolution in Physics"; while the newspapers remind us daily that the Biological Revolution is in high gear. And now, in discussions of microphysics, the cry is heard that we need a revolution in logic! We are assured that the reason we have such trouble in our intercourse with the microworld is that we are using the wrong logic in our attempts to understand it. Such talk conjures up an attractive vision: while we have been plodding along to the tune of *Principia Mathematica,* the quanta have been stepping to the music of a different drummer—or, to drop such quaint nineteenth-century metaphors and "tell it like it is": The quanta have been swinging one way, logically; while we poor squares (pardon my generation gap) are still swinging the other, unable to liberate ourselves from our fixation on old daddy Aristotle. No wonder we can't dig them quanta! They've been diffracting away through our silly old double slits, thumbing their noses at the distributive law, while Niels keeps Bohring us with all that talk about complementarity. All mysteries of the quantum world will dissolve, we are assured, if we'll only stop trying to distribute over micro-conjunctions and start harkening to what the quanta have been trying to tell us all along about *their* logic. Or, in another version of the gospel (unfortunately, more than one candidate for *the* logic of quantum mechanics has been put forward), we have to rid ourselves of our hangup on "true" and "false" and recognize that the quanta—not so hidebound as we—can swing three ways truthwise (perhaps even four, five—or an infinity of ways, for probabilistic quantum logics have also been proposed).

As a political radical, swinger *manqué*, and Jean Luc Godard movie fan I find it hard to resist the seduction of the scarlet banner, wherever raised. As a relativist in physics, if not morals, I can hardly disown the revolution in geometry, so often cited as a precedent for the revolution in logic in manifestoes of the new revolutionary party. I also cannot deny the interest and importance of much of the rapidly growing volume of work done by mathematicians and physicists under the rubric "quantum logic." Then why am I left so uneasy by the claims of some of the quantum logicians? To explain why is, of course, the purpose of this paper.

One reason must be stated at the outset. It seems to me that much of the appeal of arguments for a logic of the quanta—both to their proponents and to the public—comes from the mystique surrounding quantum mechanics. Anything touched by this formalism thereby seems to be elevated—or should it be lowered?—to a fundamental ontological status. The very words "quantum mechanics" conjure up visions of electrons, photons, baryons, mesons, neutrinos, quarks, and other exotic building blocks of the universe. But the scope of the quantum-mechanical formalism is by no means limited to such (presumed) fundamental particles. There is no restriction of principle on its application to any physical system. One could apply the formalism to sewing machines if there were any reason to do so! More to the point, quantum mechanics *has* been applied fruitfully in many cases where it is clear from the outset that the resulting quanta have nothing to do with fundamental particles, but are phenomenologically useful devices for describing low-lying states of excitation of many-body systems. I am referring to the many types of quasiparticles that have been introduced since the pioneering work of Landau.[1] In addition to the phonons and rotons he originally introduced, we now have magnons, plasmons, excitons, polarons, fluctuons, Cooper pairs—the list grows all the time. Many of these quasiparticles result from the quantization of classical field equations for matter which are clearly only phenomenological, since they may be derived from statistical-mechanical considerations. This procedure has even proved useful in electrodynamics. Phenomenological electrodynamical equations for a material medium may be derived from the vacuum Maxwell equations and some microscopic model of the medium. Quantization of these macroscopic equations for the medium gives rise to "phenomenological photons," which can be used to explain many effects.[2] Suppose tomorrow someone were to come up with some "fundamental theory," from which Maxwell's vacuum equations could be derived. Then today's "fundamental photons" would become "phenomenological photons," but this would in no way impair the usefulness of quantum electrodynamics within its range of validity. If quantum logic holds for "fundamental" photons and electrons, then it must hold for phenomenological photons and excitons. So arguments for quantum logic based on a presumed distinction between a "fundamental" quantum world and a "phenomenological" macroscopic world, the latter condemned to live by classical logic while the former ascends to quantum-logical heights, leave me unmoved. If phonons (for example) inhabit the quantum-logical world, then it cannot be such a big deal to live there!

Having cleared away this bit of phony tinsel and glitter, we come (as Leo Rosten said of Hollywood) to the real tinsel and glitter. The problem of where to begin is always difficult, especially when talking about a subject that involves consideration of such varied fields as physics, philosophy, and mathematics. Yet it seems appropriate to start by briefly stating my views on the nature of logic, a subject too often avoided in papers on quantum logic. For me, a logic always has some language as its object. The more formalized one wants the logic to be, the more formalized the language must be.[3] This implies that metaphors like the logic of events, the logic of history, the logic of the situation are just that—metaphors. Since a logic presupposes a language, the objects logic studies must be linguistic entities. I believe in the existence of other objects which are quite independent of language, but such objects, in contrast to linguistic references to them, cannot have a logic. Lest this be thought a mere verbal quibble, let me point out that someone who believed that all reality was fundamentally conceptual in nature could meaningfully and nonmetaphorically speak of the logic of the world.[4] Indeed, this is just what sets me on edge when Finkelstein states: "I think that besides mathematical logic there is now also a physical or world logic, different in principle and describing at a very deep and general level the way inanimate physical systems interact."[5]

Putnam also makes statements identifying the realms of the logical and the ontological: "The heart of the quantum-logical interpretation is that the logical relations among physical states of affairs—the relations of implication and incompatibility—are themselves an empirical matter."[6]

Bub and Demopoulos are very sparing of discussions of just what they mean by logic—or, indeed, any of the philosophical implications of their viewpoint. But it is difficult not to see a similar identification of the logical and the ontological in such phrases as: "I see quantum mechanics as a principle theory"—the term *principle theory* refers to Einstein's distinction between principle theories and constructive theories—"of logical structure: the type of structural constraint introduced concerns the ways in which the properties of a mechanical system can hang together."[7]

The word "logic" is evidently being used in the above quotations to describe relations that (nonlinguistic) objects bear to each other.[8] Now, the doctrine that all relations are fundamentally logical relations, and that there is hence no basic distinction between logic and ontology is not unknown in the philosophical literature (I use the traditional term "ontology" to make my point concisely, even though I would subject its use to extended discussion in other contexts). It has even been given a name, "panlogism," and counts some formidable philosophers among its adherents, notably Leibniz and Hegel.[9] If one wants to adhere to this doctrine, there is nothing scandalous about it—although I certainly wouldn't, since I believe there are many very good reasons not to. However, if one does adopt this position, it should be done with full awareness, and not tacitly through acceptance of a certain approach to quantum logic—particularly since I hope to demonstrate that this approach cannot do the job its adherents claim it does.

In short, I hear Hegel's Absolute Idea creeping up under cover of Finkelstein's words. At first blush, the statement "logic is empirical" sounds extremely new and daring: yet another previously sacrosanct transcendental realm is being brought down to earth for empirical investigation. Upon closer consideration of what the word "logic" can possibly *mean* in such a statement, however, the statement turns into "reality is logical," thus linking up with a hoary rationalist tradition.[10] Putnam demonstrates this process of inversion in his essay "How to Think Quantum-Logically." He starts out assuring us, "Logic is just as empirical as geometry," and ends by asserting that "physical laws have to be compatible with logic—that is to say have to be compatible with the *true* logic, which is quantum logic."[11] I doubt that Putnam would subscribe to this statement if we substitute "geometry" for "logic" in it: "Physical laws have to be compatible with geometry—that is to say have to be compatible with the *true* geometry, which is relativistic geometry." He—or at least I—would point out that physical geometry is no more than a certain aspect of, or abstraction from, the relations described by physical laws. Thus, I don't believe one can consistently uphold the "proportion" he states in the article between geometry and general relativity, and logic and quantum mechanics.[12]

None of my comments should be taken to imply that I believe there is some unique acceptable logical system. There are a variety of possible formal logical systems associated with a variety of formalized languages.[13] Indeed, ordinary (unformalized) language contains the elements of a number of possible logical systems; it is only its formalization in certain directions that results in the usual logical system. "What we must understand," as Waisman put it, "is that the choice between distinct systems of logic is not decided by logic. It may happen that one system is more suitable for this, the other for that purpose; or our decision may be guided by aesthetic preference; but neither has anything to do with any insight into the truth of a system."[14] Arthur Fine arrived at a similar conclusion: "I can find no conceptual background against which one might see the possibility of discovering that some law of logic is false. . . . There may very well be excellent reasons to explore non-classical systems of logic and even to employ them in science. It is just that, so far as I can see, the reasons for which one might thus abandon classical logic will be pragmatic; it will not be because one finds that a logic is false."[15] In other words, *what* logical systems are useful in treating *what* theories by means of *what* formal languages are questions to be decided by practice, not dogma.[16]

Emphasizing the linguistic reference of logic also does not commit one to the much narrower (and much criticized) thesis that logic is to be identified as a branch of syntax and/or semantics (often called the linguistic thesis); nor to the related doctrine of the absolute nature of the analytic-synthetic distinction.[17]

What I am arguing against, basically, is the idea that there is some sort of simple mirroring relation between the structure of language and the structure of the extralinguistic world such that logical relations mirror structural relations of "reality" in some direct, unmediated fashion. At any given state of scientific development what we confront is not "reality," but some particular theoretical structure, and its accompanying modes of experimental protocol, which enable us to understand and cope with some aspects of the world. What the assertion that "reality is logical" *really* offers us, then, is the consolations of philosophy: the assurance that things are the way they are according to some particular theory because rationally they *must* be thus and so. We are presented with a modern version of that combination of "uncritical positivism and equally uncritical idealism" that Marx stigmatised in his critique of Hegel.[18] The danger here is that such consolations may divert us from confronting tensions within the existing theoretical and experimental structure, or between that structure and other, unassimilated elements—tensions that could lead to a deeper comprehension of, changes in, or even the complete overthrow of, that structure. I will maintain later that just this danger lurks in the "logic is empirical" approach to quantum logic.

Having asserted that there is an important distinction between logical and ontological relations, the question naturally arises: what is the relation between the two? For I certainly want to emphasize that, if logic deals with language, language is meant to deal with the world. Once one starts to pose such questions, the ambiguity of the phrase "quantum logic" becomes apparent. It is open to two broadly contrasting interpretations, both of which, curiously enough, originated at almost the same time in the mid-1930s. "Quantum logic" can be taken as equivalent to "*the* logic *of* quantum mechanics," as Birkhoff and von Neumann put it; or it can be taken as equivalent to "logic[s] *for* quantum mechanics," as Martin Strauss did.[19] The first phrase suggests that the relations between certain elements of quantum mechanics (just what elements is a question to which I shall later return) are fundamentally logical; while the second suggests that the linguistic means with which quantum mechanics sets out to express certain physical relations may or may not involve nonstandard logical systems. The possibility that there may be more than one way of doing this also suggests itself. The difference between these two interpretations is nontrivial: the first point of view suggests that the logical reformulation of quantum-mechanical relations has deep explanatory power, enabling the solution of (at least some of) the problems of interpreting of the theory, or the dissipation of some of the apparent paradoxes of the subject. From the second point of view, however, it is clear from the outset that use of alternative logical systems to reformulate quantum mechanics gives us no more than alternative reformulations of the same content: if there is some mystery associated with the theory, it is not dissipated by being reformulated in a logical mode (any more than, in Molière's *Le Malade imaginaire*, the ability of opium to put one to sleep is explained by the quack doctor's reference to its "dormative powers"). Problems remain problems, no matter how reformulated.

Of course, this does not rule out the possibility that a particular interpretation of quantum mechanics may suggest, or be more compatible with, some particular approach—a point to which I shall return later. However, the considerations for or against accepting one or another interpretation of quantum mechanics—or any physical theory for that matter—are neither strengthened nor weakened by a formalization of that interpretation using some standard or nonstandard logic. The message I want to convey is that logic cannot do the work of physics—or any other natural science. So my critical comments are directed at "the logic of quantum mechanics" school of quantum logicians: those who, knowingly or unwittingly, blur the distinction between the logical and the ontological in their discussions of quantum logic. Important contemporary exponents of this approach are Finkelstein, Putnam, and Bub and Demopoulos.

I have no quarrel at the logical level with the "logic[s] for quantum mechanics" school of quantum logicians, who realize that they are dealing with problems at the linguistic level, and that logical considerations do not circumvent any difficulties in the interpretation of quantum theory. Of course, one may differ with the interpretation of quantum theory favored by a particular quantum logician of this school; but it will be clear to both parties that they are not differing (primarily) over logic.

Nor do I wish to quibble with the many mathematicians and physicists working on technical problems of quantum mechanics who hang out shingles reading "quantum logic" in front of their shops. Quantum mechanics is so full of misnomers why should one more matter?[20]

Having indicated my approach to logic and drawn the distinction between the two principal trends in quantum logic, I shall now outline the rest of this paper. First of all, I shall explore the possibility of using nonstandard logics in the discussion of physical systems. To introduce the subject in a simple and "colorful" way I set up and discuss a "color logic." Then I turn to other examples of increasing complexity and greater relevance to the problem of quantum mechanics. My aim here is threefold:

1. To introduce some of the technical machinery used in quantum logic, in order to give readers not conversant with the subject some feeling for the mathematical structures involved. Those already familiar with lattice theory may skip pages 239–43.

2. To show that many features of the various "quantum-logics" are not peculiar to quantum theory, but arise whenever one tries to formalize the treatment of systems with incompatible conditional properties, that is, properties which cannot all manifest themselves simultaneously.

3. To emphasize that it is the physical properties of the system that are of primary significance. Differing logical schemes for describing relations between these properties can be formulated, but none of these alternative logical modes contributes anything extra to our understanding of the physical relations themselves.

Finally, I turn to quantum-mechanical systems. I discuss the motivation for the introduction of Hilbert spaces in their treatment and the existence of alternative mathematical structures, such as convex sets, which are equally if not better suited to this purpose. Then, in the final section, I consider what I take to be the two major claims made by those quantum logicians who adopt the "logic of quantum mechanics" approach. First, they claim that, if one adopts this approach, one may treat a quantum-mechanical system as having properties in much the ordinary sense in which a glass of water, say, has them. Hence, one may attribute simultaneous intrinsic numerical values to the magnitude of all these properties—values, that is, not produced by measurement nor deriving their meaning fron an interaction of the quantum system with a macroscopic apparatus. Second, they claim that the application of quantum-logical laws dissipates apparent paradoxes and explicates puzzling features occurring in other interpretations of quantum mechanics. Critical examination of the first claim leads me to conclude that (up to now at least), far from something like an ordinary concept of property being applicable, *no* coherent quantum-logical concept of property has emerged. Examination of the second claim shows that the "explanations" offered for various features of quantum mechanics at best do no more than reformulate the puzzles: they assure us that things are the way they are because that's the way (quantum) things are. It should be possible for those already familiar with the subject to read this final chapter with only an occasional reference back to earlier chapters.

I dedicate this essay to the memory of Martin Strauss, whose pioneering work on quantum logic should be more widely known. Dudley Shapere's comments on an earlier draft and his generous encouragement were most helpful and much appreciated. I am also immensely grateful to Tomas Brody for a detailed critique of a portion of the earlier draft.

2. COLOR "LOGIC"[21]

One of the basic facts about perceived colors is that all of them may be produced by mixing three primary colors in the proper proportion. Consider a white screen on which beams of colored light may be projected, singly or in combination, and a (non-color-blind) person observing some color projected on the screen. The fact that all colors can be built up out of mixtures of three primaries implies that any color can be matched by a suitable mixture of only three such beams. It also means that any color can be distinguished by three characteristics: hue, brightness, and saturation, for example. Fortunately, we shall neither have to consider all colors nor go into the exact definitions of hue, saturation, and brightness. I shall only consider one set of three primary colors: red, green, and blue-violet, which I shall symbolize by *R, G,* and *B;* and the following colors, which can be obtained by appropriate mixtures of the primaries: by mixing *R* and *B* we get magenta, abbreviated *M;* by mixing *R* and *G* we get yellow, *Y;* by mixing *B* and *G* we get a blue-green named cyan, *C;* by mixing all three primaries *R, B,* and *G,* we get white, *W;* and if none of the primaries is illuminating the screen we get black, *Bl.* These relations among the colors are summarized in figure 1.
Here, straight lines connect the colors contained in a mixture: a color in the mixture appears at the lower end of a line, while the resulting color appears at the upper end. We shall say that one color *is contained* in another if there is any sequence of lines leading upward from the first

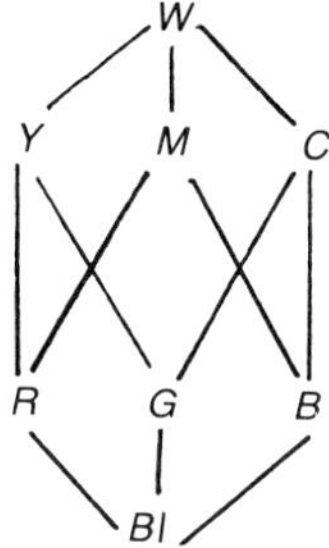

Figure 1

color to the second. For example, green is contained in white. We say that one color *covers* another if it contains it without any intermediate color lying between the two. For example, white covers yellow. Any color, then, can be obtained from a mixture of all the colors it covers. We regard black, by convention, as contained in every color; in order to make the statement about mixtures true for the primaries, we widen the containment relation slightly by saying that any color is contained in itself. We are thus led to a containment relation between two colors, X and Y (where the letters X and Y stand for any two colors on our diagram):

$$X \leqslant Y;$$

this may be read Y contains X, or X is contained in Y. Note that two colors needn't stand in this relation. Green, for example, is not contained in magenta, nor is magenta contained in green. Every color, however, by our convention, is contained within itself:

$$X \leqslant X.$$

Inspection of the diagram shows that this relation is *transitive*: that is, if one color is contained in a second, and that second in turn is contained in a third, then the first is contained in the third. Symbolically, if $X \leqslant Y$, and $Y \leqslant Z$, then $X \leqslant Z$. Mathematically, a relation with these properties (some, but not necessarily all, elements are related; an element is related to itself; the relation is transitive) is called an *ordering* relation, or a *partial ordering* in somewhat older terminology. Thus, our set of colors, related by the containment relation suggested by the properties of color mixing, is an ordered set.

But there is further mathematical structure inherent in our diagram. For any two colors in the diagram, we can define various operations that connect them with a (unique) third color. Indeed, two such "natural" operations are suggested by the containment relation. We can single out the *first* color (reading the diagram upward) which *contains* both our colors in it. For example, starting with red and blue, magenta is the first color containing them both (white also contains both, but white also contains magenta, so white is not the *first* color containing both). More abstractly, for two colors X and Y, we define the color $X \vee Y$ (which can be read "X cup Y") as the first color above the two in our diagram that contains both of them. Similarly, we can single out uniquely the *last* color *contained in* both colors. For example, starting with red and blue this would be black. More abstractly, for two colors X and Y, we define the color $X \wedge Y$ (which can read "X cap Y") as the last color contained in both (by last we mean that no higher color in the diagram is contained in both; for example, black is contained in yellow and cyan, but green is higher and contained in both, so $Y \wedge C$ is G). Mathematically, an ordered set for which such cap and cup operations are defined (relative to the ordering relations in question, of course), for every pair of elements in the set is called a *lattice*. Thus, our set of colors, related by the containment relation suggested by the facts of color mixing, is an example of a lattice. Since lattices play a major role in quantum mechanics, particularly in quantum logics, it is worthwhile looking at some other ways of picturing this color lattice, which suggest other properties that lattices may have.

First of all, we may picture the primary colors as nonoverlapping regions in the plane, in a so-called Venn diagram (fig. 2).

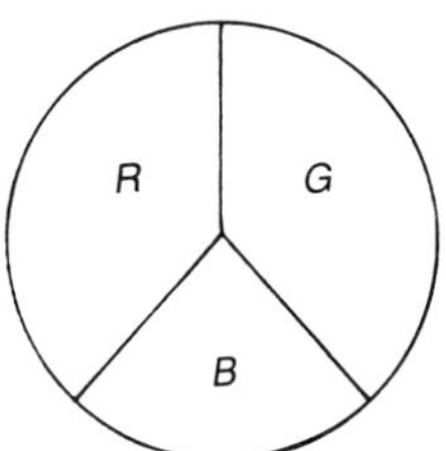

Figure 2

The other colors may be pictured as follows: black is "represented" by the null set, the set containing no points; white is represented by the entire space of the Venn diagram (or "universe" as it is often picturesquely called) consisting of the set-theoretical union of the three regions R, G and B—the entire circle, in other words. Yellow is represented by the union of the red and green regions, cyan by the union of the green and blue regions, and magenta by the union of the red and blue regions, as depicted in figure 3.

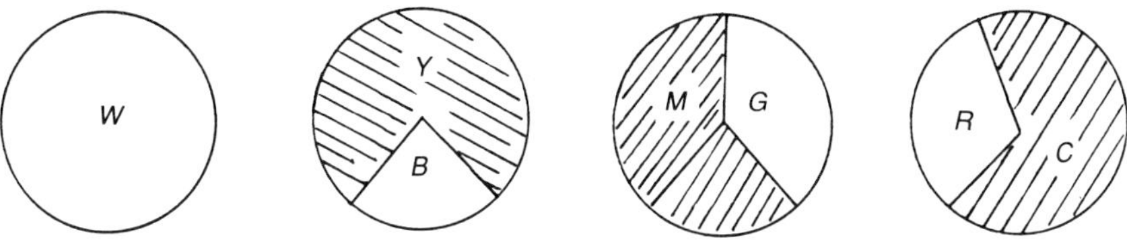

Figure 3

All the facts about color mixing, the color containment relation, and the cap and cup operations, have their counterparts in simple set-theoretical operations on these sets. The color containment relation is represented by the relation of set-theoretic inclusion: that is, one color is contained in a second if the region representing the first is contained in the region representing the second. The cup operation is represented by the set-theoretic operation of union; in other words, the color that is the cup of two colors is that color corresponding to the union of the two regions representing the two colors. Similarly, the cap operation is represented by the set-theoretic operation of intersection; that is, the color that is the cap of two colors is that color corresponding to the region common to the regions representing the two colors. Any lattice that can be represented in terms of regions in a Venn diagram (or, more generally, that can be represented by sets, with the usual set-theoretical operations of inclusion, union, and intersection representing the lattice ordering operation, the cup operation, and the cap operation, respectively) is a special type of lattice, called a *Boolean lattice.* Not all lattices are Boolean, but obviously the color lattice is.

Notice that the primary colors *R, G,* and *B* occupy a unique position in the color lattice, best brought out in the Figure 1. They are the "lowest" elements in the lattice, in the sense that they cover black, that is, no color stands between them and black, the lowest element in the lattice; furthermore, every other element in the lattice can be reached by following lines upward, starting from the primary colors. Such elements, if they exist, are called *atoms* of a lattice. A lattice built entirely out of combinations of such elements is called an *atomic lattice.* Thus, the fact that three primary colors exist, and that all other colors in our color diagram can be built up by mixtures of the primaries, may be expressed by saying that the color lattice is atomic, with three atoms.

There is another relation between elements of our lattice having its counterpart in a relation between colors that I have not mentioned. This relation is most easily seen in the Venn diagram. Notice that for any color we choose there exists another color such that the regions representing the two are nonoverlapping and, taken together, fill the circle.

For example, the red region and the cyan region are nonoverlapping and together fill the circle. Such elements of a Boolean lattice are called *complementary*; happily, colors that stand in this relation are also called complementary. They can be defined as a pair of colors containing no color in common (except black, of course) that can be mixed (in appropriate proportions, of course) to give white. We were able to *deduce* the existence of complementary colors from the Venn diagram which encoded the facts about color mixing previously mentioned. In other words, these facts already *implied* the existence of complementary colors; the Venn diagram merely helps us to see this conclusion.

You may wonder how this conclusion is pictured in Figure 1. A moment's reflection (pun intended) shows that complementary colors are symmetrical about the center point of the diagram, reproduced in figure 4 with the central point indicated by a large black dot and complementary colors indicated by a minus sign (e.g., R has $-C$ in parentheses after it to indicate that it is the complement of cyan):

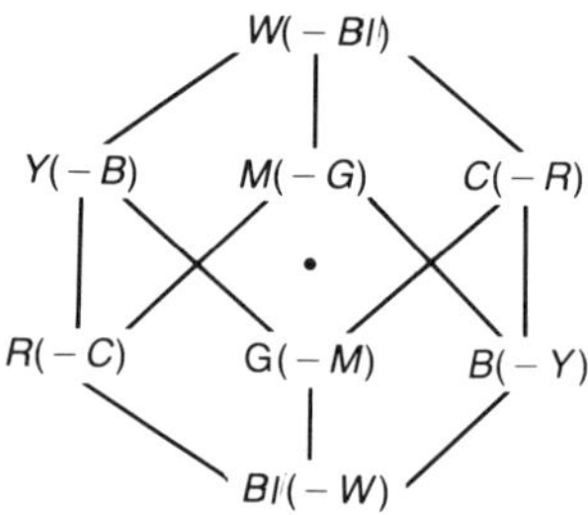

Figure 4

Two other properties of the complementary relation are clearly brought out in this diagram: first, it is *symmetric*—if one color is the complement of another, the second is the complement of the first. Secondly, complementing two colors reverses the ordering relation. For example, $G \leqslant Y$ implies that $-Y \leqslant -G$, or equivalently $B \leqslant M$.

You may also wonder about the use of a minus sign to represent color complementarity. Sometimes the minus terminology is used in the color literature, but there is another reason. The relation of complementary sets is often indicated by the use of a minus sign: if I is used to represent the universal set and X is some subset, then $I - X$ represents the complementary set. Later on, I shall give a "logical" reason for the minus sign, but before leaving the topic, remember that any Boolean lattice can be represented by a Venn diagram (or more generally by some universal set and certain subsets of it), and every set always has a complementary set relative to the universal set. It follows that every Boolean lattice is complemented.

I shall return to Boolean lattices, but first I shall describe another way of picturing colors and their relations, which is of help in visualizing the more general, non-Boolean lattices that occur in quantum mechanics. We can represent the three primary colors by three intersecting perpendicular *lines*, as in figure 5.

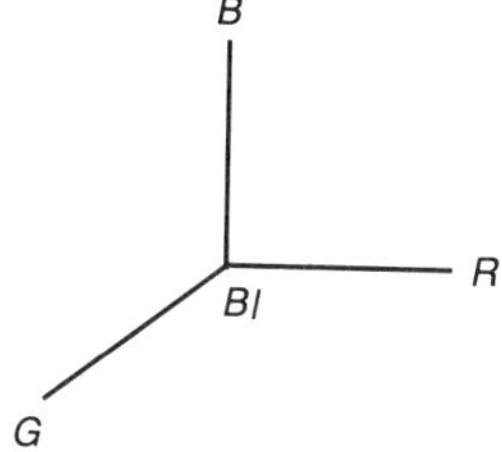

Figure 5

The colors are then represented as follows: black is represented by their *point* of intersection, which I shall call the origin. The three primaries, of course, are represented by three perpendicular *lines* through the origin; the mixture of any two primary colors will be represented by that *plane* through the origin containing the lines for the two primaries; and the *whole space* will represent white. To help visualize this, figure 6 shows the three planes with their corresponding colors:

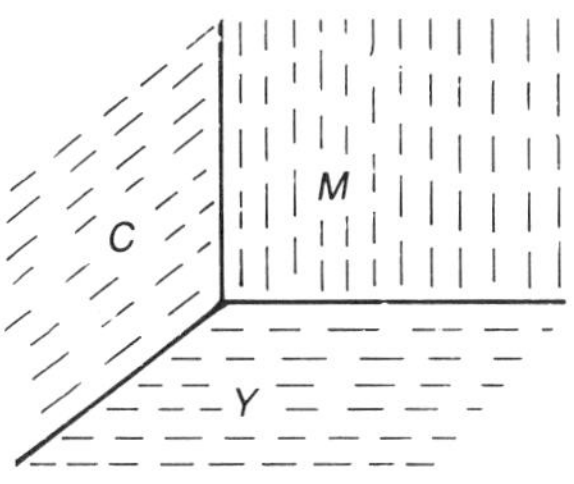

Figure 6

We can now verify that all color relations previously discussed correspond to geometrical properties or relations of the origin point, lines, planes, and the whole space representing the various colors. The atomic property of primary colors corresponds to the fact that lines are the lowest (one) dimensional entities beyond the (zero-dimensional) point representing black. That all other elements can be built from the atoms corresponds to the geometrical fact that all the planes as well as the entire space can be spanned by two or three lines. Color complementation corresponds to the *orthogonality* relation: the line representing red, for example, is orthogonal to the plane representing cyan just as red and cyan are complementary colors. The color containment relation corresponds to a point's property of lying on a line, a line's property of lying in a plane, a plane's property of lying in the whole space. For example, the green line lies in the yellow plane, and yellow contains green.

The cap relation corresponds to *intersections* of the origin point, lines, planes and the whole space. For example: the intersection of any two lines is the origin point and the cap of any two primaries is black; the intersection of the whole space with any point, line, or plane in it is that point, line, or plane just as the cap of white with any color is that color, and so on.

The cup relation is a little trickier, but not much. It is pictured as follows: each of the two colors whose cup we want is represented either by the origin point, or by some lines or planes through the origin, or by the whole space. Take the representatives of the two colors and find the point, line, plane, or whole space that just contains them (i.e., nothing of lower dimension will do); then the color corresponding to it represents the cup of the two colors. For example, two perpendicular lines just lie in the plane containing them (they also lie in the whole space, but they don't "just" lie in it—it is one dimension higher than needed). Therefore, the two primaries represented by these lines have the color represented by this plane as their cup. A line and a plane may either be contained in that plane—if the line is coplanar—or be contained only in the entire space if the line is perpendicular to the plane. As examples of these geometrical facts, we can read from the diagram that $R \vee M$ is M, while $R \vee C$ is W.

Thus, Figure 4 includes all the facts about colors represented in the other two diagrams, i.e., that follow from the fact that the color relations form an atomic Boolean lattice with three atoms. This diagram, however, suggests a question not suggested by the previous ones. We have only used one triplet of perpendicular lines through the origin and the set of planes they span. Yet there is clearly an infinity of such triplets of perpendicular lines, together with the planes they span, in space with a point picked as origin. Is there a lattice for which *every* line and plane through the origin represents an element of the lattice? There are such lattices. Indeed, they are examples of just the kind of non-Boolean lattices that occur in quantum mechanics. I shall later discuss the use of such lattices in quantum theory and quantum logic, but here I just compare them with the Boolean lattices previously considered. In some ways they are similar: they are also atomic, since lines through the origin are still the lowest-dimensional elements in the lattice (except for the origin point, of course); every plane through the origin can be spanned by a pair of lines; and the whole space can be spanned by three noncoplanar lines. There are now, of course, an infinity of atoms—or, better, a threefold infinity of atoms—corresponding to all possible triplets of orthogonal lines through the origin. In abstract lattice theory—and, indeed, for the lattices used in quantum mechanics—one introduces spaces of an arbitrary—even infinite—numbers of dimensions. There is nothing magic about the number three: it is just the highest dimension that we can easily visualize, since physical space is three-dimensional. It is just a lucky accident that only three primary colors are needed, so we can depict color space in physical space. I shall stick to such easily visualizable representations of lattices, but only for simplicity.

Another feature that these more general lattices used in quantum theory share with our Boolean lattice is the existence of a unique complementary element for every lattice element. Geometrically, this corresponds to the fact that there is a unique plane through the origin orthogonal to any line, as well as a unique line through the origin orthogonal to any plane. There are even more general lattices in which every element does not have a unique Boolean-like complement. Those lattices that do are called *orthocomplemented.*[22] Since only atomic, orthocomplemented lattices like those I have been discussing are needed in quantum mechanics, I shall consider only such lattices from now on. They can always be pictured by systems of lines, planes, and so on through the origin point of a space with some finite or infinite number of dimensions, or some generalization of such a space.[23] In such a space we can always single out a set of mutually orthogonal lines (the number needed equals the number of dimensions of the space) and generate a Boolean lattice from these lines by forming planes and the like by combining them in all possible ways, as in our three-dimensional example. We may think of the lattice as made up of an infinity of such Boolean lattices, fitted together in the special way that allows them to be pictured in a generalization of figures 5 and 6, the third type of diagram. Note that any given line through the origin can be included in an infinity of different orthogonal sets; the other sets can be obtained from any one set by rotation about the line in question. Figure 7 illustrates one such line ($\alpha_1 = \beta_1$) and two orthogonal sets that include it: (α_1, α_2, α_3) and (β_1, β_2, β_3).

I remind you that if a lattice is Boolean, its elements can be pictured in the second type of (Venn) diagram. A feature of more general lattices that is most important for the discussion of quantum logic is that the entire lattice *cannot* be so represented. That is, the elements of such a lattice *cannot* be pictured by a set of regions in a Venn diagram, with the cup operation represented by set-theoretical union and orthocomple-

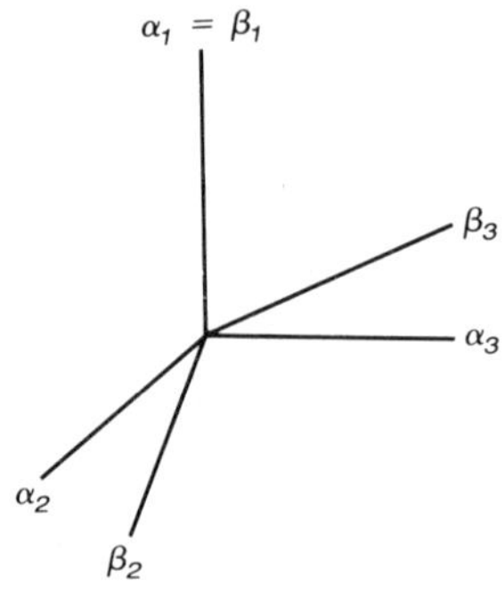

Figure 7

mentation by set-theoretical complementation of these regions. I will return to this question when discussing the role of the distributive law in quantum logic.

We now have enough mathematical machinery to begin discussing formal logic. Start with a set of *simple propositions* (for the moment taking that concept for granted) and form the set of all the *compound propositions* that result from them by iterated negation, disjunction and conjunction—that is, repeated use of the words *not*, *or*, and *and* to connect simple propositions. A relation of logical implication holds between some of the propositions in the set thus formed. For example, if "*a*" and "*b*" represent simple propositions, the proposition "*a* and *b*" implies the proposition "*a*." The set of all compound propositions then forms an atomic lattice, with simple propositions as its atoms, the implication relation as its ordering relation, disjunction (*or*) identified with the cup relation, conjuction (*and*) identified with the cap relation, and negation (*not*) identified with taking the complement. (This again demonstrates the appropriateness of the minus sign for complementation, since it is usually used for negation in logic.) Those of you who have diagramed relations between propositions on a Venn diagram will realize that the resulting lattice must be Boolean. One can also give another logical interpretation of a Boolean lattice, which is closer to my color example. Instead of a set of simple propositions, consider a set of simple properties—or rather, a set of simple predicates used to describe the properties of some object or system—and then form the set of compound predicates by iterated negation, disjunction, and conjunction of the simple predicates. There will be a relation of implication within the resulting set of predicates, and it will generate an atomic Boolean lattice of predicates. There is a simple relation between the predicate interpretation and the propositional interpretation of the lattice, of course: If we take the elementary propositions to be of the form "This object (or system) has the property *P*," they will be in one-to-one correspondence with the simple predicates *P* of the object. The compound proposition "This object has the property *P* and this object has the property *Q*" stands in one-to-one correspondence with the proposition "This object has the (compound) property $P \wedge Q$." A similar correspondence may be established between compound propositions formed with *or* and compound predicates formed with $\vee$, and between negations of simple propositions and negations of predicates.

The mathematics of atomic Boolean lattices thus serves to formalize the relations of the usual propositional logic by means of a propositional calculus or of the usual predicate logic by means of a calculus of predicates.

Now we are ready to discuss a color "logic." Suppose, for the moment, that we take the similarity of mathematical form between the lattice structure of the color diagrams and the lattice structure of the predicate calculus to imply an identity between them. That is, we interpret the complementary relation of two colors as a logical negation, the cup relation (between two colors and a third) as a logical disjunction and the cap relation as a logical conjunction. The assertion "This color is not yellow" for example, must now be regarded as equivalent to "This color is blue," since blue is the complement of yellow. "This color is red or green" must now be regarded as equivalent to "This color is yellow," since R $\vee$ G is *Y*. Similarly, "This color is yellow and cyan" must be taken as equivalent to "This color is green," since Y $\wedge$ C, is G. You will easily convince yourself—however counterintuitive such equivalences may seem—that we thereby get a perfectly self-consistent way of "negating" and "compounding" elementary color predicates (or propositions), which can never lead to any contradiction. (If it did, the original lattice would be self-contradictory.) Indeed, since we use the Boolean lattice structure of our color diagram to guide us in setting up color "logic," one might argue that we are still doing standard logic—only with a rather nonstandard interpretation of the connectives! Indeed, every law of standard predicate or propositional logic will have its counterpart here—suitably reinterpreted, of course. The law of the excluded middle ("p $\vee$ $-$p" is always true) will here mean that mixing any color with its complement always yields white. The law of contradiction ("p $\wedge$ $-$p" is always false) will here mean that any color and its complement contain only black in common. And so on.

What can one say about such a "color logic"? In the first place, note that the relations between the colors were first established empirically; only on the basis of these empirical relations were we able to set up definitions of the "color logical" connectives. We would be hard put, indeed, to explain to someone who could identify colors but knew nothing about the facts of color complementation why he or she should regard "This is not yellow" as synonymous with "This is blue." Thus our "logical" connectives are just logically sanctified versions, so to speak, of the empirical connections previously established between colors. If anyone were to take this game seriously and claim that the laws of "color logic" in some way *explained* the facts about color relations or that they deepened our understanding of these facts I think one would be justified in replying: "By baptizing empirical relations between colors as logical relations between color predicates (or between propositions about colors) and claiming that we now understand them better, you think you have done something very profound. It seems to me, however, that you are really manifesting an uncritical empiricism: You accept these relations, as they stand, asserting that they somehow have the force and necessity of logical relations among predicates or propositions. Thus, you divert attention from the problem of trying to *really* reach a deeper understanding of these relations, for example on the basis of some theory of color vision. Surely one can conceive of a world in which the relations between green, yellow, and cyan, for example, were different from those encoded in the color 'logic.' That is, one could, unless one were ready to accept that everything in the world is just as it is, and could not be otherwise, because *all* relations *in the world have the force of logical necessity*."[24]

Secondly, one might enquire into the motivation for setting up such a "color logic." I don't think one would be satisfied with a reply along the following lines: "Lattices occur in logic. Lattices occur in color relations. Therefore, the lattices occurring in color relations must be interpreted as a logic." It would require a deviant logic, indeed, for *this* to be a valid deduction![25] Stegmüller has stressed this point in his discussion of quantum logic:

> Unfortunately, discussions [of quantum logic] are partially impaired by conceptual confusions. The following hypothesis may explain how this comes about: Formalized theories are made the object of algebraic studies. This hold for logical as well as physical theories. Now, classical logic has the formal structure of a so-called Boolean algebra. If it can then be shown of a quantum-physical theory that it has a structure differing somewhat from that of a Boolean algebra, then some authors state that a "nonclassical logic of quantum physics" has thereby been discovered. . . . Whether a theory applies classical logic or a nonclassical one is not decided by the question of whether this theory can be subsumed under the same axiomatically characterizable *algebraic structure* as classical logic or under the *algebraic structure* of a different logic.[26]

One can easily forgive mathematicians who overlook this distinction and refer to the mathematical structure of quantum theory without further ado as "quantum logic," especially since the term "mathematical logic" is often taken to mean the study of those mathematical structures used in logical theories, quite independently of any logical interpretation. But I find it harder to forgive physicists, and especially philosophers, who contribute to this confusion.

Stegmüller goes on to point out that "the error is hardly ever committed in such a crude form. Rather, a concept of assertion or proposition is constructed for the theory in question, even if usually in a rather forced way, which lends at least a certain plausibility to the claim that the question of exemplification of a logic is involved."[27] This comment is especially relevant to our later discussion of quantum logic. But here we may note that, although the elementary propositions of "color logic" are acceptable, the meanings assigned to color "negation," "conjunction," and "disjunction" lead to bizarre compound propositions. We have already mentioned that, if negation is construed as color complementation, then, for example, the proposition "This color is red" must be taken as equivalent to the proposition "This color is not cyan." Clearly, "not" is being used here in quite another sense than in the usual understanding of the proposition; "This color is not cyan." The latter is generally taken as equivalent to the proposition "It is not the case that this color is cyan"—surely not equivalent to the proposition "This color is red!" Indeed, with only "color-logical" linguistic means, we have no way of expressing the content of the proposition: "It is not the case that this color is cyan." We might think we could do it by the proposition "This color is red or this color is yellow or . . . "—including in our disjunction every color but cyan. But remember that our "color-logical" disjunction *also* does not have the ordinary sense of the word "or." The suggested "color-logical" proposition must be taken as equivalent to "This color is white," since the disjunction includes all the primaries. So, to express a simple proposition like "it is not the case that this color is . . . " in color "logic," we should have to introduce a *second* negation, equivalent to the usual one. On the other hand, to express the relations encoded in "color logic" by means of ordinary logic, one need merely introduce the color-ordering relation (color inclusion). All other needed relations can be defined in terms of the latter, as we have seen.

Further troubles arise when we consider conjunction, for example. Consider the statement "This color is yellow and this (same) color is cyan." Ordinarily, we should be inclined to reject it as false, since the same object cannot simultaneously be two different colors. In other words, certain properties that an object could have are *incompatible* with each other. The assertion of one such property implies a tacit assertion that conditions prevail that allow the first property to occur, and this precludes the occurrence of those conditions that would allow the second property to manifest itself. The statement: "This color is yellow," for example, is shorthand for some such statement as, "This screen is being illuminated with a combination of electromagnetic waves of suitable frequencies which are such that the sensation of yellow is induced in the eye of a normal (non-color-blind) person looking at the screen."[28] This set of conditions is such that we *cannot* simultaneously assert that the screen is being illuminated with a set of frequencies that produce the sensation of cyan in a normal eye. Let us call a proposition which includes all the conditions for the manifestation of a property, such as color, an expanded proposition. Now consider the conjunction of the expanded propositions for yellow and cyan: "This screen is being illuminated with a combination of electromagnetic waves of suitable frequencies which are such that the sensation of yellow is induced in the eye of a normal (non-color-blind) person looking at the screen, *and* this screen is being illuminated with a combination of electromagnetic waves of suitable frequencies which are such that the sensation of cyan is induced in the eye of a normal (non-color-blind) person looking at the screen." It is either false, that is, it is the assertion that the screen is at one and the same time yellow and cyan as distinct colors; or at best it might be taken as equivalent to the assertion that "this screen is being illuminated with a *mixture* of the frequencies for yellow and the frequencies for cyan, and so on." Given the laws of color mixing, the latter assertion is equivalent to the statement "This color is white" (equivalent "color logically" to "this color is yellow *or* it is cyan"). However, our "color-logical" conjunction "This color is yellow and this color is cyan" is not equivalent to the ordinary conjunction of the expanded propositions. It is equivalent to the proposition "This color is green," or, in expanded form: "This screen is being illuminated with a combination of electromagnetic waves of suitable frequencies which are such that the sensation of green is induced in the eye of a normal person." By taking reasonable facts about color mixing encoded in the color lattice and forcing a logical interpretation upon them, we are led to construe the conjunction arbitrarily. Part of the trouble comes from insisting on attaching a meaning to the conjunction of predicates corresponding to incompatible properties—a move not unknown in quantum logic, as we shall see.

Another difficulty with giving color "conjunction" a logical interpretation is apparent if we remember that the assertion of an ordinary conjunction implies that the assertion of either conjunct separately is justified; or, to put it another way, either of the conjuncts can be validly deduced from the assertion of a conjunction. One must then contemplate a logic with the possibility of deducing that "this color is yellow" or "this color is cyan" from the assertion that "this color is green." If we interpret these statements in the expanded form given above, however, we can deduce the *falsity* of these statements.

So far, we have been trying to interpret the color connectives $-$, $\wedge$, $\vee$ as somehow related to ordinary negation, conjunction, and disjunction.

Suppose we give up that attempt and just agree to give them new names: say *nyet, et* and *aut,* letting the color lattice tell us what the meaning of these connectives is. Then, the relation of the color predicates *X* and *nyet-X* will correctly reflect the relation of color complementation, and the relation of the color predicates *X, Y,* and *X et Y* and *X aut Y* will correctly reflect the relations of color mixing. This seems perfectly acceptable as long as we realize that the formal similarity of the color lattice and a logical lattice now does not at all imply a relation between the *meaning* of the color connectives and that of the logical connectives.

To anticipate the discussion of quantum logic, Putnam has claimed on the one hand that one can "just read the [quantum] logic off from the Hilbert space,[29] and on the other hand that "a strong case could be made for the view that adopting a quantum logic is *not* changing the meaning of the logical connectives but merely changing our minds about the (distributive) law . . . (which fails in quantum logic.)"[30] The example of the color lattice shows that one could, on the one hand, read a *standard Boolean* "color logic" off from the lattice—but with no possibility of giving the connectives a standard logical meaning. On the other hand, if one reads a predicate logic off from the lattice in a reasonable way, the meaning of the color connectives has nothing to do with the standard logical connectives. Although this analogy proves nothing about quantum logic, of course, it should at least make us wary about claims that one can "just read the logic off" from any lattice, especially when it is conjoined with the claim that the meaning of the connectives is thereby unaltered.

I do not suggest my account of "color logic" provides a direct analogy to quantum logic. Rather, it should be taken as a parable, from which one can learn several lessons worth bearing in mind when discussing quantum logic. First of all, not every lattice of properties—even an atomic Boolean one—should be interpreted as corresponding to a logic of predicates or propositions. Just because one can formally associate a set of elementary predicates or propositions with the atoms of such a lattice and introduce a self-consistent association of the "negation," "conjunction," and "disjunction" of such elementary predicates or propositions with the lattice operations of complementation, cup, and cap does not mean that one has succeeded in creating a new logic.

A final lesson is the need for caution in the logical treatment of predicates or propositions referring to incompatible properties. I shall next discuss this problem at greater length, since it brings us a step closer to the problem of quantum logic.

3. INCOMPATIBLE PROPERTIES

Propositions that implicitly contain presuppositions, that is, propositions whose assertion is meaningful only when those presuppositions are fulfilled, have long been of interest to logicians. It is not always obvious how to express such presuppositions explicitly. One way is to form a conditional statement, with the presupposition(s) stated in the antecedent and the proposition as consequent. There is then no problem if the presuppositions hold. But what is to be made of such a statement if the presuppositions do not hold? Logicians call such conditionals subjunctive or counterfactual conditionals.[31]

As an example, consider the proposition "If it rains tomorrow, then I shall go out with an umbrella." If, indeed, it rains tomorrow and I go out with an umbrella, the proposition is clearly true. If it rains tomorrow but I leave my umbrella home when I go out, it is clearly false. But what if it does not rain tomorrow? Someone wedded to material implication will ruthlessly apply the truth table interpretation of "If . . . then" statements: since the antecedent is false, the statement is true, regardless of what I do. Less sophisticated persons, or those less wedded to material implication, may hesitate: One might argue that, since the presupposition was not fulfilled, the statement is in a sense annulled; perhaps it is better to attach to it no truth value at all. The statement may be interpreted as an implied promise: "If it rains tomorrow, I promise to take an umbrella out with me." The promise was not broken, so in that sense the statement is not false; but it seems excessive to say that it is therefore true.

The presupposition is often not included in the statement of the proposition. Suppose we listen to the weather forecast together one evening, and all channels unanimously predict a heavy thunderstorm the next morning. I observe, before going off to bed, "I shall take my umbrella when I leave tomorrow." You wake up the next morning, after I have left the house, to a brilliant sunlit day—not a cloud in the sky—and you find my umbrella in the stand near the door. Do you say: "There's Stachel's umbrella; he lied to me last night!" Or do you (more charitably, surely) say: "Stachel meant to say, 'If it rains I shall take my umbrella when I leave tomorrow'; so seeing a sunlit morning, he quite rightly decided not to burden himself with his umbrella—which he'd probably lose anyway without rain to remind him to keep dragging it around. So his statement just doesn't apply: it's meaningless in the context of today's weather; it wasn't really a proposition."

Blau, as expounded by Stegmüller,[32] uses the existence of such propositions with unfulfilled presuppositions as one of his main arguments for the claim that ordinary language is best treated as implicitly containing a third truth value: indeterminate (*unbestimmt*). Without pursuing this possibility, it is clear that a problem arises when we take statements with presuppositions—whether or not these are explicitly included in the statement—and try to decide how to incorporate them into a formalized logical system. When discussing the cases in which such a proposition is not clearly true, it seems important to distinguish somehow between the case in which the presuppositions are not fulfilled and the case in which they are fulfilled and nevertheless the proposition does not hold.

We move closer to the case of quantum logic by discussing predicates that can be meaningfully asserted of some object only if certain conditions are fulfilled. I shall call these conditional predicates, and also speak of the corresponding properties as conditional.[33] This is obviously a special case of the previous one: if P is such a predicate, the proposition "This object has (property) P" is an example of a proposition with implied presupposition—fulfillment of the conditions needed to make P meaningful being the presupposition.

One often attempts to define such a predicate by means of a conditional statement with the conditions in the antecedent. As an example, consider the hardness of some material, on Moh's scale, based on its ability to scratch or be scratched by various minerals.[34] The material manifests hardness only if it is in the solid phase. Under conditions (of temperature and pressure, say) in which it is liquid or gaseous, it cannot scratch anything. So we might attempt to define hardness by the statement: "If the material is in the solid phase and if it has hardness h, then it will scratch mineral m_1 and not scratch mineral m_2." Of course, the conditional here should be read as a biconditional. The converse conditional, "If the material will scratch mineral m_1 and not scratch mineral m_2, then it is in the solid phase and its hardness is h," must be true or false whenever the first conditional is. This gives a definition of hardness that is equivalent to that suggested by Strauss for reactive properties.[35] On the other hand, the biconditional, "If the material is in the solid phase, then it will scratch mineral m_1 and not scratch mineral m_2 if and only if the mineral has hardness h," might also be taken as the definition of "the material has hardness h." Indeed, this is equivalent to Carnap's partial definition of a dispositional property.[36] In case the material is in the solid phase, the two definitions are equivalent, as a look at their truth tables shows.[37] But in the counterfactual case, when the material is *not* in the solid phase, their truth values are not the same in half the cases (assuming we are justified in using material implication to interpret such conditionals.) How best to formalize the definition of conditional properties is not easy to decide. For the following discussion any reasonable definition will do.

For ordinary (nonconditional) properties[38] and propositions predicating these properties of some object, negation of the predicate can be so defined that the negation of the proposition predicating the property is equivalent to the proposition asserting the negated predicate. This equivalence is usually taken so much for granted that it may need a moment's thought to realize that the proposition "This hat is not blue" can be interpreted two ways, either as a negative proposition, "Not (this hat is blue)," or as an assertion, "This hat is (not-blue)." Things are not so simple, however, when we turn to the negation of conditional properties.

Several possibilities suggest themselves. We might be talking about a *proposition* such as "This material has hardness *h*," using one or another definition of the conditional predicate "hardness *h*." In this case, assertion of the negation "This material does not have hardness *h*" could mean that the hardness of the material, as manifested in the solid phase, actually has some other value, say *h*″. We could similarly define the *predicate* "having hardness not-*h*," in the proposition "This material has hardness not-*h*," as that predicate describing the property of having any value of the hardness in the solid phase except *h*. Then the negated proposition would again be equivalent to the assertion of the predicate "not-*h*."

On the other hand, the statement "The material has hardness *h*" might be interpreted as an abbreviated version of the expanded proposition "Under conditions of temperature *T* and pressure *P* the material has the hardness *h*." This is not equivalent to the previous interpretation, because here we place no restrictions on the range of *T* and *P* to ensure that the material is in the solid phase. The negation of this proposition could be asserted under two different sets of circumstances (remember our similar discussion of counterfactuals): (1) if the material was solid at temperature *T* and pressure *P* but had a different value of the hardness; (2) if the material was fluid at that *T* and *P*.

Even in everyday speech we often feel obliged to distinguish between these two types of negation. Suppose hardness is being measured as a function of temperature in some laboratory. At some point one researcher remarks to the other, "The hardness of this sample is *h*." The second researcher may want to deny this because he knows the hardness of the rock under the current temperature conditions to have a different value; he might, on the other hand, want to deny it because he knows that the rock has melted at the temperature reached. If he just noncommittally says, "The hardness of this sample is not *h*," he will be giving his co-worker no clue as to which meaning he intends. He may try to convey a clue by his intonation. If he says, "The hardness of this sample is *not h*," this will imply that its solidity is not in question, just the value *h*. To imply the other case he will shift emphasis: "The *hardness* of this sample is not *h*"—perhaps adding—"because the damned thing has melted, you klutz!" if he is feeling uncharitable. This is one of those cases indicating that ordinary language contains the seeds of more than one type of negation, as has been often noted.[39] Van Fraassen has discussed the distinction between two types of negation in the context of quantum logic.[40] He calls one type ("The hardness is *not h*") "choice negation"; he calls the noncommittal type ("The hardness is not *h*"—the proposition is not true for whatever reason) "exclusion negation."

Zinov'ev[41] distinguishes between negation of a predicate, which he calls "intrinsic negation," and negation of an assertion, which he calls "extrinsic negation." His intrinsic negation is equivalent to what Van Fraassen calls "choice negation," but is applied to predicates, not propositions. Zinov'ev also admits the possibility of indeterminate cases "where it is impossible to establish whether or not an object has a given attribute."[42] This indeterminate case could include the case of conditional predicates when the conditions that allow predication are not present. In that case, Zinov'ev's extrinisic negation coincides with van Fraassen's exclusion negation. Zinov'ev did not have applications to quantum logic in mind in making this distinction. He states that "In the exposition which follows we will often talk about classical or non-classical cases, systems, etc. Non-classical cases will always mean that there are two different negations and indeterminacy; the classical will mean that indeterminacy is excluded and, therefore, the negations are not distinguished (or, they are identical). From this point of view the non-classical conception of logic appears as more general than the classical. The latter is a special case. This means the usual conception of non-classical logic as a contradiction of the classical has nothing in common with our terminology."[43] The relation between the two negations has also been briefly mentioned by Martin Strauss.[44]

It is hard to see what other type of negation besides choice negation could be used to derive another predicate from a conditional predicate such as "hardness *h*." It has all the basic formal properties required of a negation: Its application twice results in the original property $(-(-h) \equiv h)$. It is also true that the conjunction "*h* and not *h*" $(h \wedge -h)$ is the absurd property ϕ, which no object ever has: $h \wedge -h \equiv \phi$ [there is no problem about defining the conjunction of compatible properties in the usual way, as we shall soon see, and h and $-h$ are compatible since the conditions for applicability of each are the same.] What about the disjunction "*h* or not *h*" $(h \vee -h)$? If we want to preserve the usual relation between disjunction, negation, and conjunction (De Morgan's law)

$$p \vee q = -(-p \wedge -\mathrm{q}),$$

we must define $\mathrm{h} \vee -\mathrm{h}$ as the trivial property which an object always has. For, if we use De Morgan's law,

$$h \vee -h = -(-h \wedge -(-h)) = -(-h \wedge h) = -\phi;$$

and the negation of the absurd property is the trivial property. At first sight, this presents a problem: We can imagine a substance that could never be solidified but stays liquid down to absolute zero. We can require of any definition of hardness, however, that it imply that *if* the material *were* solidified, it would either have hardness h or some other hardness, even though the condition is never realized. (This is the case for either of the definitions of hardness proposed earlier, for example). Then we may take $h \vee -h$ as equivalent to the trivial property I: $h \vee -h \equiv I$.

If we want to preserve the simple relation between negation of predicates and of corresponding propositions that holds for unconditional predicates, we must adopt choice negation for propositions as well. This implies, of course, that a proposition and its (choice) negation do *not* exhaust all possibilities: only a proposition and its exclusion negation would do that. We shall explore further the consequences of such a decision shortly; but it should be clear by now that the negation of a predicate and the negation of the proposition asserting that predicate are distinct logical operations.

When conditional properties are compatible, that is, when the conditions under which each can manifest itself are not mutually exclusive, we can define the (compound) conjunctive predicate so that the conjunction of two propositions that assert the object to have each of the properties is equivalent to one proposition that asserts the conjunctive predicate of the object. For example, viscosity and conductivity are compatible quantities, since both can manifest themselves in the fluid phase. Thus, the compound proposition "This sample of liquid has viscosity v and this sample of liquid has conductivity c" is equivalent to the simple proposition "This sample of liquid has (viscosity v and conductivity c)." I have put the last phrase in parentheses to indicate that it represents a single compound property. We are so used to identifying a compound sentence involving simple predicates with a simple sentence involving a compound predicate that it takes some such device (as well as a moment's thought) to even realize that it is being done. Now consider all the mutually compatible properties that a system might have at a certain time. First of all, let me emphasize that I mean *might* have and not *does* have. For example, the properties "viscosity v_1" and "viscosity v_2" are compatible according to our definition, since both require the same precondition for their applicability: that the system be in the fluid phase. This does not mean, of course, that any one liquid could possibly manifest both viscosity v_1 and viscosity v_2 at the same time. That is, "viscosity v_1" and "visocity v_2" are compatible but mutually exclusive properties. In general, there may be a relation of implication between two compatible properties whereby if a system has one property, it must have the other. For example, if it has the property "viscosity v_1," it must have the property ("viscosity v_1 or v_2"). If two properties imply each other, they are equivalent with respect to this relation (although not necessarily otherwise, of course). For example, assume electrical conductivity and thermal conductivity to be (uniquely) correlated for some material. Then the property "electrical conductivity c_1" implies the property "thermal conductivity c_2" for some value of c_2 and vice versa; so they are equivalent properties. This implication relation turns the set of all mutually compatible properties into a partially ordered set; defining conjunctions and disjunctions of two properties in the usual way turns this partially ordered set into a lattice; the negation of any property is the (unique) complementation of that property. The resulting lattice of all mutually compatible properties must then be Boolean.

If the system has two or more incompatible properties, however, the situation is not so simple. Let us take hardness and viscosity. As noted above, the quality of hardness can manifest itself only when a substance is in the solid phase. Viscosity, on the other hand, is applicable only to substances in the fluid (liquid or gaseous) phase. Their conditions of applicability cannot be realized simultaneously,[45] so the two properties are incompatible. The statements "This liquid has viscosity v" and "This solid has hardness h" will then be shorthand for "This substance in the liquid phase has viscosity v" and "This (same) substance in the solid phase has hardness h." We can thus make sense of the assertion "This material has viscosity v and hardness h" by interpreting it as just shorthand for the conjunction of these two propositions. Can it also be interpreted as a statement about a compound predicate, "(hardness h and viscosity v)"? Since the conditions for manifestation of one property are incompatible with those for the other, the condition for the manifestation of such a compound property would be an *impossible condition*—a condition that can never occur. Thus, the compound predicate is not defined for any physical circumstances. *Any* definition would be compatible with all actual physical situations. This suggests that we are under no obligation to define such a compound predicate. Similar considerations apply to the definition of a disjunctive compound predicate. Of course, if we *do* define the conjunction and already have defined negation, then—if we want De Morgan's laws relating conjunction, negation, and disjunction to hold—we *must* adopt a corresponding definition of disjunction.

One simple definition that suggests itself for the conjunction of incompatible predicates is the impossible predicate. If we adopt this definition, and the "choice" negation (for which the predicate "hardness not-h" means having some hardness other than h), then Morgan's law requires that the disjunction of hardness and viscosity be defined as the trivial predicate.

As an example, imagine a system capable of existing in only two phases (solid and liquid, say) and having only one (independent) quality in each phase (say, hardness in the solid, viscosity in the liquid phase), with only two possible values for each of the two qualities (say, h_1 and h_2

for hardness, v_1 and v_2 for viscosity). Using the definitions of compounds of incompatible predicate just discussed, we get the following diagram, (fig. 8) for the lattice of its predicates.

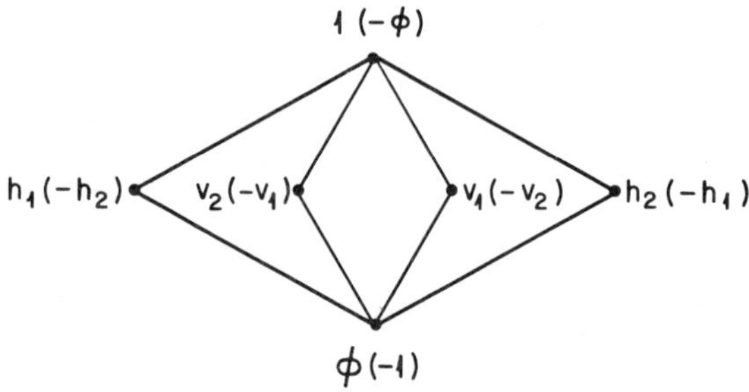

Figure 8

Finkelstein has drawn a diagram just like figure 8 in his article on "The Physics of Logic,"[46] with the caption: "The system . . . is the simplest quantumlike lattice and exhibits nondistributivity and coherence." In section 2 I noted that there are orthocomplemented atomic lattices that differ in very important ways from Boolean lattices. Since we now have an example of such a lattice, this is a good place to discuss nondistributive lattices in a little more detail. Let's start by going back to the (usual) logical interpretation of Boolean lattices, in which $\wedge$ is interpreted as conjunction and $\vee$ as disjunction.[47] If one says "*p* and (*q* or *r*)" (where *p,q,r* represent three propositions, or predicates), this is equivalent to saying "(*p* and *q*) or (*p* and *r*)," as one can see after a little thought. The corresponding relation holds for any three elements of a Boolean lattice. In other words, for any three elements *a,b,c* of a Boolean lattice, $a \wedge (b \vee c)$ is the same lattice element as $(a \wedge b) \vee (a \wedge c)$ (indeed, as discussed in appendix 2, this property can be used to define a Boolean lattice—distributivity captures the essence of Booleanness.) This distributive law can be visualized by drawing a Venn diagram, always good for picturing a Boolean lattice.

The lattice pictured in figure 8 is an atomic, orthocomplemented lattice, as we can easily check by going through the defining properties one by one (see appendix 2 for a summary of the defining properties). But the distributive law *fails* for this lattice. To see this, we take three distinct elements such as h_1, h_2 and v_1 for *a, b,* and *c,* respectively; and compare $h_1 \wedge (h_2 \vee v_1)$ with $(h_1 \wedge h_2) \vee (h_1 \wedge v_1)$ to see whether or not they define the same element of the lattice. $(h_2 \vee v_1)$ is *1*, so $h_1 \wedge (h_2 \vee v_1)$ is the same as $(h_1 \vee 1)$, which is h_1 itself. On the other hand, $(h_1 \wedge h_2)$ is ϕ, and so is $(h_1 \wedge v_1)$ Thus, $(h_1 \wedge h_2) \vee (h_1 \wedge v_1)$ is the same as $\phi \vee \phi$, which is again ϕ. Thus, the distributive law fails for this lattice of predicates of a (schematically modeled) two-phase system. Now the lattice composed of v_1, v_2, ϕ, 1 is Boolean, as is the lattice composed of h_1, h_2, ϕ, 1. What we have done is to "stitch together" these two Boolean lattices by identifying their ϕ and 1 elements. They then become sublattices of a non-Boolean lattice.

Let me emphasize two points. First of all, we got this lattice only because we adopted certain *definitions* of compound incompatible predicates. However "natural" these definitions may seem if we insist on defining such compound predicates, there is no physical content at all in their use. They are entirely dispensable: the existence of incompatible properties suggests that compounding of conditional predicates be restricted to compatible ones. Second, even granted that we introduce such compound predicates, there is no *necessary* relation between them and compound *propositions* formed from propositions asserting incompatible simple predicates of the system. We are not obliged to modify propositional logic, even if we adopt the compound predicate logic diagramed in a nondistributive lattice such as figure 8. Only if we impose the additional requirement that the negated and compounded *propositions* be so defined that their structure is isomorphic to the structure of the negated and compounded *predicates,* does a nonclassical propositional logic arise.

I am not arguing at this point that such nonstandard operations on predicates or propositions *cannot* be introduced, of course, nor even that they *should not* be. I am just emphasizing that there is no sort of *necessity* about such moves. There is certainly no *physical* necessity: physically meaningful relations between the properties of a system remain what they were, whether or not we decide to define compound predicates which would only be meaningful under conditions that can never occur[48] and whether or not we define compounding of propositions to parallel this compounding of predicates. Nor is there any *linguistic* necessity, since there are alternate ways of treating the relations of these physical properties in a formalized language—as well as the customary physicist's way of just using ordinary (nonformalized) language with a little care so as to avoid meaningless or undefined assertions. Finally, there is no *logical* necessity: as I hope the earlier discussion has persuaded you, the reason for changing a (consistent) logic cannot be because that logic is false. The concept of logical necessity is meaningful only *within* a logical system and not between logical systems.[49]

Now, consider the possibility of not defining compounds of incompatible properties for our two-phase, two-property system. There are two alternate ways to proceed, which may be depicted as follows:

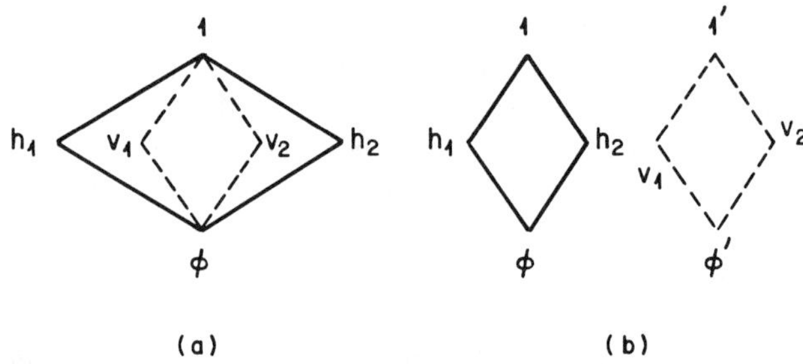

Figure 9

In the left-hand diagram, we continue to identify the ϕ and 1 elements, of both Boolean sublattices, but we do not introduce any compounding of elements from different sublattices. This is indicated by the use of different symbols to represent the ordering relation within each sublattice: straight lines for one sublattice and dotted lines for the other. Note that the two points—ϕ and *1*—which are connected by *both* types of lines form a third sublattice, which is contained in each of the other two. This diagram pictures the simplest example of a *partial Boolean algebra,* which may be characterized informally as a way of "sewing together" Boolean lattices more loosely than their previous stitching together into a lattice (fig. 8). This looser sewing together does not allow the compounding of elements which, together with their complements, do not form a Boolean lattice. (See app. 2 for a more accurate definition of a partial Boolean algebra.) It clearly represents a weakening of the previous lattice structure, which could be recovered by identifying the two types of lines. In the right-hand diagram the two Boolean lattices are entirely decoupled: the lowest and highest elements of each sublattice are distinguished as ϕ, ϕ′, and *1, 1′*, respectively.

Having noted several different types of structure we can impose on the two Boolean lattices of conditional properties of a model two-phase system, I shall now discuss the implications for logic of each of these structures. Let me start with the two separate Boolean lattices (fig. 9b). Each of these describes a phase of the system, solid and liquid respectively. The two phases are incompatible, but no difficulty arises if we restrict ourselves to predicates, or propositions about the (compatible) predicates, of each phase separately. All propositions will then, explicitly or implicitly, refer to one or the other phase. For example, "The viscosity of the system is v_1," must then be taken as shorthand for the expanded proposition "The viscosity of the system in the liquid state is v_1." The negation of that statement "The viscosity of the system is not v_1," must be expanded to "The viscosity of the system in the liquid state is not v_1." Since v_1 and v_2 are the only two values possible for the viscosity, this is equivalent to the statement "The viscosity of the system in the liquid state is v_2" or, in the shorthand form, "The viscosity of the system is v_2." Thus, the negation within each lattice is an exclusion negation (which happens to coincide with a choice negation in this simplified model because there are only two possibilities). The cup of v_1 and v_2 is the (highest) element *1* of the lattice. Expressed in propositions, this means that the statement "The system in the liquid state has either viscosity v_1 or viscosity v_2," is equivalent to "The system is in the liquid state"— the trivial proposition for the liquid state. And indeed they are equivalent under the conditions stated, in which only two values of the viscosity are possible. Similarly, the cap of v_1 and v_2 is the (lowest) element ϕ of the lattice. Expressed in propositions, this is the equivalence of "The system in the liquid state has the viscosity v_1 and it has the viscosity v_2" and the absurd statement about the liquid state, "The system is not in the liquid state." Should we regard these equivalences as logical equivalences? As is clear from the above discussion, they follow from a certain model of a physical system, and are only valid relative to that model. If we introduced a third possible value of the viscosity, for example, all these equivalences would not hold. So if we formulate the equivalences in the form "For model *M* proposition *A* is equivalent to proposition *B*," this is a logical truth that does not depend on the suitability of model *M* to represent some aspects of the behavior of a physical system. On the other hand, if we phrase the equivalence in the form "For system *S*, proposition *A* is equivalent to proposition *B*," then its truth or falsity depends on the applicability to system *S* of model *M*, and so it is not a logically necessary truth. If one just states "Proposition *A* is equivalent to proposition *B*," the statement is open to either of the above interpretations. Or, it might be given some entirely different one, such as, "The world is such that the state of affairs corresponding to proposition *A* is equivalent to the state of affairs corresponding to proposition *B*, and this equivalence has the force of logical necessity." The reader will know from section 1 that I do not accept such an interpretation. It depends on two assumptions: that language somehow directly mirrors states of affairs in the world and that the relations between states of affairs in the world have the force of logical relations. For me, if the equivalence of two propositions *A* and *B* about a model *M* has the force of logical necessity, this is clearly for reasons independent of any state of affairs in the world. It is even a metalinguistic proposition, if we want to state that equivalence as a logical truth. Of course, I do not expect everyone to agree with me on such questions, but I do wish everyone writing on "quantum logic," and meaning thereby more than a mathematical formalism, would state what he or she *does* mean by logical truth and logical necessity. Then at least one could have an informed discussion of the philosophical implications of various positions.

Returning to our model of a physical system, let us suppose we now identify ϕ and ϕ' and *1* and *1'* which converts the two separate Boolean lattices into a partial Booean algebra (that is, we move from fig. 9b to fig. 9a). By considering all elements to be part of one structure, we subtly change the meaning of the negation and disjunction. For now this one structure represents the system, regardless of the phase it is in. The common ϕ element now corresponds to the absurd proposition: "The system does not exist"; and *1,* the trivial proposition, corresponds to: "The system exists." The negations of the elements, which were *exclusion* negations with respect to the separate Boolean lattices, now become *choice* negations with respect to the partial Boolean algebra. (The exclusion negation of "The system has viscosity v_1" would now be "The system has viscosity v_2 or the system has hardness h_1 or the system has hardness h_2," where "or" is understood in the sense of set-theoretic disjunction.) Similarly, in the partial Boolean algebra $v_1 \wedge v_2$ now coincides with $h_1 \wedge h_2$, since both are equal to the *1* element of the partial Boolean algebra. Thus, "The system has viscosity v_1 or viscosity v_2" is now equivalent to "The system has hardness h_1 or hardness h_2"; and both mean no more or less than "The system exists." So the move from two disjoint Boolean lattices to one unified partial Boolean algebra has modified the meaning of the negation and disjunction, even though no compounds of incompatible properties are allowed.

Gödel proved the completeness of the propositional calculus based on Boolean lattices.[50] As Birkhoff pointed out: "One may interpret this result as follows: the classical logic of attributes cannot be *strengthened* without giving rise to absurdities; it can only be weakened."[51] Indeed, Kochen and Specker were able to show that one could construct a well-formed formula from the elements of a certain partial Boolean algebra (i.e., one that combined only elements from a single Boolean algebra at each stage of its formation) that would have been a *tautology* (i.e., true for any truth values of the elements entering the formula) if the elements had all been from a single Boolean algebra, yet was *invalid* for a particular truth valuation of the elements of the partial Boolean algebra. The tautologies of this partial Boolean algebra do represent a weakening of Boolean logic. Of course, the elements of the partial Boolean algebra entering such a formula cannot *all* come from one Boolean algebra. So as just indicated, the meanings of the negations and disjunctions in such a formula must differ from their meanings if the formula were to be given the usual Boolean interpretation. So the Kochen and Specker result does not provide any sort of "empirical refutation" of classical logic, even if such a partial Boolean algebra is used to model the behavior of some physical system.

Given the partial Boolean algebra diagram, we need merely ignore the distinction between the two types of lines to turn it into the non-distributive lattice of figure 8. Thus, the meanings of negation, and of those disjunctions of compatible elements that were already present in the partial Boolean algebra, need not be changed. The conjunctions and disjunctions of incompatible elements, however, must now be given a meaning. Since such compounds did not *exist* in the separate Boolean lattices of compatible elements or in the partial Boolean algebra, there is no possibility of their meaning remaining the same. *New* meanings must be introduced. We have previously discussed various aspects of this problem, but, in general, we can say that, like the meanings of ϕ and I in a partial Boolean algebra as such, they will have to be meanings that refer to the system without distinguishing between its solid or liquid phases.

To summarize, as long as we formulate propositions about the system with explicit or implicit reference to the phase it is in, no need arises to go beyond classical (Boolean) logical means of treating such propositions. As soon as we combine these disjoint Boolean structures into one compound structure, be it a partial Boolean algebra or an atomic, orthocomplemented lattice, we must modify the meaning of some operations and/or add new ones that must be defined. A weakening of classical logic may thereby be introduced for propositions that do not refer to phase; but no fresh insight into the physical relations of the system is obtained by doing so. The existence of alternative choices and the possibility of alternative logics indicate that this procedure may be best described as investigating "logics for" the model of the physical system in question. Any attempt to describe the use of a particular nonclassical logic as an expression of "the logic of" the physical system, in the sense that a nonclassical logic is imposed on the linguistic means of expressing the physical relations of the system by some supposed logical structure of these relations, has to account for the existence of such alternatives—including the classical one. (If, in turn, logical status is ascribed to features of the world supposed to be described by the model, there is the further problem of justifying the ascription of logical status to ontological relations.)

Putnam has tried to answer such objections based on the possibility of using alternative logics, in the quantum-mechanical context, by comparing different logics to different geometries.[52] Classical logic is then the analogue of Euclidean geometry; while quantum logic (logics) is (are) the analogue(s) of non-Euclidean geometry (geometries). In both cases, he claims, one can put the new wine into the old bottles, but only at the expense of introducing redundant hypothetical elements. One may wonder, in the light of the previous discussion, which interpretation actually introduces the redundant elements. But my objection to the analogy is more serious. Euclidean and non-Euclidean geometries are of "equal strength," in the sense that each one assigns a metrical structure to space, with all the consequences that flow from that assignment. We should expect, if the analogy is to be helpful, that we would be presented with two (or more) logical systems which are of "equal strength," but Gödel's result, mentioned earlier, shows that this cannot be the case. Both of the quantum logics Putnam has in turn adopted (atomic orthocomplemented lattice and partial Boolean algebra) as "the" quantum logic are (strict) weakenings of classical Boolean logic. We can easily show this: Suppose they were not. Then some tautology that is not a classical Boolean one would have to hold in one of these logics. But each of them contains at least one Boolean sublattice. Thus, this tautology would have to hold for a Boolean lattice, in addition to all the classical tautologies, contradicting the completeness theorem for classical Boolean logic. In other words, the set of non-Boolean tautologies must be strictly contained within the set of Boolean tautologies for *any* ordered set of elements that contains a Boolean lattice and for which the relations within the set reduce to the corresponding Boolean ones within that lattice. Some Boolean tautologies may indeed fail to hold, as Kochen and Specker showed for partial Boolean algebras and as the failure of the distributive laws shows for the orthocomplemented non-Boolean lattices. On the other hand, no tautology of such a logic can ever fail to be a Boolean tautology, for it must hold when the values of its variables are restricted to a Boolean sublattice. Thus, if there is a geometrical analogy for the relation between classical logic and these quantum logics, it is rather like that between Euclidean geometry (or hyperbolic or elliptic geometry) and Bolyai's (weaker) absolute geometry, obtained from Euclid's system by omitting the parallel postulate. Why any such weakening of classical logic should be capable of shedding light on the nature of quantum-mechanical systems is a mystery to me. At most, such a logic might shed light on why some things that are classically true are not true quantum-mechanically, but it is hard to see how it could help us to understand what *is* the case in quantum mechanics. Nor have any of the more detailed analyses of the claimed explanatory power of quantum logic, some of which will be considered in the last chapter, shed any further light on this question.

Another analogy between quantum mechanics and geometry has been proposed by Mielnik.[53] In this case, however, existing quantum mechanics is the analogue of Euclidean geometry, while some potentially interesting generalized quantum theories are the analogues of Riemannian geometries. The basis of this analogy is the convex set model of quantum states, to be briefly considered in section 3. Whatever one may think of this analogy, at least it deals with structures (convex sets) that are "equally strong," like Euclidean and Riemannian geometries, and not with systems that are all strictly weaker than one system.

Finkelstein has given an example[54] that indicates why we shouldn't expect the structural features of a nonclassical logic to give us any insight into the inner workings of those systems to which it applies. He considers a set of black boxes whose responses to certain tests may be used to order these tests. He shows that the boxes may be rigged so that the resulting ordering is not even transitive. As he says, "This models a more drastic departure from classical laws than that demanded by quantum physics." Yet he has set up his model using devices operating purely classically: "Within classical logic it is simple to set up 'malicious models' which violate any of these [classical] laws," such as transitivity of the ordering.[55] This means that the study of the "logic" of such a system, in the Finkelstein sense, gives us no insight into the inner workings of the black boxes. The question may thus always legitimately be raised, if we are confronted with such a deviant "logical" structure: What is the mechanism by which the effects described by this structure are achieved? The answer, of course, may be that there is no "black box" hidden in the system, so the "logical" structure accurately mirrors all there is to be learned about the system. It might also be the case, however, that such "black boxes" will be found to exist and their study to reveal much more about the system than is reflected in its "logic." But, baptizing that "logic," with Putnam,[56] the "true logic" of the world and insisting that "physical laws have to be compatible with . . . the *true* logic" is to foreclose such options.

I shall resist the temptation to explore more complicated models and instead summarize what has been accomplished in this section and draw some further lessons for the problem of quantum logic.

The conditional properties of a system can be divided into mutually compatible properties, which may manifest (or fail to manifest) themselves under the same set of conditions, and mutually incompatible properties, the conditions for the manifestation of one excluding the conditions for the manifestation of the other. For some simple cases I sketched how the treatment of predicates for such systems may be carried out using disjoint Boolean lattices, or combining these into one partial Boolean algebra or extending this algebra into a lattice. Analogous methods have been proposed for treating the "properties" of a quantum-mechanical system. This does not prove, of course, that quantum properties should be regarded as conditional in nature and not all compatible with each other, although this has often been suggested.[57] It does strongly indicate, however, that such an approach merits further consideration. I shall not go into this problem any further, although I think it may prove to be the most fruitful way of treating the peculiar properties of quantum-mechanical systems. My aim in this essay, however, is to deal with the claims of the "logic of quantum mechanics" advocates. One of their chief claims is that one need not treat the properties of a quantum-mechanical system as conditional or incompatible. This raises the question: Why, then, need they be treated by means of an atomic orthocomplemented semimodular lattice (Finkelstein, old Putnam)? Or by means of a partial Boolean algebra (Bub and Demopoulos, new Putnam)? An account in terms of incompatible conditional properties would at least offer some explanation for the possibility of using these structures. Whether or not the concept of conditional properties, or some refinement of it,[58] is applicable to quantum-mechanical systems, discussion of this topic has demonstrated one thing. Advocates of the "logic of quantum mechanics" approach cannot claim that partial Boolean algebras and nondistributive lattices are uniquely applicable to quantum-mechanical systems. If they have explanatory power for such systems, this must be due to some feature or features that they share with classical systems having incompatible conditional properties. If one believes that quantum-mechanical properties are *not* conditional, the problem of why this should be the case is even more puzzling.

4. QUANTUM MECHANICS

In the last section I showed by a simple example how partial Boolean algebras and nondistributive lattices may be used to model incompatible conditional properties of physical systems that everyone would agree are classical in nature. Thus, the mere occurrence of such structures cannot be characteristic of quantum-mechanical systems. Of course, in such a classical case, these algebraic structures lack many features found in the quantum-mechanical case. In this chapter I shall discuss some features characteristic of quantum systems and how they give rise to the kind of partial Boolean algebras and nondistributive lattices used to model such systems. Discussion of the example of polarized light, which may be analyzed as a classical wave or a quantum particle phenomenon,[59] will show, however, that such structures also may be used in the classical wave context. It is only the particle interpretation that distinguishes the quantum-mechanical use of these structures.

Before discussing quantum systems, I must say something about waves and particles. An important feature of the "logic of quantum mechanics" approach is that it is based entirely upon the particle picture of the physical systems considered. Putnam has emphasized this: "The world consists of particles (not of 'waves', nor of 'waveparticles.') I say this because I am quantizing a particle theory; if I were quantizing a field theory, I would say 'the world consists of fields'."[60]

All his arguments for quantum logic are based on nonrelativistic quantum mechanics in the particle picture and, in fairness, I shall similarly restrict myself in the next chapter. I must note, however, that the case for the quantum logicians becomes much weaker if we consider relativistic quantum theory. Then the inadequacy of the particle picture soon becomes clear. The one-particle Dirac equation for a free particle, for example, presents no problems, since we may consistently ignore negative energy states. As soon as we consider a particle in interaction with external fields, however—not to speak of a many-particle system proper—the possibility of transitions between positive and negative energy states arises. So, in order to preserve a "one-particle" interpretation, Dirac was forced to postulate the existence of an unobservable sea of particles filling almost all the negative energy states. The unoccupied "holes" in the sea are then interpreted as particles of opposite charge (antiparticles) with positive energy. This is sometimes a useful artifice for calculational purposes, but its necessity shows that one really does not have a consistent one-particle relativistic quantum theory. The accepted way around this difficulty is to start from a quantum field theory in which particles and antiparticles are treated symmetrically from the start as quantized states of excitation of the field. The attempt to build a fully consistent relativistic quantum field theory raises a multitude of other problems, but the point I want to stress is that one has to adopt the field, that is, wave, point of view in order to treat particles and antiparticles on the same footing in a relativistic theory.

Even restricting oneself to nonrelativistic quantum theory, one can derive all the results that follow from quantizing the classical particle picture by quantizing the corresponding classical wave equation. Since the wave equation from which one starts is the Schrödinger equation (here treated as a classical equation), this procedure is often referred to as "second quantization"—another example of a misnomer in quantum theory. This result was soon discovered by the founders of matrix mechanics. Heisenberg states:

> The fact that the particle picture and the wave picture are two forms of appearance of one and the same physical reality forms the central point of quantum theory. It is satisfying that the double nature of atomic phenomena also finds its complete expression in the mathematical apparatus of the theory . . . . Although the classical theories of the particle picture and the wave picture are absolutely different, in their mathematical as well as in their physical form, the quantum theories of both representations (*Vorstellungen*) are mathematically and physically identical.[61]

Actually, this identity only holds for situations in which the particle number is well defined and remains constant. Jordan, another coauthor of the *Driemännerarbeit* which first expounded matrix mechanics,[62] has emphasized this point. In the wave picture there is a matrix operator N_V, the integer eigenvalues of which equal the number of particles present in the volume V in the corresponding eigenstate of the field.

> The integral-valuedness of the eigenvalues of . . . N_V . . . obviously means that our theory actually leads to the occurrence of *indivisible particles* which can only occur in V in integral-valued numbers. We emphasize, however, that N_V *as a matrix* by no means always has a definite value: that is the *fundamental limitation which the quantum theory places upon the particle concept.* If we *measure* N_V, then indeed an integral value always results. But if we measure a quantity which does not commute with N_V, then the occurrence of particles can be dispensed

with. . . . In this respect the [wave] theory discussed here gives *more* than the [particle] method of a multidimensional configuration space with symmetric eigenfunctions. We are here also led to processes of destruction and creation of particles; the total number of particles present is not a prescribed number n (as in the "n-body problem"), but is taken as a variable that can assume *arbitrary* eigenvalues 0,1,2,3, . . . , and under certain circumstances can also be *undefined*.[63]

In particular, field strength and particle number are represented by operators that do not commute. Therefore, both cannot be diagonalized simultaneously. Diagonalizing the field strength operator leads to the wave picture; diagonalizing the particle number operator leads to the particle picture. Thus, Putnam's claim that in quantum mechanics one can restrict oneself to the particle picture ("The world consists of particles") just because one started out from a (classical) particle theory is much too simple-minded. If one is going to speak about "properties" of quantum systems in the way that the quantum logicians propose, they have to have wave "properties," such as field strength and phase, as well as position and momentum. The many "states" that some quantum logicians want to attribute simultaneously to a quantum system (see sec. 5) will have to include those in which field strength is sharp, as well as those in which particle number is sharp. Ignoring this question means truncating the theory.

Indeed, this truncation is manifested in the need to *postulate* either total symmetry (bosons) or antisymmetry (fermions) for the wave function of a multi-particle system. In quantum field theory, this relation between spin and statistics can be proved. As Martin Strauss noted, "In fact, the restrictive postulate 'symmetric or antisymmetric' expresses a much deeper and more fundamental property of particles than mere indiscernibility, namely the fact that like particles are *quanta of a common field* and hence—contrary to like particles of classical physics—do *not* possess the properties of independent existence and genidenity. This makes it understandable that the named postulate follows from the quantum field theory. Thus, this postulate represents a *partial anticipation of quantum field theory*."[64] The field aspect of quantum systems is of more than formal importance. It implies, for example, that electrons can transmit exchange forces giving rise to homopolar bonding between molecules, magnetic properties of matter, and the like.

Many puzzling features of quantum theory arise from exclusive reliance on the particle picture. They do not arise if one adopts the point of view of quantum field theory from the beginning, which leads one to look upon wave and particle pictures as applicable only under appropriate conditions. That a quantum field manifests particulate aspects in its interaction with a detector, for example, does not imply that it need do so in its interaction with a screen having two slits. The formalism of quantum field theory also incorporates features that reflect the differing nature of the electromagnetic and electron quantum fields. These features explain the existence of circumstances in which the electromagnetic field will manifest classical wavelike behavior, as well as of circumstances under which electron fields will manifest classical particlelike behavior. I know of no discussion of the significance of the symmetry properties of the wave function for quantum logic nor of any attempt to apply quantum-logical principles to quantum field theory by proponents of the "logic of quantum mechanics."[65]

Having made this point, I shall proceed as if one could always confine discussion to quantum mechanics in the particle picture. I shall try to make my initial exposition of the formalism as "neutral" as I can with respect to the question of its interpretation, concentrating on those features that characterize the actual use of the theory in physical applications. My aim is to provide the basis for a critical discussion of quantum logic in section 5, so I have not sought to expound my own views here. It is impossible, however, to be totally impartial in such a presentation, and I am sure some of these views will become apparent. To the extent that my presentation facilitates and does not foreclose discussion of the views of proponents of the "logic of quantum mechanics," however, it will serve its purpose.

A formalism involving a Hilbert space, Hermitean operators on that Hilbert space, and the like, has been found to be an appropriate mathematical structure to model the behavior of quantum systems. Once such a structure is postulated, the lattice of closed subspaces of the Hilbert space and the associated partial Boolean algebra of projection operators onto these subspaces follow mathematically without any further assumptions. The existence of these mathematical structures within the formalism of quantum mechanics raises but by no means answers the question of what implications follow for a logic of predicates of, or propositions about, the quantum system described by that formalism. Before I turn to such issues, it seems appropriate to discuss why Hilbert spaces are introduced in the first place. Rather than deal with this question in the abstract, I shall discuss it in the context of an example: the polarization of light.[66]

A beam of light, considered as a classical field described by Maxwell's electromagnetic theory, for example, manifests behavior that cannot be explained if we assume the beam is completely characterized by its intensity, phase, and propagation vector. This behavior can be explained by assuming that there is also a vector associated with the beam which lies in the plane perpendicular to the propagation vector. In Maxwell's theory it is a matter of convention whether one chooses the electric displacement vector **E** or the magnetic induction vector **B** to describe polarization. Both choices have been used. For the sake of definiteness I choose the **E** vector, but the discussion does not depend on the full Maxwell theory, just the existence of a polarization vector. I shall discuss only beams of light propagating in the same direction, taken as the z-axis of a coordinate system. Then the polarization vector lies in the x-y plane; in general it will vary in both magnitude and direction. For the present, consider the case where both magnitude and direction are fixed; this is called linear polarization (it is also often referred to as plane polarization, since the polarization and propagation vectors define a fixed plane). Describing polarization by a vector implies more than that it has a magnitude and direction. It implies that two polarization vectors can be composed by the vector addition law; and that it makes sense to speak of any numerical multiple of a polarization vector. (See fig. 10) In short, it implies that the linear polarization vectors form a linear vector space over the real numbers (see app. 2 for the definition of a linear vector space). This is not just a mathematical assumption about the polarization vectors. A linear vector space model of polarization phenomena has many consequences, such as predictions about interference effects for polarized light, which may be tested experimentally. (In terms of Maxwell's theory, this model may be derived from consideration of the vectorial character of Maxwell's equations.) I shall later give an example: transmission through a birefringent crystal.

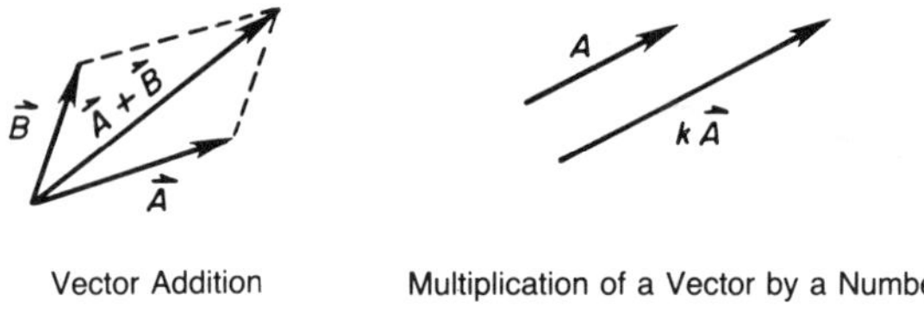

Figure 10

One must attribute further structure to the linear vector space model in order to account for the class of polarization phenomena summarized in Malus' law.[67] For a linearly polarized beam of light, there is a device called a linear polarizer which will (ideally) transmit all the incident light when its polarization vector is oriented in a certain direction perpendicular to the direction of propagation. If the polarizer is then rotated around the propagation direction, the intensity of the light is reduced until, after rotation through 90°, no light at all (ideally) is transmitted through it. The variation in intensity of the transmitted light can be modeled by assuming there exists a direction associated with the polarizer, called its axis of polarization; the transmitted intensity then varies as the cosine squared of the angle between the polarization vector of the light beam and the axis of polarization of the polarizer (Malus' law). This law is easily explained by the linear vector space model if we make the following assumptions (see fig. 11):

1. The vector space of linear polarizations, like ordinary physical space, has a Euclidean metric.
2. The length of the linear polarization vector is proportional to the square root of the intensity I of the light beam.
3. The polarization vector of the light beam transmitted by a linear polarizer is equal to the projection of the incident polarization vector onto the axis of polarization of the polarizer.

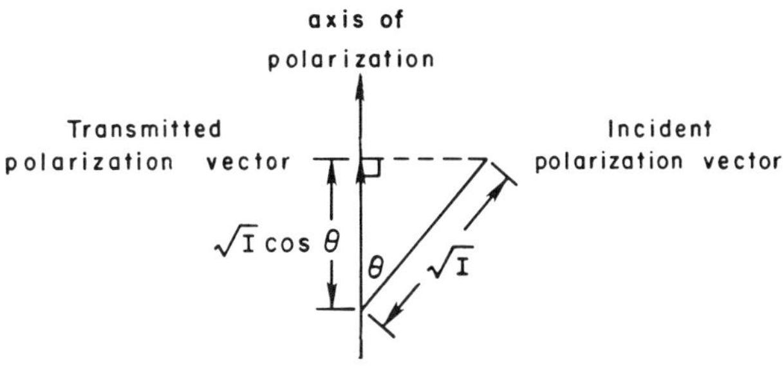

Figure 11. Malus' Law

Mathematically, one says that the two-dimensional linear vector space modeling linear polarizations has a real metric or norm, making it a real two-dimensional Hilbert space (see app. 2 for the definition of a Hilbert space). A vector in this two-dimensional real Hilbert space represents the linear polarization of a beam of light. Its direction represents the direction of polarization; its length represents the amplitude of the wave, the square of which gives the intensity of the beam. A linear polarizer may be represented in this model by a linear operator on the Hilbert space. Its effect on any polarization vector is to project out (i.e., remove) that portion of it perpendicular to the direction of the axis of polarization of the polarizer. (Perhaps it is worthwhile emphasizing that projection implies dropping a perpendicular, and so requires a metric, while taking vector components requires only drawing parallel lines, and so does not. It is an affine concept.) Such a linear operator is called a projection operator. A linear polarizer may also be represented by a direction in the Hilbert space that is, a one-dimensional subspace. Such a direction is called a ray of the Hilbert space; it is often convenient to represent rays by unit vectors (see fig. 12). The projection operator representing the polarizer acts on any vector in the space to project it onto the corresponding ray. The fraction of a light beam emerging from one linear polarizer that will be transmitted by a second is independent of the intensity of the beam: by Malus' law it is equal to the cosine squared of the angle between the two axes of polarization. Thus, to model this law, we really do not need the full Hilbert space but only the space of rays which is conveniently represented by the space of unit vectors. In two dimensions this is just the unit semicircle (see fig. 12).

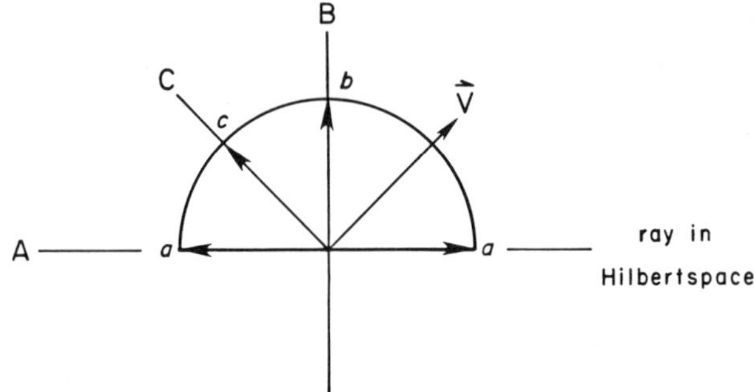

Figure 12. Rays in Hilbert space. A and B are orthogonal rays. C is a ray orthogonal to the vector v. Each point on the semicircle corresponds uniquely to a ray, if we identify the end points of the semicircle.

So far, beams of light have been treated as purely classical wave fields. Thus, the Hilbert space model of polarization could have been introduced early in the nineteenth century—indeed, some early representations of polarization came very close to this model. It is based on the possibility of modeling linear polarizations by a linear vector space. Malus' law relating incident and transmitted intensities is then modeled by introducing a norm on the linear vector space.

We may seek to justify the vector space model, on the basis of Maxwell's theory, for example—or just postulate it as a model for polarization. In either case, linear polarizers are introduced as "black boxes" which produce certain effects on a light beam. They enable us to set up a self-consistent model that has a good deal of success in treating the phenomena of polarization, but gives no explanation of how these "black boxes" work; that is, what is inside them that enables them to transform one state of polarization with a given intensity into another state with a different intensity. Maxwell's theory of light, which can validate the Hilbert space model of polarized light, is not sufficient to explain the mechanism of a linear polarizer. Some theory of matter and matter-light interactions would also be needed. It might turn out that various mechanisms inside a "black box" could produce the same effect. Since the ways of producing linearly polarized light are so varied,[68] it seems likely that several different mechanisms, sharing some basic features, must be at work. The point I want to emphasize, however, is that we are not obliged to provide a model of the inside of the "black box" in order to justify the introduction of phenomologically characterized polarizers into the model of polarization phenomena.

The discussion so far has been entirely in terms of the classical wave picture of light. But there is another set of phenomena, involving matter-light interactions, which suggests the need for a particulate picture. For example, one may observe the energy of the electrons emitted from a metal surface as a function of the intensity and frequency of the light beam illuminating it. This photoelectric emission was found to follow laws suggesting that the light beam is composed of a stream of particles or quanta, each of energy $h\nu$, where h is Planck's constant and ν the frequency of the light. This and a host of other experiments show that the emission and absorption of light by matter is explicable only on the basis of such a light-quantum hypothesis, first suggested by Einstein in 1905.[69] If the light beam is linearly polarized, the direction of emission of the photoelectrons is affected by rotation of the polarization vector of the light beam. This suggests that a polarization has to be associated with each individual quantum.

If we want to use the particle model developed for the emission and absorption of light to describe the propagation of light between emission and absorption, then we must interpret the intensity of a light beam as somehow related to the number of light quanta (photons) per unit volume of the beam. Statements about changes of intensity produced by linear polarizers must then be interpreted as statements about changes in the number of photons in the beam. This statistical interpretation correctly gives the counting rate of photons at any point where they are absorbed if the number of particles is assumed proportional to the classical intensity of the light beam (that is, to the square of its amplitude) at the point where the photon counting takes place.

A plausible assumption about the relation of beam polarization to individual photon polarization is that each photon has the beam's polarization direction. We must then conclude that a linear polarizer is capable of changing the polarization direction of those incident photons which it transmits. Statistically, it will transmit and transform the fraction of such photons given by Malus' law. What if we decrease the intensity of the beam to a point where, on the average, only one photon at a time is incident on the polarizer? There is then no way to predict from the model whether a given photon will be transmitted or not. We must here interpret the fractional intensity given by Malus' law as the probability for transmission of a photon with a certain polarization direction by a linear polarizer. If a photon *is* transmitted, however, it then *always* passes a second polarizer with axis of polarization parallel to that of the first. This implies that the photon emerging from the first polarizer must have its direction of polarization changed by that polarizer (any other assumption is inconsistent with the original one that, in a beam of linearly polarized light, each photon was polarized in the beam's direction).

We might consider an alternate assumption: each photon transmitted by a linear polarizer already has the polarization direction of the polarizer's axis of polarization *before* entering it. Then we also have to assume that a beam of linearly polarized light is composed of photons whose directions of polarization are *not* all the same, since a good number of them will pass a linear polarizer with axis of polarization oriented at *any* angle (except perpendicular) to the light beam's polarization vector. But this assumption is in contradiction with the observation that any linearly polarized light beam is transmitted, unaltered in intensity, by a linear polarizer with axis of polarization parallel to the polarization of the beam. We are forced to return to the first assumption: Each photon emerging from a polarizer has its polarization direction parallel to the axis of polarization of the polarizer.

We could assume that, *in addition* to having the polarization direction of the beam, each photon is in a definite *state* which determines whether it would or would not pass any linear polarizer on which it were incident. This hypothetical state would not be something that could be independently ascertained. The *only* way to find out if the state of a photon is such that it would pass a certain linear polarizer is to confront it with such a polarizer. If it passes the polarizer, we then know that its state *was* such that it would be transmitted by the polarizer. But since now it has been transmitted, its *new* state need not be the same as it was before transmission. So the previous answer is no help in ascertaining what its response *would* have been to any other polarizer. Thus, the "state of the photon" in this sense—a state that determines the responses to all possible linear polarization tests on the photon—may be assumed to exist. But such a "deterministic state" is unascertainable and assuming its existence has only one possible consequence that can be tested: namely, the response to *one* polarizer.

Returning to the assumption that each photon in a linearly polarized beam has the polarization direction of the beam, from it we can predict only the probability that a photon will be transmitted by a linear polarizer whose axis of polarization makes some angle with the polarization direction of the photon. We were forced to conclude, for consistency, that transmitted photons emerge with a new polarization direction parallel to the axis of polarization of the polarizer. We can ascribe a "state" to the photon in a quite different sense from that just discussed: the "state of a photon" only fixes its direction of polarization. Given this concept of "state of the photon," we can predict only the *probability* that it will be transmitted by any linear polarizer in its path (we *can* say with certainty what the new state of any transmitted photon will be). But this concept of "state" only exceptionally allows for prediction with certainty.

Generally, a probability is all that can be derived from this concept of "state of a photon," which I shall therefore call the "statistical state"of a photon.

Even someone who insisted on introducing the definition of the "deterministic state" of a single photon would have to admit that the "statistical state" was useful for a *beam* of photons. He or she might call it the statistical state of the beam or ensemble of photons, while insisting that each member of the ensemble had its own individual deterministic state. The only predictions that can be made for such ensembles of photons (except in a few special cases) will be statistical ones. For this purpose the "statistical state," whether attributed only to the beam or to the individual photons in it, would have to be used by proponents of either definition.

In the photon picture relative beam intensities (Malus' law) must be interpreted as (conditional) probabilities for photons having passed one linear polarizer to pass a second. If we adopt the concept of a "statistical state of the photon," we may describe such a probability as the transition probability for a photon in one state to go into another under the influence of a linear polarizer. The concept of conditional or transition probabilities thus arises in the Hilbert space model because we are forced to give a particle interpretation to this model, which was set up as a wave model.

Both the deterministic and the statistical definitions of the state of a quantum system are found in the literature on quantum theory. In a recent discussion of these two concepts, Freundlich has dubbed them the Popperian and Copenhagenist definitions of the state.[70] Whether these names are fully justified historically is a separate question, as Freundlich notes. What is important is that each is a coherent, self-consistent concept of the state of a system. As we shall see in the next section, some proponents of the "logic of quantum mechanics" want to give the concept of state a meaning that combines features of both the statistical and deterministic concepts. They end up without *any* coherent concept but with the claim that a quantum system has an *infinity* of states at the same time.

There are still two features of the polarization of light that remain to be mentioned. First of all, linear polarization is only one possible type of polarization. The polarization vector may vary in both length and direction as a function of time. In the most general case the tip of the polarization vector will trace out an ellipse; this is called elliptic polarization. The ellipse may degenerate into a straight line, giving linear polarization, or it may degenerate into a circle, yielding circular polarization. The rotation of the polarization vector, viewed from the direction of propagation, may be clockwise or counterclockwise. We then speak of left or right elliptic or circular polarization. All these polarizations states may be represented by vectors in a two-dimensional *complex* Hilbert space, that is, a two-dimensional vector space in which vectors can be multiplied by complex numbers. This model enables us to represent the addition of two linear polarization vectors that are out of phase with each other by using the well-known procedure of multiplying a vector by a complex exponential representing its phase. Since all possible polarizations can be produced by adding linear polarization vectors differing in phase, the two dimensional complex Hilbert space model of polarization is sufficient. The Hilbert spaces needed to model almost all quantum systems are of this complex variety. Since the real Euclidean metric was adequate for the two-dimensional Hilbert space of linear polarizations, we can visualize it easily, but, in general, the metric of a Hilbert space is complex, and so it is not easily visualizable, even apart from the higher (possibly infinite) dimensionality of the space.

The other feature is the existence of partially polarized and unpolarized light beams. Ordinary sunlight, for example, shows none of the directional features of a polarized light beam. By mixing ordinary sunlight with a completely polarized light beam, we can get mixtures of varying degrees of polarization. Can unpolarized or partially polarized light be understood by means of the classical wave model? Actually, as Lipkin puts it, "The existence of unpolarized light already gives indications of the quantum nature of light. Unpolarized light cannot be described classically as a single simple classical wave . . . or any linear combination of such waves. Unpolarized light can be described classically as a series of very rapid short bursts or pulses of light, each having a different polarization, with no correlation between the polarization of different pulses. If these pulses and the interval between them are very short compared to the characteristic times of the measuring apparatus, they will be detected as a continuous beam, and any polarization measurement will give an average of the polarization of the individual pulses."[71] Each of these pulses is a sort of "classical photon." Polarized light manifests the phenomena we associate with coherence: typical classical wave, or field effects. Unpolarized light manifests the phenomena we associate with incoherence: typical classical particle effects. If light were composed of classical particles, it would not manifest coherence effects, as does fully polarized light. If light were a classical wave, it would not manifest incoherence effects, as does totally unpolarized light. The photon model has to straddle these classical alternatives.

Since there can be incoherent mixtures of light beams, as well as coherent superpositions, we need a model for mixed light beams that allows for all possibilities from completely polarized (full coherence) through partially polarized to totally unpolarized (full incoherence). The model must depict how various mixtures are related to each other. Such a model cannot be based directly on the Hilbert space, but the set of idempotent linear Hermitean operators (often called density matrices) on the Hilbert space can be used to model mixed states. Rather than describe this somewhat abstract model, I shall describe a more visualizable equivalent model that pictures all mixed states as points of a convex set (see app. 2 for the definition of a convex set). Since the relations between mixtures do not depend on the absolute intensity of beams but only their ratios, we may restrict ourselves to beams of unit intensity (compare this with the discussion above of Malus' law and the restriction to rays in Hilbert space). Then all totally polarized states may be represented by the points of the surface of a unit sphere, called the Poincaré sphere since this model was first introduced by Poincaré.[72] Points on the equator of the sphere represent states of linear polarization. (I will describe just how in a moment.) The north and south pole represent left and right circularly polarized light; all other points represent different types of elliptically polarized light. Points inside the sphere represent partially polarized light, with the center of the sphere representing totally unpolarized light. If we take any point inside the sphere and draw any line through it, it will intersect the sphere's surface at two points. The partially polarized light represented by that interior point can then be produced by incoherently mixing beams of totally polarized light of the types represented by these two points. They must be mixed in the ratio in which the interior point cuts the chord connecting the two points on the surface (see fig. 13). For example, since the center of the sphere bisects the diameter connecting the two poles, we conclude that totally unpolarized light can be produced by incoherently mixing beams of left and right circularly polarized light in equal proportions.

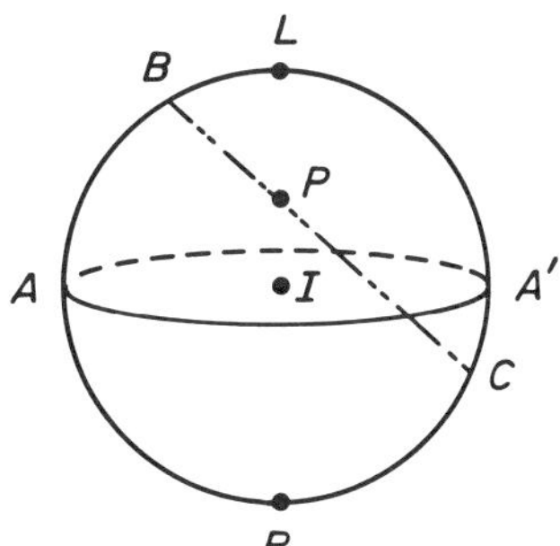

Figure 13. The Poincaré sphere. The north and south poles represent left and right circularly polarized light, respectively. The equator AA′ represents all states of linearly polarized light. The center point I represents unpolarized light. Partially polarized light represented by the point P may be obtained by mixing polarized beams corresponding to the points B and C in the proportion BP/PC.

To give a little more feeling for the Poincaré sphere, I shall describe a simple geometrical construction relating the Hilbert space model for linearly polarized light to the "Poincaré circle" (the equator of the Poincaré sphere) which also corresponds to these linearly polarized states. Inscribe the Poincaré circle inside the semicircle representing the rays of the two-dimensional Hilbert space of linearly polarized states. In this way we get a one-one correspondence between points on the Poincaré circle and those on the Hilbert space semicircle (see fig. 14). A little geometry shows that points on the Hilbert space semicircle 90° apart are represented by points at opposite ends of a diameter of the Poincaré circle. Thus, a pair of such diametrically opposite points represent orthogonal (linearly) polarized states.

Malus' law may also be represented simply in the Poincaré sphere model and, indeed, in a generalized form. A device that selects any specific state of polarization is called an analyzer (a linear polarizer is just one example). Any two states of (complete) polarization, will be represented by two points on the Poincaré sphere. The fractional intensity of a beam in one state of polarization transmitted by an analyzer that selects the other state of polarization is given by the cosine squared of half the angle subtended at the center of the sphere by the two points (see fig. 15)[73] Thus, all the information about (incoherent) mixtures of polarizations and (coherent) transition probabilities between polarizations is represented in the Poincaré sphere model.

One might wonder if all the other information encoded in the Hilbert space model can also be represented in the Poincaré sphere model. It can, sometimes with great elegance. For example, consider a transparent doubly refracting (birefringent) crystal. Such a crystal splits a beam of polarized light into two beams traveling with different speeds.

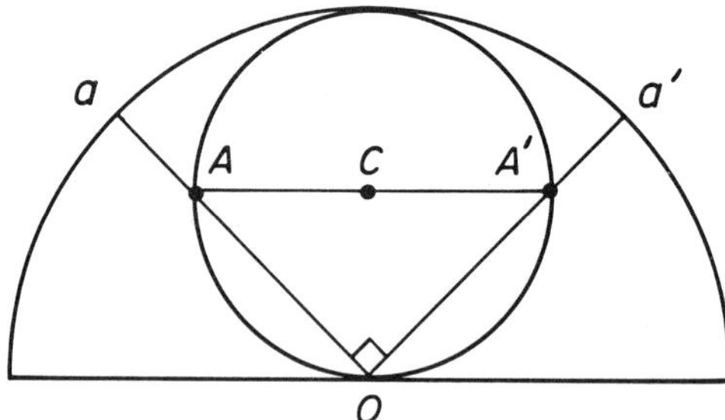

Figure 14. The Poincaré circle inscribed in the Hilbert space ray semicircle: *a* and a′ correspond to orthogonal polarization directions in the Hilbert space. *a* is mapped into *A* on the Poincaré circle by a radius of the semicircle, as is a′ into A′. *A* and A′ are diametrically opposite points of the Poincaré circle which thus correspond to orthogonal polarization directions.

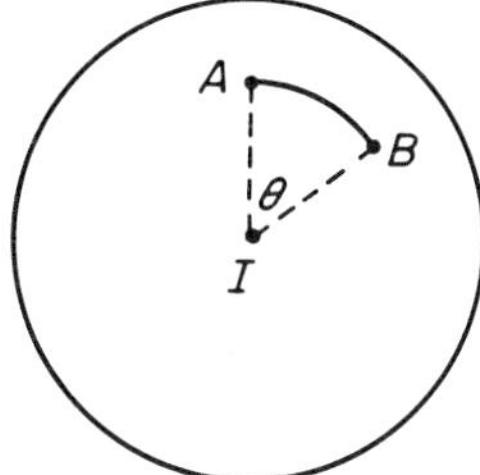

Figure 15. Generalized Malus' law. The probability for a photon with polarization state A to pass an analyzer for polarization state B is given by $\cos^2 \frac{1}{2}\Theta$.

Each of these beams is polarized in a way characteristic of the material (usually linearly) and orthogonal to the polarization of the other beam.[74] Since the two beams travel at different speeds, the crystal introduces a phase difference between the two that depends on the length of the crystal traversed by each beam. Suppose a beam of polarized light is incident on such a crystal and the emergent beams are allowed to recombine. What is the polarization of the resulting beam? The Poincaré sphere provides a simple geometrical construction to get the answer. The two orthogonal polarizations define a diameter of the Poincaré sphere. If the Poincaré sphere is rotated about this diameter through an angle equal to the phase difference introduced by the particular crystal, the point of the sphere representing the polarization of the incident beam is rotated into the point of the sphere representing the polarization of the recombined beam (see fig. 16).[75] Other examples of the superposition principle can also be worked out with the help of the Poincaré sphere.

It is worth dwelling a little longer on the birefringent crystal, since it provides a polarization analogue of the notorious double-slit experiment

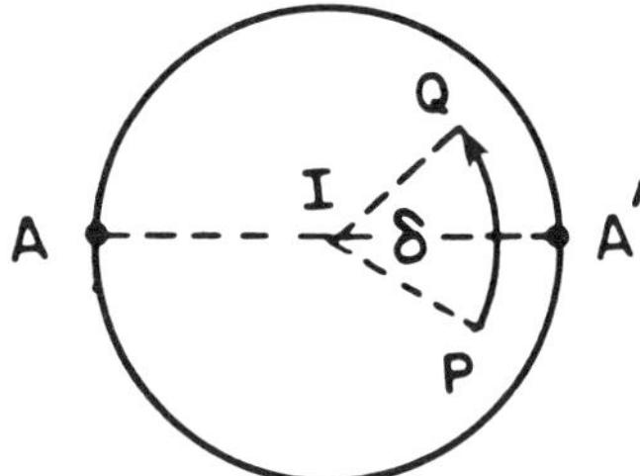

Figure 16. A birefringent crystal picking out orthogonal polarization states A, A′ introduces a phase difference δ between the two. An incident beam of polarization P will emerge with polarization Q.

for electrons discussed in the next section. Suppose the incident light beam is weakened in intensity until on the average (in the particle picture) only one photon at a time is in the crystal. If we allow the two beams to rejoin coherently, as envisaged above, the resulting photons will always be found to be in the state predicted by the construction using the Poincaré sphere. A photon enters the crystal with one polarization and one emerges having another polarization. What happens in between? If we try to imagine the photon following the path of one beam or the other, we run into difficulty. It would then have to have one or the other of the two orthogonal polarizations associated with each of these paths, rather than the final polarization that it does show. If we insist in using the photon picture to describe propagation through the crystal, we have to say that the photon "interferes with itself": it somehow divides and passes along both paths before recombining. Yet we know that a photon of a given frequency could not divide without changng frequency, since energy E and frequency ν are related by the Planck formula $E = h\nu$. On the other hand, if we place a photon counter along the path of each of the two beams before they can recombine, one or the other counter will always signal the passage of a photon of frequency ν after one is incident on the crystal. The polarization of the photon triggering the counter will always be found to be the one appropriate to the path along which the counter is located. After passing through these counters (indeed, a counter on one beam would suffice), the two beams will be found not to combine coherently but incoherently, to produce a mixture of the two orthogonal types of polarized light associated with the two beams. Those familiar with the double-slit experiment will recognize the close analogy.[76]

Another feature of the Hilbert space model is important for any quantum logic: All the linear subspaces of a Hilbert space form a special type of lattice which is characteristic of that Hilbert space. (See app. 2 for these results). There is also a lattice inherent in the Poincaré sphere, which is equivalent to the Hilbert space lattice (I shall not go into further detail here but will allude to this lattice later in the course of a more general discussion of lattices associated with convex sets). In short, the Poincaré sphere model can do as much as the Hilbert space model. Indeed, the Poincaré sphere model represents incoherent mixtures directly, while the Hilbert space model can represent mixtures only indirectly in terms of linear operators on the Hilbert space.

Now, drawing on the polarization example, I shall discuss the general structure of nonrelativistic quantum mechanics in the particle picture.

First of all, how does one characterize a particular type of quantum system—electrons or pi mesons, for example? There are a number of parameters, such as mass, charge, spin, and so on, the values of which serve this purpose. Their values may be called *characteristic properties* of the quantum system. It seems safe to regard these as properties of the individual system in much the ordinary sense of the word. Their existence has been overlooked by those who claim that the standard interpretation deprives particles of any properties at all, in the ordinary sense of the word "property." Paul Davies, for example, has recently asserted that "in response to Einstein's challenge, Bohr maintained that microscopic systems have no intrinsic properties whatever."[77] Bohr however, described quantum theory, as "a rational generalization of classical physics, allowing for the existence of the quantum but retaining the unambiguous interpretation of the experimental evidence defining the inertial mass and electric charge of the electron and the nucleus."[78]

A typical quantum-mechanical experiment involves the preparation of a beam (or ensemble) of quantum systems, some manipulations performed on the beam, and, finally, the registration of the possible outcome(s) of this process, often called a measurement. Even if only one system is explicitly considered, the preparation procedure inherently defines the conditions for an ensemble of systems. (An example from polarization would be preparation of a beam of light with some definite polarization, its passage through a birefringent crystal, followed by feeding of the resulting beam into a set of appropriate analyzers, using photon counters to record the passage of a photon through an analyzer.) The ensemble (beam) initially prepared may be coherent or partially or totally incoherent. (Note the contrast with the classical situation: A classical beam of particles could not be prepared with coherence-properties; only a classical field could be.) Once prepared, manipulation of the ensemble may preserve its degree of coherence or it may produce a decrease in it. The final registration procedure *must* result in a net decrease in the degree of coherence of the initial ensemble. (Of course, after registration only a portion of the ensemble may be selected for further manipulation. The degree of coherence of that portion may be greater than that of the ensemble incident on the registration apparatus.)

One may thus speak of coherence-preserving processes or devices and of processes or devices partially or maximally destroying the coherence of an ensemble. (The latter class of devices are often called measuring apparatuses.)[79] It seems that all devices that actually produce a decrease in the degree of coherence of a quantum ensemble must be macroscopic and themselves undergo some irreversible change in the registration process. This change constitutes a record of the process, which is finished at this point whether or not any human observer ever looks at that record. At any rate, devices that may be used to manipulate ensembles coherently or incoherently, and thus produce transitions between initial and final ensembles, must be postulated as part of the theoretical structure of quantum mechanics.[80] (Polarizers played this role in our example.) How models of such devices may be constructed, whose internal constitution and mode of operation are such that they produce (exactly or to arbitrarily good approximation) the effects postulated for these devices in quantum mechanics, is a separate question. The situation is analogous to the employment of ideal rods and clocks with postulated characteristics (or similar devices) in setting up the special theory of relativity. The existence of the postulated entities must be consistent with the resulting theory. For example, postulating an absolutely rigid body would not be consistent with the special theory of relativity, nor would postulation of a device capable of simultaneously registering precise values for both members of a pair of conjugate variables (such as position and momentum) be consistent with quantum mechanics. Special relativity is a principle theory and one would have to invoke constructive theories to build models of rods and clocks with the postulated properties.[81]

If one regards quantum mechanics as a principle theory (as Bub and Demopoulos do), there is no reason at all why it should explain the construction of the devices it postulates. Even if one does not regard it as a principle theory, the question of whether quantum mechanics *alone* is sufficient to construct such entities remains open. Maxwell's theory of electromagnetism is a constructive theory, yet it is not able by itself to construct models of rods and clocks. The quantum theory of systems with an infinite number of degrees of freedom can go a long way toward constructing models of entities that behave like registration devices;[82] but even if it could not, this would not cast doubt on the consistency of the nonrelativistic quantum theory of systems with a relatively small number of degrees of freedom.

One must try, with the help of quantum theory, and various auxiliary theories, to understand the functioning of those devices actually used in the laboratory to exemplify measuring apparatuses. Such apparatuses will rarely function in just the way postulated for the devices used in setting up quantum mechanics. This is no more surprising or disturbing than the fact that the functioning of devices actually used to measure spatial and temporal intervals in the laboratory involves correcting for the many ways in which the functioning of such devices may deviate from the behavior postulated for ideal rods and clocks in setting up special relativity.

Great confusion results from an immediate identification of the postulated entities with actual laboratory devices. The theory seems to have been given an anthropomorphic or even subjective cast. But devices postulated by the theory imply no immediate relation to human agency—any more than do other entities postulated by the theory, such as quantum systems themselves. Quantum mechanics may be applied to understanding what goes on in a laboratory; it may just as well be applied to what goes on in an unmanned spaceship with servo-mechanisms preprogramed to carry out some experiment and record the results on a magnetic tape or to the explanation of interactions that took place on earth before life arose that have left a record in the rocks.[83] One might, of course, want to argue that *all* theory involves a human element. But such an argument would clearly apply as well to classical as to quantum theory, so here we may leave such questions aside.

The quantum-mechanical formalism enables us to compute the relative intensities of the various subensembles corresponding to the possible results of a registration procedure (relative to the initial ensemble intensity, of course). As discussed above for polarization, if we regard the ensemble as made up of particles (photons, in that example), we have to interpret these intensity ratios as the probabilities that the individual quantum systems (particles) making up the initial ensemble will give a particular registration result. These probabilities are *transition probabilities:* the probability of B following A—as opposed to the probability of B, *tout court.* Thus, they resemble the conditional probabilities of classical probability theory rather than absolute probabilities. A classical conditional probability, however, may be interpreted as the probability that a member of an ensemble, each member of which has property *A,* also has property *B.* To give such an interpretation to quantum-mechanical probabilities would be to prejudge the issue of whether we may attribute intrinsic properties (other than its characteristic ones) to a quantum system. So I shall use the words "transition probability" without any such implication.[84]

Indeed, one could just regard such transition probabilities as relations between initially prepared and finally registered ensembles (as Malus' law could be interpreted as a relation between beams transmitted by linear polarizers) without ever referring to the state of individual quantum systems. This is the point of view of proponents of the statistical interpretation of quantum mechanics.[85] Or one could introduce the concept of the "statistical state of the (individual) quantum system," regarding the outcome of a preparation as an ensemble of quantum systems, each of which is characterized by the same statistical state. Then transition probabilities may be regarded as giving the probability for a transition of a quantum system from one such state to another. This is what Freundlich calls the Copenhagenist position,[86] and I use this name as a shorthand way of referring to this viewpoint without prejudicing the question of whether it adequately represents Bohr's views. Another possibility is to accept the concept of the "deterministic state of the quantum system," according to which the state of a system fixes the results of any future registration of that system (maximal or not). As discussed for polarization, the deterministic state cannot be ascertained by a preparation or any other manipulation of the system. This is what Freundlich calls the Popperian position,[87] and I also use this as a shorthand name, without necessarily implying that it correctly represents the views of Sir Karl. Proponents of each of these viewpoints will use essentially equivalent methods to compute transition probabilities, but the names and interpretations they give to various elements of the formalism will naturally differ.

The discussion of the concept of state so far has been based on the (customary) assumption that the ensemble to be studied is initially defined by a preparation.[88] The statistical state (whether only applied to ensembles or to individual systems) is then used to compute transition probabilities to possible future states. This customary procedure is called "prediction." It is quite possible, however, to consider ensembles defined by some registration result and to use a statistical state defined for such an ensemble to calculate the probabilities that the system had been prepared earlier in various ways. This procedure is called "retrodiction." In the prediction case the statistical state of an ensemble or individual system characterizes it just *after* a nondestructive registration used as a preparation; in the retrodiction case the statistical state characterizes the ensemble or individual system just *before* a registration. Thus, it is quite possible for an individual system undergoing some sequence of manipulations to have *one* statistical state at a certain time as a member of one (predictive) ensemble and *another* statistical state as a member of another (retrodictive) ensemble, This is a clear indication of the relative, or conditional, significance of the statistical state vector.[89] As David Finkelstein once remarked: Having a wave function is like having a haircut; if you are not careful you start to wonder where your haircut is when you are not having it.

This conditional significance of the quantum state of an individual system may be compared with the classical case, where the (unique) state of an individual system, defined by specification of the values of a complete set of kinematical and dynamical variables (such as position and momentum for a particle), may be used for both prediction of its future and retrodiction of its past states. Any uncertainty about the behavior of an individual member of an ensemble of classical systems, such as the ensemble corresponding to a complete solution to the Hamilton-Jacobi equation, arises only from an incomplete specification of the state of an individual member of the ensemble; for example, specification of the parameters characterizing a complete solution fixes only half the complete set of variables needed to specify an individual system's state (with a certain distribution of values for the conjugate variables). In many respects such an ensemble is the closest classical analogue to a quantum ensemble.[90] Classically, a probabilistic element enters only because of an incomplete specification of what could be specified. In the quantum case, a probabilistic element characterizes the most complete possible specification of the system.

From now on I shall confine myself (as the quantum logicians tacitly do) to ensembles defined by a preparation. In that case, one may use the concept of (predictive) statistical state without ambiguity. When I use the word "state" without further qualification in the sequel, it is always to be taken in this sense.

There are three key elements that must be represented in any mathematical structure serving as a model for a quantum-mechanical system or ensemble:

1. Coherent or pure states, and the possibility of (coherent) superposition of such pure states.
2. Incoherent or partially coherent mixed states, and the possibility of (incoherent) mixing of pure states to produce all possible such mixed states.
3. Transitions between states, and a method of calculating transition probabilities for such transitions.

Quantum-mechanics differ from classical mechanics with respect to each of these three points:

1. Classically, coherent states and their superposition are meaningful only for fields obeying linear wave equations. Quantum mechanically, these concepts apply to all systems, including particles.
2. Classically, particle ensembles are always in a totally incoherent mixed state. The decomposition of such a state into "pure states" for the individual particles is unique. Quantum mechanically, mixed states of all degrees of coherence are possible. The decomposition of such a state into pure states is generally not unique.[91]
3. Classically, transition probabilities between mixed states of particle ensembles can be derived from a deterministic treatment of transitions between pure states of individual particles. Quantum mechanically, even transitions between pure states must be treated probabilistically.

For photon polarization, as we saw, a two-dimensional complex Hilbert space may be used to model all three features. More complicated cases require Hilbert spaces of higher dimension. Indeed, any case involving the motion of a particle requires an infinite-dimensional, separable Hilbert space (see app. 2 for definitions), but the basic structure of such Hilbert spaces remains similar. Use of the complex number field has proved sufficient for all known quantum systems. However, a generalization to a third number field, quaternions, is possible[92] and, indeed, cannot be ruled out by the representation theorems for quantum-mechanical lattices (see app. 2 for the representations of quantum lattices).

The Poincaré sphere, the alternate model I introduced to represent these three features of a quantum system in the case of photon polarization, can be generalized, giving rise to an interesting new possibility. Convex sets are the most natural generalization of the Poincaré sphere (see app. 2 for a definition) and every known quantum system may be modeled by an appropriate convex set.[93] While all Hilbert spaces are basically similar in structure, convex sets exhibit much greater variety. All convex sets give representations of relations between mixtures of coherent and incoherent states, with associated transition probabilities, much like those of known quantum systems. But they do not all incorporate the relations between coherent states in a way corresponding to a linear vector space (with the associated superposition principle). Since, on the classical level, such relations reflect the properties of waves obeying linear wave equations, Mielnik has suggested that convex sets might provide models for the quantization of systems corresponding to classical nonlinear wave equations.[94]

To say something further about the relative merits of convex set and Hilbert space models, I must discuss the structure of Hilbert space in more detail. Since the lattice of closed linear subspaces of a Hilbert space, or the corresponding lattice of projection operators acting on that Hilbert space, are used in many treatments of quantum logic, this discussion is also of direct relevance to the next section, on quantum logic. The reader may want to refer back to section 2, on "color logic," where many of the concepts were first introduced, or to appendix 2, where more formal definitions of these concepts can be found.

The set of closed linear subspaces of a Hilbert space forms a lattice, with set-theoretical inclusion as the ordering relation between lattice elements. The zero-dimensional origin is the lowest or ϕ element, the entire Hilbert space is the highest or *1* element of the lattice. The one-dimensional linear subspaces or rays are the atoms of the lattice. In the two-dimensional real or complex Hilbert spaces used to model polarization, there are no other elements. The atoms form orthocomplementary pairs, corresponding physically to orthogonal polarizations. Since the meet of any two lattice elements corresponds to the set-theoretical intersection of the corresponding subspaces, the cap of any two atoms is the ϕ element (as it must be in any lattice). Since the join of any two elements corresponds to the linear subspace spanned by the two linear subspaces corresponding to the two elements (i.e., the *smallest* linear subspace containing them both), the cup of any two atoms is the *1* element (this is a peculiarity of the two-dimensional case). The resulting polarization lattice is very much like the lattice of figure 8 in section 3 for the incompatible predicates, except that there are now an infinite number of atoms—one for each angle between 0° and 180°.

Once one goes to Hilbert spaces of three or more dimensions (an infinite number of dimensions is required for any quantum system that includes spatial position), there will be linear subspaces of two, three—

indeed, any dimension below that of the entire Hilbert space. In the case of an infinite-dimensional Hilbert space, there are mathematical complications involving the existence of limits (we should really speak of separable Hilbert spaces, for example) and the requirement that linear subspaces be closed becomes nontrivial. But for purposes of general orientation, the real three-dimensional Hilbert space corresponding to ordinary Euclidean space with a point picked as origin serves quite well and I shall often use this example. All straight lines through the origin are one-dimensional linear subspaces (atoms); all planes through the origin are two-dimensional linear subspaces. In this case, the element orthogonal to any atom is not an atom but is the orthogonal plane through the origin. Capping (∧) and cupping (∨), or meeting and joining, operations are easily pictured (see figures in sec. 2).

Physically, vectors of the Hilbert space correspond to completely coherent or pure states of the quantum system. Since each such vector lies in one and only one ray (one-dimensional subspace), it is often loosely said that rays correspond to pure states; but rays cannot be added, whereas vectors can be; vectorial addition corresponds to the coherent superposition of pure states. Partially or totally incoherent states or mixtures, cannot be represented by elements of the Hilbert space. They are represented by a certain class of linear operators on vectors of the Hilbert space, called density operators (I shall not bother with an exact characterization of such operators). Pure states, as special cases of mixtures, are also representable by a special type of density operator.

One type of coherent manipulation of an ensemble of quantum systems in a pure state is represented by the action of another class of linear operators, called unitary, on the corresponding state vector. This action may be visualized as a "rotation" of the state vector in Hilbert space (quotes around rotation, to remind you that the metric of Hilbert space is usually complex). Free or inertial motion is an example of such a coherent manipulation, as is coherent scattering by some force field, and the like. The effect of one class of coherent manipulators is to project the incident ensemble coherently onto a set of orthogonal subspaces of the Hilbert space. If the set of subspaces onto which projection occurs is complete (i.e., if they span the entire Hilbert space), we shall call this a complete coherent projection. (This is the inverse of a coherent superposition.) If part of the incident ensemble is destroyed (or completely removed from further consideration), we shall call this an incomplete coherent projection. In principle, one should be able to produce the effect of a complete coherent projection by postulating some interaction corresponding to the coherent manipulator, the effect of this interaction on the state of the incident ensemble producing the desired projection effect. For example, a birefringent crystal has the effect of coherently dividing an incident ensemble into two orthogonal subensembles, so it acts as a complete coherent projector. By introducing a suitable model of the crystal's charges and currents into Maxwell's equations, the evolution of an incident light wave should lead to the postulated division of the beam. Often, however, the existence of some device with the desired projection effect is just postulated.

The effects of registration devices are also represented by projection operators onto linear subspaces of the Hilbert space. But, in this case, the subensembles projected onto each orthogonal subspace are postulated to be mutually incoherent. An incident pure state is almost always turned into a mixture by registration. Registration devices may be maximal, that is, maximally decreasing the coherence of the incident ensembles, or nonmaximal, producing some lesser degree of incoherence. The maximal effect is produced by a device whose effect is represented by a set of projection operators onto a complete, mutually orthogonal set of one-dimensional subspaces or rays. In real three-dimensional Hilbert space, for example, this corresponds to projections onto three mutually perpendicular axes. A nonmaximal registration corresponds to a set of projection operators onto mutually orthogonal subspaces, not all of which are one-dimensional.[95] In three dimensions the only such case that can occur is projections onto a ray and the orthogonal plane, but in higher-dimensional cases many more possibilities exist.

Transition probabilities between an incident state and a set of pure states, corresponding to all the possible results of a registration are represented in the Hilbert space model by the "cosine" squared of the angle between the vectors (or, loosely, the rays) representing the initial and each final pure state. I put "cosine" in quotes as a reminder that we are dealing with quantities in a complex Hilbert space, so this "cosine" will actually be a complex number of modulus less than one; it is the square of this modulus that represents the transition probability.

The (pure) state vector characterizing the subensemble of systems "transmitted" for a certain registration result, (assuming that the incident ensemble was pure) is given by the so-called projection postulate: The final state vector is the projection of the incident state vector onto the subspace of Hilbert space corresponding to the particular registration result[96] (see fig. 17). This projection postulate is based on the assumption that the registration device has no further effect on the subensemble it picks out, such as its destruction or some additional coherent effect, which would then also have to be taken into account.

For mixed states, the transition probabilities, like the states themselves, cannot be directly represented in the Hilbert space model. One must work with linear operators on the Hilbert space. Without going into further detail here, let me say there are algorithms for computing such transition probabilities that generalize the "cosine squared" rule and projection postulate for pure states.

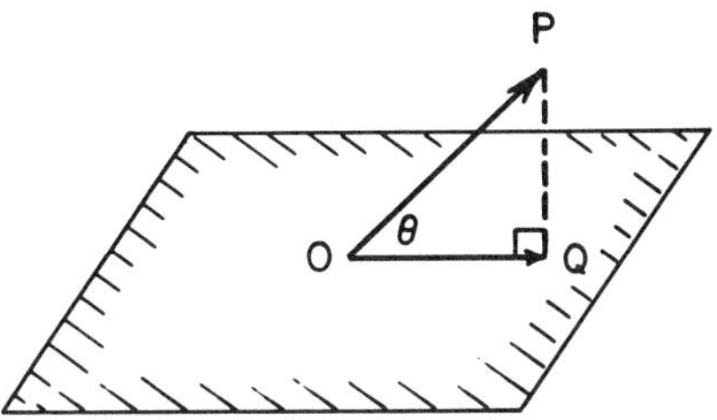

Figure 17. Projection postulate. Let $\overrightarrow{OP}$ represent the state vector before a registration which may have a result corresponding to projection onto the subspace pictured. Then $\cos^2 \Theta$ represents the transition probability and $\overrightarrow{OQ}$ represents the state vector of the transmitted portion of the ensemble after the registrations.

Now I shall return to the question of the relative merits of convex set and Hilbert space models. For polarization, the Hilbert space and Poincaré sphere models were about equal in the ease with which they could represent transition probabilities between states, because of the extreme simplicity of both models, in which (as we have seen) only atoms lie between the ϕ and *1* elements of the lattice. In general, the convex set model is much superior in its ability to represent transition probabilities. To start with a point valid even for polarization, convex set models represent transition probabilities directly. Complex Hilbert space models yield a complex number (of modulus less than one) as the "cosine" factor, which then has to be "squared" (i.e., its modulus has to be squared) to get a transition probability. This means that a phase factor (complex number of modulus unity) appears in the Hilbert space model which has no physical significance, whereas all information in the convex set model is physically relevant. Let me recall Hertz's classic words about the criteria for models in physics: "the consequences of the models [*Bilder*] should in turn be models of the consequences. . . . Of two models of the same object, that one will be the more suitable which represents more essential relations of the object than the other. In the case of equal scope, that one of the two models will be more suitable which, in addition to the essential features, contains the lesser number of superfluous or empty relations, which is therefore the simpler."[97] On the basis of the last criterion, the convex set model is superior.

Even more impressive is the ease with which the convex set model represents transition probabilities for mixtures. In this model, all states, pure and mixed, are represented by points. Linear subspaces of a Hilbert space (or the corresponding projection operators) are represented by certain linear subsets of the convex set, called walls (see app. 2 for definitions). The transition probability for *any* state, corresponding to any possible outcome of a registration (maximal or nonmaximal), is given by a certain geometric invariant (called the affine ratio) defined by the point representing the state and the wall representing the linear subspace. Thus, a unified representation of transition probabilities is realized in the convex set model.

The walls of any convex set form a lattice which, for convex sets corresponding to Hilbert spaces, is isomorphic to the lattice of subspaces of that Hilbert space. As mentioned above, there are convex sets that do not correspond to *any* Hilbert space. The lattice of their walls is thus a "non-Hilbertian lattice." Should quantum systems modeled by such convex sets ever prove to be of physical importance, Hilbert space lattices would prove too narrow a basis for quantum theory.

Color mixing (see sec. 2) can also be pictured by a convex set diagram. The three primary colors, which I took as red, green, and blue-violet, are represented by the three extremal elements of a convex set, that is, the vertices of a color triangle. Color mixing corresponds to generating all other points of the triangle by linear conbinations of the vertices. For the color triangle, the empty set, the vertices, the sides, and the entire triangle are the only walls. A partial ordering of the walls is given by set-theoretical inclusion. The relation between complementary colors, defined in section 2, is represented by the relation between a vertex of the triangle and the opposite side (see fig. 18). Sides are given the name of the color of their midpoint, and the whole triangle the name of the color (white) of its central point. Incidentally, the color triangle representation goes back to Maxwell.[98]

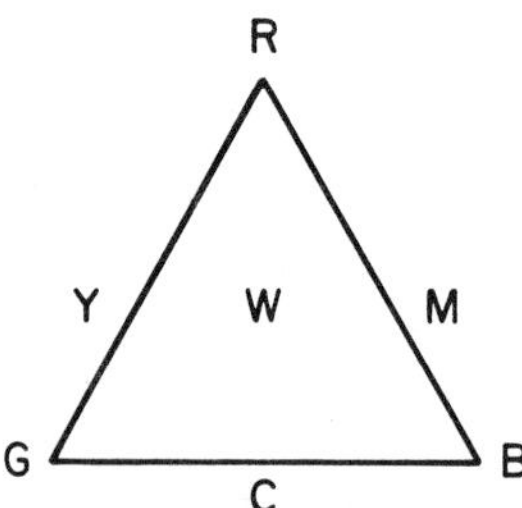

Figure 18. The color triangle

5. QUANTUM LOGIC[99]

When discussing elementary particles, one should remember that, at the deepest level of understanding so far reached, they should be regarded as quanta of a relativistic field. This implies such profound differences between these quanta and classical particles that perhaps they are not yet all fully understood or appreciated.[100] As stated in the last section, however, I shall ignore this point from now on in fairness to the quantum logicians, who take nonrelativistic quantum mechanics in the particle version as their point of departure (usually further confining themselves to single-particle systems).

There is a related difficulty, however, that cannot be so easily ignored. This is the question of how to treat an individual quantum system. As discussed in the previous section, quantum mechanics is set up in a way that treats such a system as one member of an ensemble. This ensemble is often not explicity mentioned, and the ensemble to which a system belongs may change in the course of its treatment. But an initial ensemble, inherently defined by some explicit or implicit preparation procedure and whose fate is determined by further manipulations and the final registration procedure, is always involved in application of the quantum-mechanical formalism to any concrete problem. One may admire or detest this feature of the theory, one may regard it as highlighting some ultimate limitation of human ability to comprehend the microworld; or one may regard it as a symptom that quantum mechanics is a sort of phenomenological "black box" theory, in need of supplementation by some more profound explanation of the microworld—but one cannot evade this feature. Quantum mechanics as a working theory does not treat a quantum system "in itself" but as part of an ensemble defined and manipulated by certain ideal devices,[101] although the verbal treatment may often disguise this feature. Many discussions start with a phrase such as: "Consider an electron in the state. . . . ," or, "Consider an electron with momentum p. . . . "Behind such phrases, of course, lies some preparation procedure—itself an element of the theoretical structure—capable of preparing an unlimited number of samples of the quantum system in question, all in the same "state." If we unpack the shorthand (discussed in section 4), "state" here means: "statistical state relative to the ensemble associated with that preparation procedure."

So, if one wants to talk about individual systems in the quantum-mechanical context, two questions arise: First, to what extent does the theory *require* or *mandate* application of particular concepts of the theory to the individual quantum system? Second, to what extent does it *permit* such applications, even if they are not mandated? After ascertaining what the theory mandates and what it (merely) allows, one may adopt either of two attitudes: the relaxed (whatever is not forbidden is allowed); and the strict (whatever is not mandatory is forbidden). Let us see how these attitudes apply to the concepts of statistical and deterministic states of individual quantum systems. If it is agreed that quantum theory deals basically with ensembles of systems, then the concept of the statistical state of an ensemble is mandatory; the concept of the statistical state of an individual system is then allowed but not mandatory. The concept of an ascertainable deterministic state of the individual system (determining the outcome of any registration procedure to which the system might be subjected) is forbidden. The concept of a deterministic state is allowed, however, so long as it is *not* assumed to be ascertainable. The statistical interpretation can thus be seen to be strict on both questions. The Copenhagenist is relaxed on the first question but strict on the second, whereas the Popperian is relaxed on both questions.

Of course, one may quite legitimately raise related questions outside the quantum-mechanical formalism. One may devise *Gedankenexperimente,* for example, to show that, under certain conditions, microsystems *must* possess certain attributes or conversely, to show conditions under which they *cannot* possess certain others. Heisenberg's analysis of the uncertainty principle[102] and the Bohr-Rosenfeld analysis of the measurability problem for the electromagnetic field[103] were designed to show that the results of such *Gedankenexperimente* were always consistent with the formal analysis given by quantum mechanics. One might well try (as Einstein did at one time) to design *Gedankenexperimente* with the contrary aim, but all such efforts are attempts to relate the quantum formalism to something outside it and, thus, to test or criticize the theory. One is also free to try to develop alternative theoretical formulations, outside the quantum-mechanical framework, aimed at explaining the same empirical material on a quite different theoretical basis (Einstein's ultimate goal). If one's aim is to understand and elucidate the consequences of the quantum-mechanical framework, however, naturally one must remain within that framework. One cannot claim to be working within this conceptual framework and then introduce criteria from outside it. Since the quantum logicians I am discussing claim to be elucidating features inherent in the (correct) interpretation of the quantum-mechanical formalism, I assume we agree on the need to remain within the quantum-mechanical framework for the discussion of quantum logic.

I intend to discuss the question of whether and under what conditions we can ascribe such properties as definite values of position and momentum to individual quantum systems. Before directly confronting this question, however, I have to discuss in some detail a separate but closely related question: In characterizing quantum systems, what is mandated and what is allowed if some result is obtained from a registration apparatus? Using a registration apparatus is commonly called making a measurement. The word "measurement," however, conveys anthropomorphic, and even subjective, overtones to many people (such as Wigner's friend), so I often use the term "registration," introduced in section 4, even at the expense of occasional awkwardness of expression, to emphasize the objective nature of the processes.[104] Such acts of registration are represented in the quantum-mechanical formalism by projection operators acting on the Hilbert space representing the system. The mathematical structure attributed to the set of projection operators plays an important role in distinguishing various approaches to quantum logic, so I must start with a discussion of this question. Then I shall consider the use of projection operators to represent various types of (ideal) registration devices, illustrating the general discussion by a detailed consideration of the double-slit experiment. After this long excursion, I shall take up the problem of the ascription of states and properties to individual quantum systems and the question of whether quantum logic really offers a better understanding of quantum-mechanical paradoxes such as those associated with the double-slit experiment.

5.1 The Algebra of Projection Operators

As mentioned in section 4, there is a unique correspondence between a linear subspace of a Hilbert space and the projection operator onto that subspace. This correspondence allows us to ascribe a lattice structure to the projection operators just by taking over the lattice structure of the subspaces. However, a more limited algebraic structure is inherent in the projection operators themselves. Two projection operators P_1 and P_2 may commute: $P_1P_2 = P_2P_1$. This implies that their effect on any vector of the Hilbert space is independent of the order in which they are applied to that vector. It is easy to show that this happens if and only if either the linear subspaces onto which P_1 and P_2 project are orthogonal, or the subspace onto which one of them projects lies within the subspace onto which the other projects. But commutation is defined directly for the two operators, and thus does not involve any reference to subspaces of a Hilbert space on which they act. Cap and cup operations may be defined for two commuting projection operators as follows:

$$P_1 \wedge P_2 = P_1P_2; \quad P_1 \vee P_2 = P_1 + P_2 - P_1P_2.$$

(The sum of two linear operators is always defined and is independent of the order in which the sum is taken.) Introducing Φ as the symbol for the projection operator onto the zero-dimensional subspace (which takes any vector into the zero vector), and I as the symbol for the projection operator onto the entire Hilbert space (which takes any vector into itself), one can easily show that the set of all projection operators forms a partial Boolean algebra (see app. 2 for a complete definition). Because it is easier, I shall explain what this means with the help of the corresponding subspaces of Hilbert space. Remember, though, this is only a crutch: all algebraic properties are defined for the set of projection operators independently of this correspondence. If we take a complete set of orthogonal rays in Hilbert space, they generate an atomic Boolean lattice (see app. 2 for definition) with the rays as its atoms. In real three-dimensional Hilbert space, for example, the x,y, and z axes of any rectangular coordinate system are the atoms of a Boolean lattice. The xy, yz, and xz planes, the origin, and the entire space are the remaining elements of the lattice (See figs. 5 and 6, sec. 2). The corresponding projection operators all commute with each other, forming the elements of a Boolean algebra (just another way of looking at a Boolean lattice). I have defined the cap and cup operations above. If P is the projection onto a certain subspace, then $I - P$ is the projection onto the orthogonal subspace; thus P', the complement of P in the Boolean algebra ($P \wedge P' = \Phi$, $P \vee P' = I$), is given by $P' = I - P$. Since *any* set of orthogonal axes can be used, there are an infinite number of Boolean lattices (or algebras) contained in the Hilbert space lattice. Two or more of them may share an element in common. For example, rotation of two of the axes about the third in any orthogonal triplet of axes yields another set of orthogonal axes that has one axis in common (see fig. 7, sec. 2). These Boolean algebras are "interwoven" in a very tight fashion. For example, the set of *all* elements that two such Boolean algebras have in common always forms a Boolean algebra. Appendix 2 gives a definition of the way in which a set of Boolean algebras must be interwoven to form a partial Boolean algebra. The set of projection operators on a Hilbert space satisfies all these conditions. Although (as mentioned) the set of projection operators also can be made to form a lattice, they do so only because they "inherit" this structure from their correspondence with the subspaces. Since the product of noncommuting projection operators is not a projection operator, there is no way to define cap and cup operations for noncommuting projection operators without somehow going outside the set of these operators. So cap and cup are not "natural" operations in this case.

5.2 *Registration and the Double-Slit Experiment*

As discussed in section 4, a maximal registration of a quantum ensemble is represented by the set of projection operators onto a complete set of orthogonal rays. We have just seen that such a set generates a Boolean algebra. Thus, a Boolean algebra of projection operators, or the corresponding rays of Hilbert space, is picked out by any maximal registration. We are interested primarily in their effect on quantum ensembles, so we regard two devices that effect the same maximal registration as equivalent; or, to put it the other way around, two such devices are regarded as different exemplifications of the same ideal maximal registration procedure.

Consider a pair of mechanical quantities that are canonically conjugate in classical mechanics—for example, position and momentum. An individual trajectory of a classical mechanical system is characterized, at any time, by the values of a complete set of such canonically conjugate variables. The Hilbert space of the corresponding quantum system is spanned by a complete set of orthogonal rays. Such a set, however, corresponds to only *half* of the classical complete set: to position variables only, or momentum variables only, for example. That is, *one* such mechanical quantity corresponds to a complete set of orthogonal rays in quantum mechanics.[105] If the quantum-mechanical formalism provides an adequate description of quantum systems, this implies that a maximal registration device acting on a quantum system can yield values that correspond to only half the classical quantities; for example, values corresponding to the classical position variables or to the classical momentum variables—but not values corresponding to both. I say "values corresponding to the classical quantities," because the interpretation of a registration result on a quantum system as related to a quantity of the corresponding classical system is not immediate and straightforward; it requires application of some correspondence principle relating the quantum and classical systems. I shall give examples of such correspondence-principle arguments in discussing the double-slit experiment.[106]

Let us focus attention on the subensemble selected by the registration of a particular value of position, for example. By the projection postulate, this subensemble will be in a pure state, with state vector lying in the ray corresponding to this value of the position. Now suppose we immediately carry out a maximal registration corresponding to a momentum measurement on this pure ensemble. The "cosine squared" rule indicates that all values are equally likely to be registered.[107] Classically, of course, we could carry out position and momentum measurements simultaneously, but if we choose to carry out the momentum measurement immediately after the position one, the result will not differ appreciably. Thus, classical mechanics *mandates* attribution of definite values of *all* members of a complete set of canonically conjugate mechanical variables to a system. Quantum mechanics, on the other hand, *forbids* the attribution of unique values to the corresponding set of variables in a pure ensemble—the closest we can get to homogeneity of a quantum ensemble. But it *permits* attribution of such values to *half* of such a set in a pure ensemble. These considerations set the limits within which one may maneuver in trying to attribute properties to a quantum system. Whether such values are to be associated with a pure quantum ensemble just *before* or just *after* the registration is not decided a priori. Classically, of course, such values must be associated with both times. Quantum mechanically, as our discussion in section 4 showed, the choice depends on whether we define the ensemble by a preparation or a registration.

As emphasized earlier, these conclusions are compelling only to the extent that one regards the quantum-mechanical formalism as providing an adequate account of quantum systems. A whole corpus of *Gedankenex-*

perimente, starting with Heisenberg,[108] has been devised to demonstrate the consistency and adequacy of the formalism—or disprove it. I believe that none of the proposed counterexamples has been based on a correct understanding of the implications of the formalism. More to the point, the quantum logicians accept the adequacy of formalism, so it must be taken as the basis for a discussion of quantum logic.

So far I have discussed only maximal registration procedures, but one must often deal with nonmaximal ones. A nonmaximal registration procedure (also called an incomplete measurement) corresponds to some set of projection operators onto a complete set of orthogonal subspaces, not all of which are rays. The orthogonality and completeness of a set of subspaces (completeness, remember, means that the set spans the whole Hilbert space) is reflected in the algebra of the projection operators by the fact that any two of them have Φ as product, while the sum of all of them equals I:

$$P_iP_j = \Phi, \ \Sigma P_i = I.$$

A registration, maximal or nonmaximal, may be destructive in the sense that all or part of the incident ensemble is destroyed, or at least omitted from consideration of the subsequent evolution of the rest of the registered ensemble.

Since I shall soon consider quantum-logical discussions of the double-slit experiment, it provides a good example to illustrate some possible types of registration devices. Consider a screen with two small holes in it (these considerations could be extended to any number of holes). Suppose the screen is a mirror (perfect reflector) on one side, and let a beam of light fall on the screen from the mirrored side. Part of the beam will be transmitted through the holes; the rest will be reflected from the mirror. So far, only a coherent manipulation of the beam has taken place: In principle, by cleverly reflecting both the transmitted and the reflected beams, we could get them to (coherently) interfere. So the screen is an example of a complete coherent projector (see sec. 4).

Now, place a small (localized) photon counter just behind each hole in the screen. Then a signal from either counter is a registration, to be interpreted as the passage of a photon through the corresponding hole with an accuracy depending on the size of the counter, the distance between counter and hole, and so on. In principle, this registration may be carried out with arbitrary accuracy. If the screen is in the x-y plane, we may say, in correspondence with the behavior of a classical particle, that we have registered (or measured) the x and y coordinates of the photon position[109] at the moment (given by the time of the counter signal) it passed through the hole. The screen and pair of photon counters thus constitute a maximal registration device for that part of the incident ensemble transmitted by the screen. (The nontransmitted part is usually not mentioned in discussion of the double-slit experiment, but I shall return to it in a moment). Suppose we want to make a nonmaximal registration of this part of the ensemble. This requires a record of the emergence of a photon from the screen, without any possibility of associating the record with either of the two holes. We might use an absorber placed beyond the two holes. Absorption of a photon by the absorber is detectible in principle. For example, the energy of the photon could be thermally redistributed throughout the absorber, causing a change in its temperature (or entropy) which (assuming the energy of the photon to be large enough) could be measured classically. If the absorber is rigidly fixed to the same framework as the screen, there is no possibility of using its recoil momentum to associate absorption of a photon with passage through one hole or the other. Thus, the screen with absorber constitutes a registration device for a nonmaximal registration of the portion of the incident ensemble transmitted by the screen. How shall we interpret such a nonmaximal registration physically? One may be tempted to interpret it as a measurement of the *alternatives* for the position of the photon in the *x-y* plane at the time it supposedly passed through one of the two holes: *either* its *x-y* position was that of one of the holes *or* it was that of the other hole. But that is precisely what the *first* device (screen plus two localized counters) registers if we just *add together* the signals from the two counters, disregarding the distinction between thier signals. The second device (screen plus absorber) is *not* equivalent to the first, so it *cannot* be registering the same thing.[110] Using the particle picture, it is hard to interpret what is being registered, except to say that it is the emergence of a photon into the region beyond the screen. We might perhaps say that the position of the photon is thereby fixed to an extent measured by the distance between the two holes. The quantum logicians, of course, claim that we *can* say that the second device registers the alternatives for the position of the photon in the *x-y* plane, if only we reinterpret alternation ("or") in a quantum-logical way. I shall return to the validity of this claim later.

The double-hole (or double-slit) experiment is usually considered with a different registration device. A set of small (localized) photon counters is distributed uniformly over a plane (The emulsion grains of a large photographic plate are usually taken as the counters) which is placed well beyond the screen (i.e., many wavelengths of the incident light away from the screen). Then a signal from one of the counters (sensitization and development of a grain) indicates that a photon has emerged beyond the screen without indicating through which hole. This arrangement also constitutes a nonmaximal registration, equivalent in this respect to the second device just discussed; but it accomplishes much more. The screen is a coherent projector, and the space between the screen and the plane of the photographic plate constitutes a region in which a coherent ensemble emerging from the two holes evolves under the influence of the unitary operator representing force-free evolution in time. In the classical wave picture, this corresponds to the evolution of a diffraction pattern at the plane of the plate. The statistical distribution of a large number of photon registrations on the photographic plate produces this diffraction pattern, which enables us to evaluate the wavelength of the incident light beam, treated as a classical wave traveling in the z-direction. (assumed to have been monochromatic). Using the Einstein-de Broglie relation between the momentum of the photons of the beam in the particle picture and the wavelength in the wave picture ($p = h/\lambda$), we can interpret this as a registration of (the magnitude of) the momentum of the photons in the incident ensemble.

The photons in the ensemble striking the plate all come from a region of the screen of size given by the distance between the two holes. We may take this distance as a measure of the spread or "uncertainty" in the definition of the position in the x-y plane of the photons in the ensemble when they emerge into the region beyond the holes. There is a consequent angular spread or "uncertainty" in the direction of photons in the ensemble striking the photographic plate at a given point, which entails a similar spread or "uncertainty" in the component of their momentum in the x-y plane. We can easily prove that the product of the spread or "uncertainty" in the x-y components of position and momentum is of the order h. This "uncertainty relation" applies in the first instance to the ensemble of photons transmitted by the screen. Considerations like this serve to elucidate the physical significance of the mathematical uncertainty relations derived from the quantum-mechanical formalism. Whether and in what sense these relations are applicable to individual quantum systems is a separate question which we shall soon discuss.

In speaking of "the position of the photon," one is actually using a "correspondence principle" argument: if a classical particle triggered a small, localized counter, this would be because the position of the particle at the moment the counter clicked was inside the counter. What we are doing is interpreting the triggering of the quantum registration device in classical terms; that is, taking a set of concepts developed to describe the state of a classical particle and applying them to a quantum system. If the two-hole experiment is interpreted as allowing us to ascertain momentum, this involves an even more indirect chain of reasoning. A sufficient number of photons must register their positions on the photographic plate in order to establish a diffraction pattern. If this pattern has the appropriate form for a monochromatic classical wave, its wavelength λ may be computed. The classical wave picture for the ensemble of photons has thus been used. λ is then associated with a momentum p using the Einstein–de Broglie relation $p = h/\lambda$, and this momentum is attributed to each of the photons in the initial ensemble incident on the screen. No "position measurement" of one photon could give us enough data to ascertain a momentum.

Up to now we have assumed that the source of the light beam, the screen, and the counters are fixed in one rigid (inertial) frame of reference. There was hence no problem about the relation between the position of the holes in the screen and points in this inertial frame. An alternative way to get information about the incident photons would be to let the screen move freely in its own plane. Then, passage of a photon through one hole or the other imparts a certain transverse momentum to the screen. The subsequent motion of the screen—either up or down—indicates through which hole the photon has gone. Thus, this arrangement constitutes a device for registering the x and y components of the position of a photon at the screen, but the screen is now functioning as a registration device, whose mobility "smears out" the diffraction pattern: the ensemble of transmitted photons no longer forms an interference pattern on the photographic plate. So no conclusion about the momentum of photons incident on the screen can be drawn.

This pair of *Gedankenexperimente* exemplify a general conclusion that follows from the quantum-mechanical formalism. It is inconsistent with the theory to assume the existence of a device that will enable simultaneous (sharp) registration of both the position and the momentum of a quantum system—or, indeed, of any other pair of quantities that are represented classically by canonically conjugate variables.[111]

This statement does not, of course, settle the question of whether we can consistently *ascribe* simultaneous values of position *and* momentum (for example) to a quantum system. Still less does it settle the question of whether we can ascribe a "deterministic state" to a system which determines the outcome of all possible registrations to which the system might be submitted. I shall come to these questions. But first I will briefly discuss the fate of that part of the incident ensemble not transmitted by the screen. As noted above, if the other side of the screen is a perfect reflector, the screen functions as a complete coherent projector, projecting onto the subspace spanned by the rays corresponding to the positions of the two holes in the screen and onto the (orthogonal) subspace spanned by the rays corresponding to all other positions on the screen. If absorbers rigidly attached to the screen frame of reference are added on each side of the screen, this forms a nonmaximal registration device that (incoherently) projects onto these two subspaces—but also destroys the incident photons. If we replace the mirror by a photographic plate, it will act as a maximal (destructive) registration device for the part of the ensemble that does not emerge behind the screen. If no further interaction of the two parts into which the incident ensemble is split by the screen ever occurs, one can neglect what happens on the other side of the screen as far as the fate of the transmitted part of the beam is concerned. (If the entire screen were of relatively small size, this would not be possible).

None of these complications of the double-split experiment need arise in the polarization analogue discussed in the last chapter. As long as the birefringent crystal is large enough to receive the entire incident ensemble, it acts as a complete coherent manipulator of the incident beam.

5.3 Filters, Projections, States, and Properties

There is a class of ideal devices for manipulation of quantum systems much in favor with quantum logicians as well as others concerned with an axiomatic foundation for quantum mechanics. The devices are variously called filters[112] or post controls,[113] yes-no experiments[114] or decision effects.[115] A filter is usually taken to be the sort of device that effects an incomplete coherent projection: it destroys or removes from further consideration all the incident ensemble except the part that is projected onto one subspace of the Hilbert space. There is an ambiguity about the analogue of a "yes-no experiment" in my terminology. It might be a complete coherent projection that projects onto two orthogonal subspaces of Hilbert space and takes one of them as the "yes" alternative; or it might be a nonmaximal registration device, incoherently projecting onto the two orthogonal subspaces with the subensemble corresponding to the "yes" alternative treated nondestructively, so that it is available for further manipulation. Since nothing is usually said about the fate of the subensemble corresponding to the "no" alternative in a yes-no experiment, it is hard to know which of these two interpretations is intended.[116] Of course, if there is never any further interaction of the "yes" and "no" subensembles, it makes no difference. My previous discussion of the double-slit experiment provides examples of both filters and yes-no experiments. A linear polarizer is another example of a filter, a birefringent crystal of a yes-no experiment.

We can set up equivalence classes of filtering devices, regarding two filters as equivalent if the ensemble transmitted by one is always transmitted unchanged by the other, in whichever order the filters are placed. Taking "filter" to mean such an equivalence class, there is a unique projection operator corresponding to each filter. If we assume, conversely, that one can correlate at least one (ideal) filtering device to each projection operator—a rather large assumption—then we may say there is one-one correspondence between filters and projection operators.

The partial Boolean algebra of projection operators may be paralleled by physical relations between filters. For example, if we have constructed a "filter" from a "yes-no experiment" by discarding the "no" output, then the ensemble that corresponds to keeping the "no" output (and discarding the "yes" output) may be taken as the "*yes*" output of a *second* filter. These two filters correspond to a pair of orthogonal projection operators (P and $I - P$). (But Mielnik has questioned whether there need always be a "yes-no experiment" corresponding to every filter)[117] If there is never any output from a sequence of two filters in either order, no matter what the input, the filters are *orthogonal.* Filters may be partially ordered: Place one in front of the other. If the output of the first is *always* transmitted by the second, we say the first filter *precedes* the second: if two filters are either orthogonal or one precedes the other, they are *compatible* (this corresponds to commutation for operators). The *product* of two compatible filters is the filter resulting from placing one beside the other (in either order). the sum of two filters is that physical arrangement that coherently superposes the output of the two when the input is coherent. For example, a screen with two holes may be looked upon as the "sum" of two screens, each having one hole in a place corresponding to one of the two original holes. It is not always so easy, however, to see how to physically realize the sum of two filters. Putting aside such doubts about the physical possibility of realizing all these operations, we have defined a partial Boolean algebra of filters paralleling that of the projection operators.

If one wants to use physical relations among filters as the basis for an axiomatic treatment of quantum mechanics, however, this is not enough. The full lattice structure of the closed linear subspaces of a Hilbert space is needed. One could just transfer this structure from the subspaces to the projection operators by fiat, and then from the projection operators to the filters. This is more or less what Finkelstein does, for example.[118] He uses the definition of partial ordering of filters mentioned above and turns the partially ordered set of filters into a lattice by *assuming* the existence of filters corresponding to the cap and cup operations defined on the lattice of subspaces. (See app. 2 for the definition of a lattice, and the relation between lattices and Hilbert spaces). Of course, if one wants to *base* a construction of the Hilbert space structure on the structure of the set of filters, this method will not serve. So Jauch and Piron try to show how filters with the required properties of the cap and cup operations may be *constructed* from suitable combinations of two filters, even if they are not compatible. They then impose sufficient postulates on the resulting filter structure to guarantee the desired lattice structure,[119] trying to give a "physical" motivation for the postulates whenever possible. Mielnik has pointed out that this lattice structure by no means follows from intuitively reasonable physical conditions one might be tempted to impose on filters. He has suggested other ways of defining the properties of filters that lead to more general convex set structures.[120]

Suppose (for the sake of argument at least) that a successful axiomatization of quantum mechanics has been given that allows the construction of the lattice of subspaces of a Hilbert space, whether or not the motivation for the axioms is given by postulated properties of a set of filters, yes-no experiments or what have you. This does not entail the probabilistic interpretation of the Hilbert space. That is, it does not entail the (generalized) "cosine squared" rule, nor the projection postulate. On the other hand, the Hilbert space model of transition probabilities *does* entail the interpretation of subspaces of Hilbert space as related to registration results and hence the lattice structure. Indeed in many ways this transition probability structure seems the most fundamental physical feature of quantum mechanics embodied in the Hilbert space model. If this observation is correct, it undermines the usual quantum logic program, with its emphasis on the truth or falsity of propositions. Von Neumann, in some of his later comments on quantum logic, seems to have adopted this viewpoint. Consequently, he started to work on a probabilistic foundation for the theory. In comments made in a talk in Warsaw in 1937, he said, "A complete derivation of quantum mechanics is only possible if the propositional calculus of logics is so extended, as to include probabilities, in harmony with the ideas of J. M. Keynes. In the quantum mechanical terminology: the notion of 'transition probability' from a to b, to be denoted by $P(a,b)$, must be introduced. $P(a,b)$ is the probability of b, if a is known to be true. $P(a,b)$ can be used to define $a \leqslant b$ and $-a$: $P(a,b) = 1$ means $a \leqslant b$, $P(a,b) = 0$ means $a \leqslant -b$. But $P(a,b) = \phi$, with a $\phi > 0, < 1$ is a new 'sui generis' statement, only understandable in terms of probabilities."[121] Landé has emphasized this point and attempted to derive quantum mechanics starting from postulating the existence of such $P(a,b)$. After imposing certain conditions on the $P(a,b)$'s, he tries to show that these conditions could be satisfied only if the transition probabilities were derived from quantum-mechanical probability amplitudes, but I do not regard this attempt as successful.[122] Martin Strauss developed a general theory of transition probabilities or, as he put it, "the probability of induced stochastic transitions between states of the same object." He calls this "prob_2 theory," defined over a partial Boolean algebra, to distinguish it from classical "prob_1 theory," defined over a Boolean algebra. Quantum-mechanical probabilities are a special case of his general theory,[123] but he did not succeed in giving a categorical (i.e., unique) characterization of this special case. The work of Mielnik and others on the convex set model has defined a generalized class of transition probability structures, of which the usual quantum-mechanical transition probabilities form a special case.[124] This whole approach, starting from a generalized probability structure and working toward quantum mechanics, is potentially very fruitful. If nothing else, it shows that the quantum logic approach is only one possible way, perhaps not the best, of searching for a deeper understanding of the formal structure of the theory.

Returning to the partial Boolean algebra and lattice structures of a Hilbert space, however obtained, what have they got to do with logic? Some quantum logicians propose that filters be regarded as tests for corresponding properties of quantum systems: If an ensemble of quantum systems in some pure state is incident on a certain filter and each member passes it, then each system in the ensemble is said to have the property corresponding to that filter. The structure of the filters, on this interpretation, reflects the structure of the properties of the quantum system. Whether that structure is taken to be a partial Boolean algebra or a lattice depends on the quantum logician.

Two questions immediately arise: Is it justified to regard filters as tests for properties inhering in individual quantum systems? And, granted that it is, is it further justified to regard the relations between these properties as *logical* relations, to be assimilated to the negation, conjunction, and disjunction of classical properties, for instance? The further claim is sometimes made that at any moment an individual quantum system either does or does not have each and every property testable by such filters. For example, it has a particular value for it position and also a particular value for its momentum. In this case a third question arises: Can such a claim be justified?

All quantum logicians regard the ascription of at least some properties to individual systems as justified. All, however, do not relate these properties to filters. Some associate properties with closed linear subspaces of Hilbert space: If the state vector of a quantum system lies in a certain subspace of Hilbert space, then the system has the corresponding property. Suitably rephrased in terms of states and subspaces, however, our questions still apply.

Jauch, Piron, and many other quantum logicians answer the first question affirmatively but answer the second by denying the logical significance of the structure of filters or subspaces.[125] Strauss,[126] Mittelstaedt,[127] van Fraassen,[128] dalla Chiara,[129] and others grant a logical significance to one or the other of these mathematical structures at the level of properties or of corresponding propositions about these properties. They differ over *which* structure. just *what* logical significance, and whether use of a particular quantum logic is *necessary* or is just one of several alternatives, including the use of standard logic. But they all use, explicitly or implicitly, a concept of logic that applies to linguistic structures, rather than directly to the structure of the world.

While differing in their answer to the second question, both groups agree in answering the third question: they do not ascribe each property or its complementary property to a quantum system at the same time. The advocates of what I call "the logic of quantum mechanics" viewpoint, however, including Finkelstein, Putnam, and Bub and Demopoulos, answer each of the three questions affirmatively. In particular, they agree on simultaneously ascribing a complete set of classical properties to a quantum system. (Such a view is not confined to these quantum logicians. Popper subscribes to this view of properties while opposing all quantum logics.[130]) As noted above, the natural structure of the projection operators is a partial Boolean algebra, while the natural structure of the linear subspaces of a Hilbert space is a lattice. Curiously enough, Finkelstein favors the filter approach, yet opts for the full lattice structure for properties, while Putnam, who favors the linear subspaces approach, has retreated from the lattice to the partial Boolean algebra structure for properties. He has joined Bub and Demopoulos, who also favor the subspace approach but who opted for a partial Boolean algebra structure for properties from the start.

From now on I shall focus exclusively on the "logic of quantum mechanics" school, and a critical examination of two claims made by its adherents which—if justified—would make a strong case for this approach as an interpretation of the quantum-mechanical formalism:

1. Thanks to the quantum-logical interpretation, the properties of a quantum-mechanical system may be treated just like properties in the ordinary sense of the word. All apparently paradoxical features of quantum-mechanical properties that occur in other interpretations, this school claims, may be removed by the quantum-logical interpretation of the logical connectives—or at most of the connectives plus the logical quantifiers. Once these are understood quantum logically, properties per se, they suggest, present no problems.

2. Similarly, they claim that the apparently paradoxical features of the particle interpretation, when applied to *Gedankenexperimente* such as the double-slit experiment, in which interference effects manifest themselves, are dissipated by the application of one or another quantum-logical approach. (I say "one or another" because, unfortunately, quantum logicians do not agree on a unique quantum logic.) I shall focus on Putnam, both because he is probably the best-known exponent of this viewpoint and because his odyssey from the lattice to the partial Boolean algebra approach offers an opportunity to discuss both, but I shall also discuss some of Bub's views.

5.4 Quantum Logic and Properties

An example of a quantum-logical explanation is offered by Putnam at the end of his paper "How to Think Quantum-Logically," written at a time when he accepted the idea that one should "just read the logic off from the Hilbert space."[131] He is discussing how one can maintain that an electron, which he calls Oscar, can have a position and a momentum at the same time and nevertheless not fall foul of complementarity and the uncertainty relations. First of all, he identifies state vectors, such as eigenvectors of position and momentum, as properties that Oscar can have. The statement "Oscar has a position," for example, is identified with the statement "$(E\psi)$ (ψ is an eigenvector of position. Oscar has ψ)." (This may be expanded to read: "There is a state vector ψ such that ψ is an eigenvector of position and Oscar has the state vector ψ.") The existential quantifier ("There is," symbolized by "*E*") is then identified with quantum-logical alternation over all position eigenvectors, that is, with "or" interpreted as the Hilbert space "V", so that the above statement is equivalent to: "$(\psi_1, \vee\ \psi_2 \vee \ldots \vee \psi_n)$." (Putnam restricts himself to the case of a finite dimensional Hilbert space, so the alternation only includes a finite number of alternatives). Presumably this statement ("Oscar has $[\psi_1, \vee \psi_2\ \vee \ldots \vee \psi_n]$") is supposed to be equivalent to: "Oscar has ψ_1 V Oscar has ψ_2 V . . . V Oscar has ψ_n." Now, the quantum-logical interpretation of "V" is spanning of the subspaces involved, so $\psi_1 \vee \psi_2 \vee \ldots \vee \psi_n$ is equivalent to the entire Hilbert space. How are we to translate this into a proposition? Presumably by: "Oscar has a Hilbert space": or, more fully,"(EH). (H is a Hilbert space and Oscar has H)." Perhaps we may express this more colloquially by some other version of the trivially true proposition, such as "Oscar exists." Putnam's quantum-logical assertion "Oscar has a position" is entirely equivalent—quantum logically—to the statement "Oscar exists." If "Oscar exists" is true, then so is "Oscar has a position," because they are quantum-tautologically equivalent statements. The trouble is, our ordinary intuitive feeling about the meaning of the statement "Oscar has a position" tends to be subtly carried along when we accept the truth of the proposition that really means no more than "Oscar exists," as explicated above (following Putnam). A similar analysis, of course, can be made for Putnam's "Oscar has a momentum," with momentum eigenstates just replacing position eigenstates everywhere in this discussion. So Putnam's "Oscar has a momentum" is also tautologically equivalent to "Oscar exists." Hence, "Oscar has a position" is tautologically equivalent to "Oscar has a momentum." Our ordinary intuitive feeling about the meaning of each of the two statements must not be carried along when considering this quantum-logical tautology. Since the two propositions are equivalent, it is no surprise that Putnam is able to assert: "Then . . . the conjunction . . . 'Oscar has a position. Oscar has a momentum' is true." (Putnam uses a dot for conjunction, which I have represented by $\wedge$.) Of course, it is also the case that, on Putnam's reading, the disjunction "Oscar has a position *or* Oscar has a momentum" is also tautologically *equivalent* to the conjunction "Oscar has a position *and* Oscar has a momentum." Putnam does not mention this, and it certainly does not bolster his case for the explanatory power of the quantum-logical interpretation of propositions. This case is based, I think, on tacitly allowing the standard logical interpretation of the meaning of these propositions to hover about their discussion, while proving them by a strictly quantum-logical interpretation of their meaning. One can pinpoint the place where the elision between the standard and the quantum-mechanical sense of alternation is effected. Putnam silently makes the transition from the compound proposition "Oscar has $\psi_1 \vee$ Oscar has ψ_2" to a simple proposition about the compound property: "Oscar has $(\psi_1 \vee \psi_2)$." As noted in section 4, if alternation is interpreted in its usual, ordinary (Boolean) sense this transition is unproblematical. But the first (compound) proposition always carries overtones of ordinary alternation, whereas the second *must not,* if $\vee$ is to be interpreted quantum logically. Their equivalence is, thus, by no means obvious.

But let us proceed with Putnam's argument. If ψ_1 is an eigenvector of the position and ϕ_1 an eigenvector of momentum, "then the statement 'Oscar has ψ_1. Oscar has ϕ_1,' is a logical contradiction! For this is just another way of saying that Oscar has a definite position r and a definite momentum r' in violation of complementarity." Quantum logically, the reasoning is impeccable. $\psi_1 \wedge \phi_1$ is the empty set, which corresponds to the absurd or tautogically false proposition. The reader may feel disappointed, having been led to believe that quantum logic was going to allow us to treat a particle as having all its properties at the same time. Yet we seem to have shown, on the basis of quantum logic itself, that it does not. But it appears we were too hasty in drawing this conclusion: "A system has a position *and* it has a momentum. But if you *know* the position (say, r) you cannot *know* the momentum. For if you did, say, know that the momentum was r', you would know 'Oscar has the position r. Oscar has the momentum r',' which is a logical contradiction." How human *knowledge* suddenly got into the act is not explained further, nor why it has anything to do with the assertion of certain propositions. Putnam objects strongly to bringing *measurement* into the interpretation of quantum mechanics as anything more than the verification of already existing properties of a system, and taxes the Copenhagen interpretation for violating this canon (quotations, from Putnam will be given later). Yet *knowledge* is apparently acceptable as an explanatory device in explicating the meaning of the absurdity of a quantum-logical proposition. Even more confusing, measurement is next introduced into the explication of the role of knowledge. Immediately after the sentence just quoted comes the following final paragraph of the article: "The logic by itself does not say exactly how any particular position measurement will make momentum uncertain. But we know that in each case the physical laws will in some way have to say that the position measurement makes the momentum uncertain, because physical laws have to be compatible with logic—that is to say, have to be compatible with the *true* logic, which is quantum logic."

To summarize the nature of the quantum-logical explanation being offered: the logic is read off the Hilbert space. The physical laws have to be compatible with this logic. Measurement has to be compatible with these physical laws. Our knowledge has to be compatible with the measurements.

If you fail to see in what sense a deeper understanding of quantum theory is achieved than follows from simply accepting the rules of quantum mechanics, I can only join you in puzzlement. I also worry about the ontological status of Hilbert space and its logic—or rather the many Hilbert spaces needed for all the quantum systems in the world. They seem to be the unmoved movers of a whole process culminating in human knowledge—but I have already commented on the danger of panlogism I see inherent in this approach (see sec. 1 and app. 1).

The example just used to discuss the claim that quantum logic deepens our understanding of quantum theory can also be used to discuss the claim that quantum logic helps us avoid paradoxes arising in the particle interpretation of the theory. "Oscar has a position and Oscar has a momentum" may be rewritten:

$$(\psi_1 \vee \psi_2 \vee \ldots \vee \psi_n) \wedge (\phi_1 \vee \phi_2 \vee \ldots \vee \phi_n)$$

in the state vector shorthand form used by Putnam. Application of the distributive law (see app. 2) yields:

$$(\psi_1 \wedge \phi_1) \vee (\psi_1 \wedge \phi_2) \ldots \vee (\psi_2 \wedge \phi_1) \vee (\psi_2 \wedge \phi_2) \vee \ldots \vee (\psi_n \wedge \phi_n).$$

This is just the quantum-logical disjunction of all the quantum-logical conjunction that we have seen to be tautologically false: "Oscar has position r_i and Oscar has momentum r_j'." But the distributive law fails for quantum logic. Thus, Putnam claims quantum logic saves us from the classical logical paradox of asserting that "Oscar has a position and Oscar has a momentum," yet denying all the individual assertions "Oscar has position r_i and momentum r_j'." Quantum logically, the reasoning is again impeccable: a statement that is tautologically true *would* be equivalent to a statement that is tautologically false *if* the distributive law held; but it does *not* hold, so there is no problem. A problem arises only if we carry along any of our ordinary intuitive feelings about the meaning of the various propositions; that is, if we imagine that the words "or" and "and," as used quantum logically in the above statements, have the significance of ordinary alternatives and conjunctives, respectively. If they did carry this significance, then, of course, the distributive law *could not* fail, as Kripke has emphasized.[132] The greatest caution, then, must be exercised in quantum-logically construing such statements as "Oscar has position r_1 or Oscar has position r_2 or . . . Oscar has position r_n" not to let any hint of the ordinary significance of "or" (enumeration of alternatives) slip into the meaning. (One cannot have one's cake and eat it!) Then Putnam's explanations of the quantum paradoxes, which all boil down to the failure of the distributive law, don't really help. Our puzzlement comes from considering a series of alternatives in the ordinary meaning of the term. The explanation offered is based on the quantum-logical join operation. As an example, take Putnam's discussion of the two-slit experiment.[133] Consider a photon emitted from a source and confronted with a two-slit screen on its way to a photographic plate. If we keep entirely to the particle picture (as Putnam tells us we must) and want to assign a definite position to the photon at each moment between its emission and absorption (as Putnam tells us we may), then the photon has to pass through one slit or the other on its way to the plate. "Let A_1 be the statement 'the photon passes through slit 1' and let A_2 be the statement 'the photon passes through slit 2.'" R is the statement that "the photon hits a tiny region R on the photographic plate." Thus, the statement that the photon passes through one of the two slits is represented by "A_1 or A_2," where "or" has its ordinary, alternative significance. Putnam, of course, represents it by "$A_1 \vee A_2$," where "$\vee$" has its quantum-logical significance of the join operation. The statement that the photon passed through one of the two slits and hit the screen at R, as ordinarily understood, is represented by "(A_1 or A_2) and R." Putnam, of course, represents it by "$(A_1 \vee A_2) \wedge R$" (he actually uses the dot for the meet operation). This statement is tautologically equivalent to "R" for Putnam, since "$A_1 \vee A_2$" is tautologically equivalent to the trivially true statement (Putnam ignores photons not transmitted through the screen). Thus, *on the quantum-logical interpretation,* "The photon passed through one of the two slits and hit R" is *strictly* equivalent to the statement "The photon hit R." Now, if we were to apply the distributive law to the Putnam statement "$(A_1 \vee A_2) \wedge R$" as we surely can to the *ordinary* statement "(A_1 or A_2) and R"—we would get "$(A_1 \wedge R) \vee (A_2 \wedge R)$." Then, if we were to use this equivalence in the ordinary probability laws, *interpretating* "$\vee$" *and* "$\wedge$" *with the ordinary meaning of* "*or*" *and* "*and*" *respectively,* we would get the (incorrect) classical prediction for the relation between the probability distribution of photons on the plate when both slits are open and that with only one slit open at a time (as Putnam shows). What conclusion are we supposed to draw from Putnam's discussion? Putnam does not claim here that it helps us to understand the actual interference pattern that quantum theory predicts. (He does discuss this question, in a paper with Friedman, after changing his quantum-logical approach. I shall discuss this paper soon). Its merit is supposed to be that it allows us to maintain that the particle did, indeed, pass through one of the two slits and yet avoid the conclusion that we should therefore get the classical probability distribution. Yet, the proposition in which Putnam formalizes the assertion that the photon passed through one of the two slits and then hit the plate at R ("$[A_1 \vee A_2] \wedge R$") has no hint of the ordinary meaning of an alternative in it. As discussed above, it means no more (and no less) than "The photon hit the plate at R." Thus, it seems to serve neither as evidence for nor as evidence against the claim that one may think of the photon as having a definite position at each moment, and in particular for the claim that it passed through one slit or the other. Further, in a situation where one may interpret "or" as involving an alternative in the usual sense and "and" as involving conjunction in the usual sense, then the distributive law *must* hold, as we shall see in a moment. What I want to emphasize is that the *explanation* is provided with the help of the quantum-logical join, while the *problem* arises in the first place because one intends "or" in the ordinary sense of an alternative. We do not get an answer to the question that troubled us.

Up to now, with Putnam, it has been tacitly assumed that the photographic plate (P) on which the photons are registered is far from the screen(S) with the double slits. To investigate the distributive law in the context of the double-slit experiment, it is helpful to think of the distance (D) between S and P as variable. If P is placed just behind S ($D = 0$) then the experimental arrangement constitutes a maximal registration of the position of each photon in the ensemble emerging from the slits in S (see section 4 for a discussion of the entire ensemble emerging from the source). Suppose a photon is registered at a small region R on P. If R is just behind A_1, the photon passed through the first slit; if R is just behind A_2, the photon passed through the second slit. Indeed, the rest of P is redundant for $D = 0$. It could be removed, leaving only two small pieces just behind A_1 and A_2, without affecting the results in any way. The distributive law is certainly valid in this case.

Proof. The two subspaces of the Hilbert space that correspond to A_1 and A_2 respectively span the entire Hilbert space of the ensemble emerging from S. Symbolically, $A_1 \vee A_2 = 1$. Thus, $(A_1 \vee A_2) \wedge R = R$. (Indeed, this holds true no matter what the value of D, a result we shall need later). If R lies behind A_1, $A_1 \wedge R = R$ and $A_2 \wedge R = \phi$; if R lies behind A_2, $A_2 \wedge R = R$ and $A_1 \wedge R = \phi$ (The rest of the plate is redundant, as pointed out above; points not behind A_1 or A_2 do not correspond to any operators on this Hilbert space). So $(A_1 \wedge R) \vee (A_2 \vee R) = R \vee \phi = R$, if R is behind either A_1 or A_2. *QED.*

Now imagine D to be increased continuously. The nonredundant portions of P gradually increase in size (the diffracted image of each slit on the plate increases in size). But, until a value of D is reached at which the nonredundant portions start to overlap (the diffracted images start to interfere), nothing basic changes in our analysis above. As long as R is within the nonredundant part of the plate behind on slit or the other, the distributive law still holds. It is only when the nonredundant portions of the plate overlap that this argument—and the distributive law—breaks down. For intermediate values of D, in which the overlap region does not effectively cover the entire plate P, the experimental arrangement still gives partial information about which slit photons passed through (i.e., if the photon is registered at a point outside the overlap region). The distributive law breaks down for points in the overlap region. In the limit, where D is very large (i.e., compared to the wavelength of the light used), the overlap region covers P. Registration of a photon cannot give any information about which slit it passed through, and the distributive law fails for any R on the plate. In this sense, failure of the distributive law is a measure of the degree of interference (superposition principle) between the images of the two slits on the plate.[134]

What implications does this discussion have for quantum logic? In the cases where we can give " $\wedge$ " and "$\vee$" the ordinary logical interpretation of "and" and "or", respectively—that is, where passage through one slit *or* the other has its usual meaning and not a Pickwickian, quantum-logical one—the distributive law does not break down. Where the distributive law does break down, the quantum-logical connectives cannot be given the ordinary meaning: passage through "$A_1 \vee A_2$" then means no more than appearance somewhere on the screen. It is no wonder that the classical probability argument fails! At the least, this example clearly serves to invalidate a claim that the quantum-logical connectives preserve their ordinary meaning, while only certain laws of logic need be changed. But there is also a problem with the claim that adoption of the quantum-logical approach frees us from the need to introduce dependence on experimental context into the interpretation of the theory. Without knowing the value of D, one cannot tell whether or not the distributive law fails for the double-slit experiment. As photons pass through S, they have to know how far away P is located before deciding whether or not to obey the distributive law. Is this really an improvement on Copenhagen contextualism?

Another modification of the double-slit experiment (see sec. 4) is to leave P far away from S but put a photon counter behind on of the two slits. Then, for each photon reaching P, we always register through which slit it passed. The distributive law always holds, the classical probability argument applies, and the classical statistical pattern for the photons appears on P. So a photon passing through the slit *without* a counter behind it has to know that a counter is lurking behind the *other* slit and adjust its behavior accordingly by obeying the distributive law and contributing to the classical pattern. Again, this seems to be an inescapable contextual feature of the particle interpretation.

Part of the discussion of Putnam's position is obsolete, for a new asceticism has recently come over him. Formerly, he was liberal with connectives, allowing them to range over the entire lattice of subspaces of Hilbert space. Now, under the influence of Kochen and Specker and Bub and Demopoulos, he has become parsimonious: only compatible propositions may be connected. He has retreated from the full lattice fortress to the partial Boolean bunker. Since he still needs the Hilbert space (where would he be without the projection postulate, for example?), it is not clear why we can no longer "just read the logic off from the Hilbert space"; nor does Putnam, in his recent article with Friedman,[135] offer any explanation beyond the bald assertion that "incompatible propositions can be true simultaneously, but their propositional combinations simply don't exist (states of affairs are not closed under propositional combination)!" His move to partial Boolean algebras entails certain consequences for Putnam's previous arguments. In a partial Boolean algebra the distributive law for three elements cannot fail to hold whenever each side is meaningful.[136] So failure of the distributive law cannot be used to explain the two-slit paradox. We can no longer even assert that a statement like "Oscar has a position *r* and Oscar has a momentum *r′*" is tautologically false. It cannot be formed. So there cannot be any question of applying the distributive law to such statements. The distributive law for the two-slit paradox is inoperative, as they used to say in Washington in similar cases.

But if he is currently chary of connectives, Putnam is still spendthrift with states. For both the old and the new Putnam a particle is not content with just one state at a time, but must have an infinite number! Old Putman wrote: "A system has a position state *and* it has a momentum state . . . and a system has many other 'states' besides (one for each 'nondegenerate' magnitude)."[137] Lest one be in doubt that by "state" he means "state vector," I quote "How to Think Quantum Logically": "Thus, on the quantum logical interpretation, contrary to what is often maintained, a system has *many* state vectors; it has a state vector of each nondegenerate observable, in fact."

The new Putman reaffirms this position in his paper with Friedman. Speaking of "the quantum logical interpretation," they remark: "(Of course, on such 'realistic' interpretations, a system has more than one state at any given time!)." In trying to ascertain the concept of "state" Putnam employs that allows this infinite multiplicity, the only explanation I could find is in "Is Logic Empirical?" Just after the sentences from that paper quoted above, he says: "These are 'states' in the sense of *logically strongest consistent statements,* but not in the sense of 'the statement which implies every true physical proposition about [the system].'" So it is clear why he states in "How to Think Quantum Logically" that "a system *has* more than one state vector, on the quantum logical interpretation, but one can never *assign* more than one state vector" (p. 61). For if the "state" is the logically strongest consistent statement we can make about the system, to assign (or assert) two or more of them would be to assert a logical *inconsistency*—presumably a quantum-logical one, so that no fiddling with the distributive law or anything else could palliate it. But then one is left to puzzle out what it means to "*have*" a state (with or without italics) that cannot be assigned or asserted. This is only one aspect of the question: What does it mean in quantum logic for a system to have a property in general? For Putnam treats a state—or a state vector—as if it were something that a system has, just as it has all its other quantum-logical properties.

Before passing to that question, however, let me recall a few things about states already mentioned in Section 4. As Freundlich recently noted:[138] one can always introduce a unique state that connects any input with any output. He has shown how to explicate the Popperian concept of state in this way and thereby make sense of Popper's claim to be able simultaneously to assign values to all observables of a system. So it is possible to introduce a *unique* state which, in Putnam's words quoted above, "implies every true proposition about [the system]." Of course, this approach has nothing to do with quantum logic, managing to get by with ordinary logical resources.

Secondly, speaking of the state vector of a system without qualification is a dangerous game. The state vector is a relative entity which, as Groenewold[139] and Aharonov, Bergmann, and Liebowitz[140] have noted, may be elminated entirely from quantum mechanics. A system may also have different state vectors assigned to it at some time, depending on whether they are to be used for prediction or retrodiction. Putnam himself admits that "talk of 'state vectors' . . . tends to be highly misleading" ("How to Think Quantum-Logically," p. 61).

Now I shall turn to the claim that, on the quantum-logical interpretation, quantum systems can be treated as particle systems that simultaneously have values for all of their classical attributes (properties, magnitudes, observables, or whatever synonym you prefer)—as well as, presumably, for such nonclassical attributes as spin, isospin, and so on. Measurement, Putnam asserts, only ascertains the preexisting values of such attributes without (for an ideal measurement, at any rate) changing them. Both the old and the new Putnam are quite explicit on this point. I shall quote primarily the new Putnam, writing with Friedman. Together they make the following assertions:

(1) The quantum logical interpretation is "realistic" in that it assumes, contrary to the Copenhagen interpretation, that quantities *always* have well-defined values. Such measurement-independent values are perfectly precise . . . There is no relativity to 'experimental arrangement' and no need, therefore, to use the term 'measuring apparatus' in formulating the basic laws of the theory. Moreover,

whether or not incompatible projection operators can be simultaneously measured, they do have simultaneous values. Incompatible propositions can be true simultaneously. (pp. 310–11)

The old Putnam added: "Measurement only determines what is already the case: it does not bring into existence the observable measured, or caused it to 'take on a sharp value' which it did not already possess." ("How to Think Quantum-Logically," p. 57). Friedman and Putnam then assert the new (for Putnam) asceticism, which I quoted previously: "(2) Incompatible proposition can be true simultaneously, but their propositional combinations simply don't exist (states of affairs are not closed under propositional combination)!" (p. 311) But now comes the real surprise: "(3) Third, not only will we refrain from asserting propositional combinations of incompatible propositions, we will not simultaneously assert individual incompatible propositions . . . .Thus the quantum logical interpretation *agrees* with the Copenhagen interpretation that incompatible propositions cannot be simultaneously asserted" (p. 312). The reason given for this restriction is "Gleason's theorem [which] states that *all* (generalized) probability functions on the partial Boolean algebra of projection operators are generated by quantum mechanical pure (or mixed) states. It follows that incompatible propositions cannot simultaneously have probability 1" (p. 312). Since Friedman and Putnam "idealize assertion [of a proposition] as the assignment of probability 1" to that proposition, this restriction is a forced move. No further explanation is given of the meaning of "assertibility of a proposition," or of its identification with assignment of probability one, which seems to introduce a probabilistic element into the foundations of the quantum-logical interpretation.[141] If assertibility is somehow related to cognizability, this would also introduce a subjective element into the foundations, but perhaps this is not intended. At any rate, the upshot of the argument so far seems to be the claim that at any time a particle has a position, it also has a momentum; but that we cannot *assert* it has a position and also *assert* that it has a momentum.

But even this is not the end of the story: "(4) Finally, although incompatible propositions cannot be simultaneously asserted on the quantum logical interpretation, there exist incompatible propositions that are simultaneously *true*—and we can assert *that*" (p. 313). One hardly knows what to make of this last claim; let us try to see just what it *really* means. The quotation continues: "Although we have to formulate it carefully: e.g., as ⌜ $\exists_{\alpha_i}$ and $\exists_{\beta_j}$,⌝ but not ⌜$\propto$; and β_j!⌝ (p. 313)"

If we expand the last two expressions, they are just the propositions "Oscar has a position and Oscar has a momentum" and "Oscar has a position and a momentum." I have discussed the quantum-logical meaning of both propositions at length above, so need remind the reader only that the quantum-logical meaning of the first (true) proposition is tautologically equivalent to the trivially true proposition "Oscar exists." I do not understand why Friedman and Putnam call its assertion an assertion of "incompatible propositions that are simultaneously *true.*" As noted above, "Oscar has a position" and "Oscar has a momentum," far from being incompatible, are strictly equivalent quantum logically. Both mean no more than their (quantum) conjunction "Oscar exists." The second proposition was previously meaningful, but false. It is now strictly meaningless in new Putnamese. Aside from this difference, the discussion here of these two propositions suggests that Friedman and Putnam are still carrying over some of the ordinary, intuitive connotations of the connectives into their quantum-logical discussion.

Since the propositions discussed here are all statements about the values of attributes, Friedman's and Putnam's final three theses leave me at a loss to understand what they mean by the assertion in their first thesis "that quantities always have well-defined values." Apparently, I am not alone in this puzzlement. Recently, Bub has given an excellent summary of his view of the current status of the quantum logic program.[142] He formulates four conditions that the quantum-logical view of physical properties must satisfy:

(a) For any system *S*, at every time *t*, every magnitude has a value.
(b) The values of the magnitudes of *S* preserve the characteristic algebraic structure of the magnitudes of a quantum mechanical system, i.e., if the value of *M* for *S* at time *t* is *m*, and the value of *N* for *S* at time *t* is *n*, and $M = f(N)$, then $m = f(n)$.
(c) Ideal measurements merely reveal the values of these magnitudes . . . without disturbing *S*; i.e., the measurement of *M*, say, with result *m* on a statistical ensemble of systems is understood as the selection of those systems with the value *m* of *M* from the ensemble, without altering the value of *M* or the values of any other magnitude.
(d) The statistical states of quantum mechanics represent all possible probability distributions over the values of the magnitudes for *S*, and transformations of probability distributions on measurements are derived by conditionalising the initial probability distribution in accordance with the measurement result, where a measurement is regarded as in principle ideal (i.e., no additional disturbance transformation is evoked). (p. 36)

He points out "that, even on the assumption of a non-Boolean possibility structure for a quantum mechanical system, maintaining the [above] four views faces formidable difficulties" (p. 35).

Bub starts his paper with one example illustrating these difficulties: successive measurements of the polarization of a photon. But an even more striking example involves only classical properties of a particle, and its conclusion involves only ensembles. If (a) is true, every quantum system has a position and a momentum. If (c) is true, a momentum measurement selects a subensemble, all having the same momentum, without changing the positions of the members of the subensemble. Then a subsequent position measurement yields a sub-subensemble whose members all have the same position and momentum, without altering these values. In other words, a dispersion-free ensemble would result. This could be verified by subsequent measurements of the position of half the members if the ensemble and the momentum of the other half. This certainly violates the predictions of quantum mechanics (uncertainly relations for ensembles). It is hard to see how any construction could be put on (c), which speaks of selections of subensembles from a statistical ensemble, that could avoid this conclusion—no matter how one defines "property," "proposition," and the like.

Popper has seen the force of such an argument, at least for the case of polarization of photons. While insisting on (a), he is willing to grant that (c) cannot be true.[143] Why does Bub insist on (c)? He lumps (c) together with (b) and (d) in his answer, but I think the point is clear; "If we drop requirement (a), then the quantum logical interpretation incorporates the essential feature of the Copenhagen interpretation, and cannot avoid the measurement problem. . . . Requirements (b), (c) and (d) cannot be dropped without reducing quantum logic to the status of an algebra of measurement operations" (p. 37). In other words, quantum logic would be demoted thereby from a theory of the inherent structural properties of a quantum system to an algebra of measuring devices (filters). As we saw, this algebra can be given a non-Boolean structure, but Popper's example shows that the logic used in such an approach can be classical.

Bub states that the only possible way out of these difficulties is "a non-Boolean analysis of the semantic notion of a property 'obtaining'," since the classical (Boolean) notion of a property obtaining is clearly incompatible with his four criteria. He admits that no such non-Boolean analysis now exists: "The open problem of the quantum logical interpretation . . . is a non-classical theory of properties, i.e., a theory of the 'obtaining' relation which does not require the existence of a 2-valued homomorphism assigning 1 to every property which obtains" (p. 38). This last is the ordinary notion of a property either holding or not holding—which leads straight to a Boolean "analysis of the semantic notion of a property 'obtaining'." Bub can only speculate on what such a non-Boolean analysis might look like: "Such a theory might be expected to yield a theory of sets without points, in a sense analogous to von Neumann's generalisation of projective geometry to continuous geometry, which he conceived as a geometry without points." (p. 38).

Until the day this hope is realized, or at least turned into a precise research program, I think one must conclude that—far from allowing us to think of properties in more or less the ordinary way—the "logic of quantum mechanics" approach has no consistent concept of property at its disposal. This is one reason why criticism of some arguments based on this approach is so difficult. The word "property" or one of its synonyms ("attribute, magnitude," etc.) is used with what appear to be the usual connotations, but if one proceeds on this assumption, one is told that this is not what is meant. (Recall the discussion of properties by Friedman and Putnam, for example.)

5.5 Quantum Logic and the Projection Postulate

Finally, let us examine in detail the new Putnam's claim that: "The quantum logical interpretation of quantum mechanics gives us an *explanation* of interference that the Copenhagen interpretation cannot supply."[144] Leaving "the Copenhagen interpretation" to fend for itself (discussion of it would take us too far afield), I shall examine the profferred quantum-logical explanation. Great stress is laid upon its ability to understand Lüders's rule for the state of a system after an incomplete measurement.[145] Lüders showed that the projection postulate, originally introduced by von Neumann for a complete measurement, also applies to an incomplete measurement. Prepare a system in some state represented by the state vector ϕ, and make a nondestructive measurement (incomplete or not) on it. If E is the projection operator corresponding to the subspace of Hilbert space singled out by the measurement, then the state of the system after the measurement is represented by the state vector ϕ_E, where ϕ_E is the projection of the state vector ϕ onto the subspace singled out by E. Friedman and Putnam correctly note that: "(as Bub first pointed out) . . . (1) [Lüders' rule] is responsible for the phenomenon of interference." Neither Lüders nor Bub, (nor anyone else, as far as I know) have noted that Lüders' rule is just a special case of Feynman's rules for computing probability amplitudes which, as is well known, include quantum interference effects.[146] Feynman's approach is based on the contrast between processes that are *distinguishable* within a given physical context and those that are *indistinguishable* within that context. A process is distinguishable if some record of whether or not it has been realized results from the process in question; if no record results, the process is indistinguishable from other alternative processes leading to the same end result. In my terminology, a registration of the realization of a process must exist for it to be a distinguishable alternative. In the two-slit experiment, for example, passage through one slit or the other is only a distinguishable alternative if a counter is placed behind one of the slits; without such a counter, these are indistinguishable alternatives. Classical probability rules apply to distinguishable processes. Nonclassical probability amplitude rules apply to indistinguishable processes. These (Feynman) rules may be formulated as follows:

1. There is a complex probability amplitude for each distinguishable

process leading from an initial preparation to a final (registered) result. The probability for that process is equal to the modulus squared of its amplitude, which must be a complex number of modular $\leqslant 1$.

2. If several alternative subprocesses, indistinguishable within the given physical arrangement, lead from the initial preparation to the final (registered) result, then the amplitudes for all the indistinguishable subprocesses must be added to get the total amplitude for the entire (distinguishable) process (quantum law of superposition of amplitudes).

3. If several distinguishable alternative processes lead from the initial preparation to the same final result, then the probabilities for all these processes must be added to get the total probability for the final result (classical law of addition of probabilities).

4. If an indistinguishable process consists of a sequence of steps, the amplitudes for all the steps must be multiplied to get the total amplitude for that process (quantum law of multiplication of amplitudes).

5. If a distinguishable process consists of a sequence of steps, the probabilities for all the steps must be multiplied to get the total probability for that process (classical law of multiplication of probabilities).

The Feynman rules do not constitute an explanation of quantum-mechanical interference, rather, they are a codification of rules for computing probability amplitudes which include interference among their consequences. Friedman and Putnam claim that "the Copenhagen interpretation" requires two separate postulates to explain quantum interference: the eigenvalue-eigenstate link for maximal registrations and the projection postulate for nonmaximal registrations. Far be it from me to try to decree just what is "the Copenhagen interpretation," although I think Feynman's approach should be included, broadly speaking. In any case, if one bases quantum mechanics on Feynman's rules, both the eigenvalue-eigenstate link and the projection postulate follow without further assumptions (see app. 3). So whatever advantage in explanatory power Friedman and Putnam claim for their approach over "the Copenhagen interpretation" (and we shall examine this claim next), there does not appear to be any advantages over an approach based on the Feynman rules.

As Feynman and Hibbs said, "The concept of interfering alternatives is fundamental to all of quantum mechanics."[147] If the quantum logicians could really explain the Lüders rule, they would indeed have contributed to a deeper understanding of the theory! Before exulting prematurely, let us examine the profferred "*explanation* of the interference." Friedman and Putnam pick as an example the spin-one system discussed earlier by Kochen and Specker.[148] By modifying the two-slit experiment discussed previously, a simpler example can be given. Consider a screen with three slits in it and a photon counter directly behind one of the slits which registers the passage of a photon through that hole but does not otherwise interfere with passage of the photon to the plate. With a plate far from the screen, this constitutes an incomplete measurement of the position of the photon ensemble emerging from the screen. If the counter clicks, the photon passed through the first slit; if it does not click, the photon did not pass through the first slit. One might be tempted to say this means it passed through one or the other of the remaining two slits, but this, of course, begs the question of what "or" means in the phase "passing through one *or* the other slit." At any rate, if we interpret α_i as representing passage through the ith hole ($i = 1,2,3$) then ($\alpha_2 \vee \alpha_3$) represents nonpassage through the first hole. If φ represents the state of the photons incident on the screen, this example may be used to help visualize the following abstract discussion.

Consider a three-dimensional Hilbert space, and let α_1, α_2, α_3 be projection operators onto one set of orthogonal axes in this space. Then the α_i commute, so even the new Putnam can take their conjunctions and disjunctions. Assume that the system is in the state φ, represented by a unit vector in the Hilbert space. Now make the incomplete measurement on the system corresponding to the projection operator onto the $\alpha_2 - \alpha_3$ plane. So E. in our formulation of the projection postulate, corresponds to $\alpha_2 \vee \alpha_3$. The projection postulate asserts that, in computing probabilities for the result of any subsequent measurement on the system after the incomplete measurement symbolized by $E = \alpha_2 \vee \alpha_3$, we should use the state vector $\phi_E = \phi_{\alpha_2 \vee \alpha_3}$ to represent the system. Friedman and Putnam claim that "the quantum logical interpretation [can] account for [the use of this state vector] without invoking an additional ad hoc postulate like [the projection postulate]." Their reasoning (somewhat simplified) is as follows.

If the measurement is maximal, that is, if it corresponds to a ray (one-dimensional subspace) of the Hilbert space, they see no problem. "By Gleason's theorem there is one and only one probability measure on the algebra of propositions consistent with this information," namely, the information that the probability assigned to a given pure state (one-dimensional subspace) is one. So, in going from one pure state to another after a maximal measurement, "all we can do is adjust our probability function to the new information. . . . But there is one and only one conditional probability . . . such that our new probability function results from the old one by *conditionalization,* and that is given by (1) [the Lüders rule]." In the language of the Feynman rules for probability amplitudes (which I find clearer), this amounts to saying: If a complete registration takes place in between the initial preparation and final registration, then only Feynman Rules 3 and 5 come into play, that is, only classical compounding of probabilities is involved. The "only" remaining question is: "Why is each classically compounded probability itself derived from an amplitude? Friedman and Putnam are content to invoke Gleason's theorem: "*all* (generalized) probability functions on the partial Boolean algebra of projection operators are generated by quantum mechanical (pure or mixed) states." Presumably if you don't feel immediately enlightened by this explanation, your knowledge of quantum logic is defective. Feynman, of course, took a different path, attempting to understand the form of the probability amplitude for a process through his space-time, sum-over-all-paths interpretation of the wave function. He could have spared himself the trouble if he had only known quantum logic and seen there was no problem!

After this triumph of insight, Friedman and Putnam turn to "the really interesting case," incomplete measurement. In this case, since "there are an infinite number of distinct probability functions such that $Pr\ (\alpha_2 \vee \alpha_3) = 1$," Gleason's theorem won't help. So some other argument is needed to help select the correct function. Quantum logic to the rescue!

They argue as follows: "Suppose we are given φ and then given $\alpha_2 \vee \alpha_3$. . . . [O]n the quantum logical interpretation we have some additional information: namely, that if φ is true then $\alpha_2 \vee \alpha_3$ and $\phi_{(\alpha_2 \vee \alpha_3)}$ are equivalent!" (p. 314). They then proceed to "define equivalence between quantum-mechanical propositions in the usual way ($a \sim b$) if (a and b) or (not-a and not-b)," where "and," "not," and "or" are interpreted quantum logically, so that this may be written

$$a \sim b = df\ (a \wedge b) \vee (a' \wedge b')$$

in my notation. Then they "observe that

$$\phi \leqslant [\phi_{(\alpha_2 \vee \alpha_3)} \sim (\alpha_2 \vee \alpha_3)],"$$

and conclude "Since $\phi_{\alpha_2 \vee \alpha_3}$ and $\alpha_2 \vee \alpha_3$, are equivalent on the assumption of ϕ, we can conclude that $\mathrm{Prob}_\phi(F, \alpha_2 \vee \alpha_3) = \mathrm{Prob}_\phi(F, \phi_{(\alpha_2 \vee \alpha_3)})$" (all quotations from p. 314).

Let us try to make (quantum-logical) sense out of this argument. "We are given ϕ and then $\alpha_2 \vee \alpha_3$." What does "then" mean in this phrase? Either we are always free to use both premises, in which case "then" has no logical force; or "then" implies some temporal logic in which, at the time we want to assert ϕ, $\alpha_2 \vee \alpha_3$ does not yet hold; and, at the time we want to assert $\alpha_2 \vee \alpha_3$, ϕ no longer may be asserted. Friedman and Putnam ruled out the first alternative in their third assertion listed earlier): "Not only will we refrain from asserting propositional combinations of incompatible propositions, we will not simultaneously assert individual incompatible propositions." They reiterate this point a few lines later: "Thus the quantum logical interpretation *agrees* with the Copenhagen interpretation that incompatible propositions cannot be simultaneously asserted" (p. 312). Since ϕ and $\alpha_2 \vee \alpha_3$ are incompatible (except in trivial cases where the initial state lies in the α_2-α_3 plane, that is, $\phi = \phi_{\alpha_2 V \alpha_3}$), it seems that they cannot be asserted simultaneously. We are not dealing with any simple assertorial propositional logic in which propositions once used as premises may be used again at any later stage of the argument. Quantum-logical arguments must then involve some sort of temporal logic, in which the *sequence* of assertions plays a crucial role. This is reminiscent of Mittelstaedt's dialogic interpretation of quantum logic in which the validity of an argument depends on the sequence in which registrations of a system occur, but I fail to see how it can be said to be compatible with a "realistic" interpretation of quantum logic, in which logic is supposed to tell us something about the structure of a system, independent of any sequence of registrations. At any rate, this issue is unexplored: Friedman and Putnam offer no rules of quantum-logical deduction, nor even a fully developed example. Let us suppose that we accept such a temporal logic and assert only ϕ initially. From ϕ and $\phi \leqslant [\phi_{\alpha_2 V \alpha_3} \sim (\alpha_2 V \alpha_3)]$, using *modus ponens,* we draw the conclusion $\phi_{\alpha_2 V \alpha_3} \sim (\alpha_2 V \alpha_3)$ (note that $\phi \leqslant [\phi_{\alpha_2 V \alpha_3} \sim (\alpha_2 V \alpha_3)]$ is a statement in the metalanguage: $\leqslant$ is not a propositional connective). Presumably, I am now free to introduce the premise $\alpha_2 V \alpha_3$, having somehow got rid of ϕ. What conclusion can I draw from $\alpha_2 V \alpha_3$ and $(\alpha_2 V \alpha_3) \sim \phi_{\alpha_2 V \alpha_3}$? Since they are compatible, I may assert them simultaneously and draw the conclusion $(\alpha_2 V \alpha_3) \wedge [(\alpha_2 V \alpha_3) \sim \phi_{\alpha_2 V \alpha_3}]$. But this last expression is just $\phi_{\alpha_2 V \alpha_3}$. A little geometry makes the last two steps clear (see fig. 19):

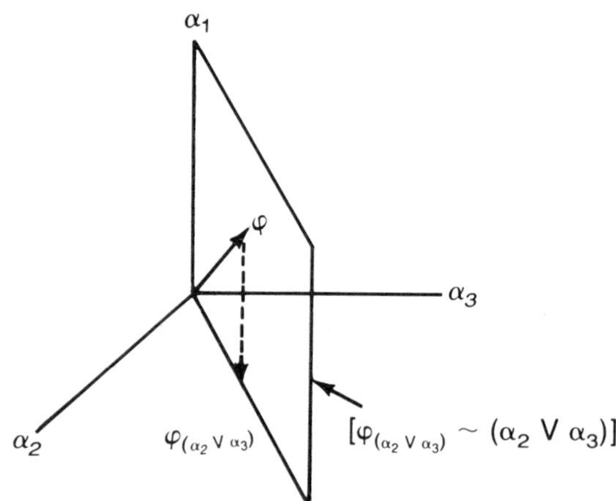

Figure 19

$(\alpha_2 V \alpha_3) \sim \phi_{\alpha_2 V \alpha_3}$ is just the $\alpha_1 - \phi_{\alpha_2 V \alpha_3}$ plane; $\alpha_2 V \alpha_3$ is the $\alpha_2 - \alpha_3$ plane. These two planes are orthogonal and so correspond to compatible propositions. Their intersection is $\phi_{\alpha_2 V \alpha_3}$; hence; the last step. One is still left with the problem of establishing that $\mathrm{Prob}_\phi(F, \alpha_2 V \alpha_3) = \mathrm{Prob}_\phi$

(F, $\phi_{\alpha_2 \vee \alpha_3}$). I don't see how such an equality of probabilities follows from the fact that one can deduce $\phi_{\alpha_2 \vee \alpha_3}$ from the premises ϕ and $\alpha_2 \vee \alpha_3$— no matter what the deductive argument. This is an independent assumption, which we formulate as: Whenever a proposition involving a nonatomic lattice element *X* occurs in an expression of the form "Prob_ϕ(F,X)," then *X* is to be replaced by the equivalent atom, the equivalence following from the assertion ϕ.

It may be Friedman and Putnam have in mind not the use of *modus ponens* to deduce $\phi_{\alpha_2 \vee \alpha_3}$, but a substitution rule: Once ϕ has been used to deduce $(\alpha_2 \vee \alpha_3) \sim \phi_{\alpha_2 \vee \alpha_3}$, the assertion of $\alpha_2 \vee \alpha_3$ allows us to substitute "$\phi_{\alpha_2 \vee \alpha_3}$" for "$\alpha_2 \vee \alpha_3$" wherever it occurs, and hence when it appears as an argument in "$\text{Prob}_\phi(\text{F}, \alpha_2 \vee \alpha_3)$." Such a universal substitution rule surely cannot be correct. Hellman has shown that Friedman and Putnam's equivalence relation can be reduced to the Sasaki hook, an implicational connective which may be defined on the lattice of subspaces of a Hilbert space.[149] This implication relation, however, does not even satisfy the strong transitive law or the law of contraposition,[150] so no equivalence relation based on it could possibly satisfy such a substitution rule. Informally, this is obvious from our example. From $\psi \leq (\alpha_2 \vee \alpha_3)$, where ψ is any ray in the α_2-α_3 plane, we certainly cannot deduce by substitution $\psi \leq \phi_{(\alpha_2 \vee \alpha_3)}$: the first is true, the second is false (unless $\phi = \phi_{(\alpha_2 \vee \alpha_3)}$). Thus, any such substitution rule will have to specify under what conditions we may replace "$\alpha_2 \vee \alpha_3$," by "$\phi_{(\alpha_2 \vee \alpha_3)}$." In other words, it still requires some *independent* assumption to assert that in "$\text{Prob}_\phi(F, \alpha_2 \vee \alpha_3)$" such a substitution is allowed. So no advantage over the Copenhagen approach may be claimed in this respect.

At best, Friedman and Putnam have succeeded in finding a way of giving certain formal prescriptions that are equivalent to postulating the projection postulate. In what sense does this yield any deeper understanding of quantum mechanics? I claim that it gives no more—and feel that it gives considerably less—insight than we get from the Feynman formulation (See app. 3). The latter at least points a finger at the sore point—superposition of probability amplitudes—instead of tranquilizing us with dubious assurances of having dissipated a major mystery. The Feynman approach makes it clear where classical probability arguments apply (distinguishable alternatives, Rules 3 and 5) and where they do not (indistinguishable alternatives, Rules 2 and 4). The distinction is certainly context-dependent: experimental arrangements that differ only in the presence or absence of a registration device involve different probability amplitudes (e.g., the double slit with or without a counter behind one slit). Thus, the Feynman approach may be broadly classed as a variant of the Copenhagen approach.

As demonstrated in app. 3, to find a conditional probability involving a nonmaximal measurement, both the classical and nonclassical rules are needed. Friedman and Putnam's statement that "if E and F do not commute we have no *conditional probability* for F given E in the usual sense—where $\text{Prob}_\phi(F,E) = \text{Prob}_\phi(F \wedge E)/\text{Prob}_\phi(E)$" is not correct. (Or, rather, is only trivially correct because Friedman and Putnam choose to restrict themselves to a partial Boolean algebra approach in which $F \wedge E$ and hence $\text{Prob}_\phi(F \wedge E)$ is not defined. If one thinks in terms of transition probabilities from an initial state, via an intermediate nonmaximal registration result E, to a final (maximal) registration result F, one may write the following table of correspondences:

$$P_{\phi \to E \to F} \Leftrightarrow \text{Prob}_\phi(F \wedge E)$$

$$P_{\phi \to E} \Leftrightarrow \text{Prob}_\phi(E)$$

$$P_{E \to F} \Leftrightarrow \text{Prob}_\phi(F, E).$$

The notation $P_{a \to b}$ stands for the probability that a system in the state a is registered in the state b, and so forth (see app. 3). Defining the probabilities on the right written by Friedman and Putnam according to this table, the equation they consider nonexistent actually holds. In my notation, it is:

$$P_{E \to F} = P_{\phi \to E \to F} / P_{\phi \to E},$$

proved in Appendix 3 (with the use of a,b,c instead of ϕ, E, F). It is only the use of Rules 2 and 4 (superposition of probability amplitudes) to calculate $P_{\phi \to E \to F}$ that introduces the peculiarly quantal aspect of the problem—as well as the introduction of probability amplitudes in the first place, of course. $P_{\phi \to E}$ is then derived from this by the classical Rule 3: $P_{\phi \to E} = \Sigma P_{\phi \to E \to F}$. The final result is the Lüders rule (the projection postulate). See app. 3 for details.

Friedman and Putnam do not discuss the double-slit experiment. Presumably, having explained interference in general, they feel no need to discuss it in particular. Bub, who adopted the partial Boolean algebra approach upon recruitment to the quantum-logical camp, goes into more detail. He states: "In general, then, insofar as problems of interpretation in quantum mechanics have their source in probability relations which are anomalous classically, these problems are resolved by recognizing the projection postulate in its corrected Lüders version as the appropriate conditionalization rule on the non-Boolean possibility structure of a quantum-mechanical system. If the logical interpretation of quantum mechanics can make sense of truth, it can make sense of probability."[151]

A bit earlier he had made a comparison with the classical (Boolean) case: "My point is that the projection postulate in the corrected Lüders version is properly understood as mere *conditionalization on a non-Boolean possibility structure*, since it is the analogue of mere conditionalization on new information in the Boolean case" (p. 389).

His analysis of the two-slit experiment consists of a demonstration that application of the Lüders rule correctly gives the quantum interference pattern if the incident state is coherent and the screen sufficiently far from the two slits: "Thus, the 'paradox' involved in the 2-slit experiment is resolved by showing precisely how the assumption of a non-Boolean possibility structure explains the existence of the 'anomalous' interference effect" (p. 389).

What evidence does he offer "that the Lüders rule . . . is the appropriate rule for conditionalizing probabilities in the non-Boolean possibility structure of quantum mechanics" (p. 383)? After showing how to rewrite the classical rule for conditionalizing probabilities in terms of a classical statistical operator, he remarks: "Bearing in mind the possibility of non-commutative algebras of magnitudes as in quantum mechanics (i.e., non-Boolean possibility structures), it seems appropriate to represent the transition corresponding to conditionalization with respect to an event α_i by the symmetrical expression [equation omitted]. . . . *Now this is just Lüders' version of von Neumann's projection postulate*" (p. 385).

He thus uses a formal analogy with the classical case to guide him in postulating an expression he already knows to be correct because it works in such applications as the double-slit experiment. Why replacement of the addition of probabilities by the addition of probability amplitudes is called "conditionalization on a non-Boolean possibility structure" is not clear. Bub's own discussion of the double-slit experiment (pp. 387–89), for example, shows that the possibility of going through one slit and the possibility of going through the other both belong to the same Boolean subalgebra. He states that "when some physical device is incorporated into the experimental arrangement for detecting the passage of an individual photon through either slit A or slit B . . . [there is] no interference" (p. 390, fn. 5). So the Lüders rule and the classical rule will yield the same result in this case. Why the first situation corresponds to a "non-Boolean possibility structure," whereas the second (presumably) corresponds to a Boolean possibility structure is not explained. Both situations involve the same two subspaces of Hilbert space. It seems that Bub just accepts the probability structure of quantum mechanics, regarding the phrase "non-Boolean possibility structure" as sufficient explanation of how "the logical interpretation of quantum mechanics can make sense . . . of probability." Again, I fail to see how this constitutes an increase in explanatory power over, say, Feynman's procedure of postulating rules for probability amplitudes. Indeed, the two methods are quite equivalent: Feynman works with amplitudes to set up his formal analogy, while Bub works with conditional or transition probabilities, which are just the "squares" of Feynman's amplitudes, to set up his. Recalling Bub's remark quoted earlier, one may wonder: if the logical interpretation of quantum mechanics calls *this* making sense of probabilities, what does it call making sense of truth?

By now the reader can appreciate Simon Kochen's perceptive comments on the explanatory power of the quantum-logical approach:

> Now quite apart from the cogency and self-consistency of this approach, one has only to look at the explanations given by quantum logic to see how unsatisfactory an account of the paradoxes that this affords. One is left with the same feeling of mystery after the explanation that one had originally. The reason for this is not hard to find. The trouble lies in this, that maintaining that the logic of quantum mechanics is given by the lattice of projection operators on Hilbert space is equivalent, via the spectral theorem, to assuming that observables correspond to self-adjoint operators. One could as well "explain" the Uncertainty Principle by pointing to the non-cummutativity of the relevant operators. One is in essence reduced to explaining the paradoxes by simply saying that this is how the formalism works. To base an interpretation of quantum mechanics on this logic is to beg the question.[152]

It is this tranquilizing element in the "logic of quantum mechanics" approach which I find so disturbing. The combination of uncritical idealism (panlogism) and uncritical positivism (things are thus and so in quantum mechanics because that is the way things are) is not unprecedented in the history of philosophy.[153] Acceptance of such a viewpoint by physicists and philosophers of science could divert them from the attempt to reach a deeper understanding of quantum theory.

APPENDIX 1: PANLOGISM

In an earlier paper on quantum logic[154] I stated that those quantum logicians who look to logical principles to provide some sort of explanation or resolution of otherwise paradoxical features of the quantum theory were in danger of committing the "sin that has been called panlogism"[155] Howard Stein, in private comments on that paper (for which I thank him) stated that he did not understand my reference to panlogism and asked me to elaborate. I have done so in Section 1 and will discuss the term itself here. By panlogism I mean the philosophical tendency to obliterate the distinction between logical and ontological principles. The word was coined by Erdmann in the midnineteenth century to describe Hegel's philosophy; but the tendency is much older, and Coutourat used the term around the turn of the century in discussing Leibniz's metaphysics.

A recent book on Leibniz, entitled *Leibniz: Philosophie des Panlogismus,* offers several relevant characterizations.[156] In the Introduction, Gurwitsch gives a definition of panlogism as "the doctrine that in the universe as a whole as well as in all its parts, that is in everything which exists and happens in it, a logic is reflected [*niedergeschlagen*] and realized. The universe is understood through and through as an incarnation of logic.

According to the principle of logical-ontological equivalence, every logical structure may be translated into an ontological one, and inversely every ontological one into a logical" (p. 3). In chapter 1, "Meaning and Presuppositions of Panlogism," in the first section, on "logical-ontological equivalence," he elaborates:

> We define Leibniz's panlogism as the thesis of the thoroughgoing reasonableness [*Vernunftgemässheit*] and logical lawfulness of the universe, whose basis—as Kabitz puts it—is that "logical necessity in the coherence of thought . . . " is transferred "into the coherence of the world of things." *Logic is precipitated and incorporated in the universe; it is inscribed in the structure and total constitution of the latter as inherent in it. The universe is realized and incarnated logic.* (P. 15)

Next, he makes the point that I stressed in Section 1: "*All ontological relations,* of whatever type or level, *bear logical relations within them as incorporated within them, as immanent and determining moments.* Therefore all relations between realities, phenomenal as well as metaphysical, may be expressed in logical form; all have a logical pendant." Naturally, such a doctrine about the relation between logic and ontology implies a doctrine about the nature of reality. Gurwitsch makes this explicit in his discussion of Leibniz's views: "Things are realizations of concepts of reason. It is not sufficient to maintain that the logical and ontological viewpoints can never be fully distinguished from each other, or that no separation, no abyss exists between reason and reality. One seems most faithful to the situation, if one speaks of an identity, or better: of an *equivalence of the logical and the ontological*" (p. 16). Elsewhere, he makes the point in these words: "The conformity between thought and being is based, not only on the circumstance that being is constituted by thought (the thought of God), but—however paradoxical it may sound—being is precisely realised and incarnated thought" (p. 15). I am not claiming everyone who espouses the "logic of quantum mechanics" viewpoint accepts these conclusions, let alone is motivated by them. But, once one has accepted that logical relations must be given a fundamentally ontological status, it is hard to see how one can avoid slipping into some form of panlogism—unless one just avoids altogether the question of the philosophical implications of the position taken.

APPENDIX 2
SOME MATHEMATICAL CONCEPTS USED IN QUANTUM LOGIC

Partially Ordered Set (poset)

A set S of elements $a,b, \ldots x,y, \ldots$ is a partially ordered set if there exists a two-place ordering relation between some of its elements, symbolized by $a \leqslant b$, which is (1) reflexive, (2) transitive, and (3) antisymmetric:

1. $a \leqslant a$,
2. $a \leqslant b$, and $b \leqslant c$ implies $a \leqslant c$.
3. $a \leqslant b$ and $b \leqslant a$ implies $a = b$.

One speaks of *least* or *greatest* elements, satisfying certain conditions, meaning thereby least or greatest with respect to the ordering relation. One also speaks of a as contained in b if $a \leqslant b$.

Lattice

A partially ordered set S is a lattice if:

1. a *meet* or *cap* operation is defined for every pair of elements a,b in S resulting in an element $a \wedge b$ which is the greatest element of the set contained in both a and b:

$a \wedge b \leqslant a,b$; if $x \leqslant a$, then $x \leqslant a \wedge b$.

2. A *join* or *cup* operation is defined for every pair of elements a,b in S resulting in an element $a \vee b$, which is the least element containing both a and b:

$a,b \leqslant a \vee b$; if $a,b \leqslant x$, then $a \vee b \leqslant x$.

3. S contains a least element ϕ and a greatest element 1:

$$\phi \leqslant x,\ x \leqslant 1 \text{ for all } x \in S.$$

A lattice may also be considered as an abstract algebra.

Complete lattice

A lattice for which meet and join elements are defined for *any* subset of elements of the lattice is called *complete.* Every lattice with a finite number of elements is complete; a lattice with an infinite number of elements is complete if meet and join are defined for all infinite subsets of elements.

Complement (not unique)

An element b of a lattice S is *a complement* of an element a of S if:

$$a \vee b = I \text{ and } a \wedge b = \phi.$$

In general, there may be more than one complement of a given element. A lattice is called *complemented* if every element has a complement.

Orthocomplement (unique)

A lattice S is *orthocomplemented* if there is a one-one mapping of S into itself such that, for every x element of the lattice:

1. It has a complement x'
2. $(x')' = x$
3. If $x \leqslant y$ then $y' \leqslant x'$.

Such an orthocomplementation is unique, if it exists.

Atoms

An element a of a lattice is called an *atom* if there are no elements between it and the least element of the lattice:

$$\text{If } \phi \leqslant x \leqslant a, \text{ then } x = \phi \text{ or } x = a.$$

Atomic lattice

A lattice S is called atomic if every element x of the lattice except ϕ contains an atom:

If $x \neq \phi$, then there exists an atom a such that $a \leqslant x$.

Distributive laws

If some three elements a,b,c of a lattice S obey one of the following two equations:

$$a \vee (b \wedge c) = (a \vee b) \wedge (a \vee c)$$
$$a \wedge (b \vee c) = (a \wedge b) \vee (a \wedge c),$$

(which are dual to each other under interchange of $\vee$ and $\wedge$), then we say that, for these elements in that order, joining is distributive over meeting, or vice versa in the second case.

Distributive or Boolean lattice

A complemented lattice for which the distributive laws hold for each triple of elements in any permutation is called a *distributive* or *Boolean* lattice. In this case, complementation is unique and is also orthocomplementation. Conversely, if complementation is unique for an orthocomplemented lattice, it is distributive. Considered as an algebra, a Boolean lattice is called a Boolean algebra.

Modular lattice

An orthocomplemented lattice is *modular* if $a \leqslant b$ for two elements a,b of the lattice implies that the distributive laws hold for a,b,x in any permutation, where x is any element of the lattice. Since, under these assumptions the distributive laws are automatically satisfied except for

$$a \vee (x \wedge b) = (a \vee x) \wedge b,$$

the latter condition is often used to define a modular lattice.

Orthomodular lattice

An orthocomplemented lattice is *orthomodular* if $a \leqslant b$ for two elements a,b of the lattice implies that the distributive laws hold for a,b,a' in any permutation. Since the distributive laws are automatically satisfied under these assumptions, except for

$$a \vee (b \wedge a') = b,$$

the latter condition is often used to define an orthomodular lattice.

Sublattice

a subset of elements of a lattice which themselves form a lattice is called a *sublattice*.

Compatible elements

Two elements of an orthocomplemented lattice are *compatible* if, together with their orthocomplements, they generate a Boolean sublattice under iterated application of the join and meet operations on the four elements.

Representation of Boolean Lattices

Every Boolean lattice can be represented by a Boolean lattice of subsets of some fundamental set I. The order relation is represented by set-theoretical inclusion. The meet $\wedge$ is represented by set-theoretical intersection $\cap$; the join $\vee$ is represented by set-theoretical union $\cup$; and the complement is represented by set-theoretical complementation relative to the fundamental set. ϕ is represented by the null set 0, and 1 is represented by the fundamental I: Let $a \leftrightarrow A$ $b \leftrightarrow B$, where a,b are elements of the abstract lattice; A,B subsets of some set I. Then

$a \leq b \leftrightarrow a \subseteq B$	$\phi \leftrightarrow O$	$1 \leftrightarrow I$
$a \wedge b \leftrightarrow A \cap B$	$a \vee b \leftrightarrow A \cup B$	$a' \leftrightarrow I - A$

Irreducible Lattice

An orthocomplemented lattice is called *irreducible* if ϕ and 1 are the only elements of the lattice compatible with every element of the lattice.

Linear Vector Space

A set of elements V, called vectors, and a field of elements F, called scalars, form a *linear vector space* if:

1. for any two vectors a,b in V there is defined another element c in V, called the *vector sum* of a and b: $c = a + b$. This sum is associative and commutative: $(a + b) + d = a + (b + d)$; $a + b = b + a$.

2. There is an element 0 in V such that, for all a in V, $a + 0 = a$. This element is called the *zero vector*.

3. For any a in V, there exists an element $-a$ in V such that $a + (-a) = 0$. This element $-a$ is called the *inverse* of a.

4. For any vector a in V and scalar α in F, there is defined a unique vector $\alpha\, a$, called the *scalar multiple* of a by α, such that scalar multiplication is distributive over vector addition: for any α in F and any a,b in V

$$\alpha\,(a + b) = \alpha a + \alpha b;$$

scalar multiplication is associative: for any α, β in F and a in V

$$(\alpha\beta)\, a = \alpha(\beta\, a);$$

addition of scalars is distributive over scalar multiplication of a vector: form any α, β in F and any a in V,

$$(\alpha + \beta)\, a = \alpha a + \beta a.$$

5. For the unit element of the field 1, and any vector a in V

$$1\, a = a.$$

The field of scalars is usually the real or complex numbers in physical applications, although quaternions have also been considered. For real or complex numbers, the scalar field is also *commutative:* for any α,β in F $\alpha\, \beta = \beta\, \alpha$.

Inner Product Space

A linear vector space over the complex numbers is called an *inner product space* if to any pair elements a,b in V there corresponds a complex number α in F, symbolized by $\alpha = (a,b)$ and called the *inner product* of the two vectors, such that:

1. For any a in V, (a,a) is real and $\geqslant 0$; $(a,a) = 0$ implies $a = 0$.
2. (a,b) is the complex conjugate of (b,a): $(a,b) = (\overline{b,a})$.
3. The inner product is linear in the second vector: for any a,b,c in V

$$(a,b + c) = (a,c) + (b,c).$$

Together with (2) this implies that the inner product is also linear in the first vector

$$(a + b,\, c) = (a,c) + (b,c).$$

4. For any α in F and any a,b in V: $(a,\alpha b) = \alpha(a,b)$.

Together with (2) this implies that

$$(\alpha a,b) = \bar{\alpha}(a,b).$$

Orthogonality

Two vectors a and b of an inner product space are called *orthogonal* if their inner product vanishes: $(a,b) = 0$.

Orthonormal Set

A set of vectors of an inner product space is called *orthonormal* if any two vectors of the set a_i,a_j are orthogonal if different and have inner product one if they are the same:

$$(a_i,a_j) = \delta_{ij} = \begin{matrix} 1, \text{ if } i = j \\ 0, \text{ if } i \neq j \end{matrix}\,.$$

Completeness

An inner product space V is called *complete* if, for any sequence for vectors a_i ($i = 1,2, \ldots .$) which converges—that is, such that for any number $\delta > 0$ there is an integer N for which $(a_i, - a_j, a_i - a_j) < \delta$ for all $i,j > N$—there is a vector a in V which is the limit of that sequence, that is, for any number δ there is an integer N for which $(a_i - a, a_i - a) < \delta$ for all $i > N$.

Hilbert Space

A complete inner product space is called a *Hilbert space* over the complex field.

Orthonormal Basis

If S is an orthonormal set, and no other orthonormal set contains S as a proper subset (i.e., S is maximal), then it is called an *orthonormal basis* or *complete orthonormal* set. Every Hilbert space has an orthonormal basis.

Dense Subsets

A subset of vectors in an inner product space is called *dense* if every vector of the space can be expressed as the limit of a sequence of vectors chosen from that subset.

Separable Hilbert Space

A Hilbert space with a countable dense subset is called *separable.* A separable Hilbert space has a countable orthonormal basis.

Linear Operator

Any function f which maps a vector a of a linear vector space V into another vector $f(a)$ of that space is called a *linear operator* if, for any vectors a,b in V and any scalars α,β in F:

$$f(\alpha a + \beta b) = \alpha f(a) + \beta f(b).$$

Bounded Operator

A linear operator f on a Hilbert space is *bounded* if there is a number $C > 0$, such that for every vector a in the Hilbert space

$$(f(a), f(a)) < C\ (a,a).$$

Self-Adjoint Operator

A bounded linear operator f on a Hilbert space is *self-adjoint* if for any two vectors a,b in the Hilbert space $(f(a),b) = (a,f(b))$.

Projection Operator

A bounded self-adjoint operator P on a Hilbert space is a *projection operator* if it is idempotent: $P^2 = P$.

Closed Linear Subspace

A linear subspace of a Hilbert space is *closed* if it is itself a separable Hilbert space. This means that the subspace contains the limit vector of any convergent sequence of vectors within it. There is a one-one correspondence between the projection operators and the closed linear subspaces of a Hilbert space: any vector in the subspace is mapped into itself by the corresponding projection operator, and any vector orthogonal to the subspace is mapped into the zero vector.

Representation of a Complete, Orthocomplemented, Atomic, Modular Lattice

Every such lattice which is irreducible can be represented by the lattice of all linear subspaces of a finite-dimensional Hilbert space. The order relation is represented by set-theoretical inclusion. The meet relation is represented by set-theoretical intersection of subspaces. The join relation is represented by the linear subspace spanning the subspaces representing the elements being joined. The orthocomplement a' is represented by the linear subspace orthogonal to the subspace which represents a. ϕ is represented by the zero-dimensional linear subspace; *1* is represented by the entire Hilbert space. If the lattice is not irreducible it can be represented by a direct sum of such Hilbert spaces.

Let $a \leftrightarrow \alpha$, $b \leftrightarrow \beta$, where a,b are elements of the abstract lattice and α,β linear subspaces of some finite-dimensional Hilbert space H. Then

$a \leqslant b \leftrightarrow \alpha \subseteq \beta$ $\phi \leftrightarrow \{o\}$ $1 \leftrightarrow H$

$a \wedge b \leftrightarrow \alpha \cap \beta$, $a \vee b \leftrightarrow \gamma$, where γ is the linear subspace spanning α and β

$a' \leftrightarrow \alpha^{\perp}$, where $\alpha^{\perp}$ is the lattice subspace orthogonal to α.

Representation of a Complete, Orthocomplemented, Atomic, Orthomodular Lattice

Every such lattice which is irreducible can be represented by the lattice of all closed linear subspaces of a (possible infinite-dimensional) Hilbert space. The representation is as in the previous definition, except that the word "closed" must be added when the operation does not automatically lead to a closed subspace. For example, the join of two elements must be put into correspondence with the *closed* linear subspace spanning the subspaces representing the two elements to be joined.

In both the previous definitions, the Hilbert space inner product field is restricted to be isomorphic to the real numbers, the complex numbers, or the quaternions.

Partial Boolean Algebra (pBa)

A family of Boolean algebras B_i are said to form a *partial Boolean algebra* (*pBa*) if the following two conditions hold:

1. If B_i and B_j are members of the family, then their set-theoretical

intersection forms a third member of the family: $B_i \cap B_j = B_k$. This implies that there are unique ϕ and *1*, elements which belong to each member of the family.

2. If *a,b,c* are three elements of the Boolean algebras B_i such that any two of them belong to some Boolean algebra of the family, then there is one Boolean algebra of the family to which all three belong.

Complement

The *complement* of any element of *pBa* is defined as its (unique) complement with respect to any Boolean algebra of the family to which the element belongs. It then follows from the definition of a *pBa* that the complement of an element is unique and belongs to every Boolean algebra to which the element does.

Compatibility

Two elements of a partial Boolean algebra are said to be *compatible* if there is a Boolean algebra of the family to which both belong. It follows that, if two elements are compatible, they and their complements all belong to a common Boolean algebra.

Meet and Join

The *meet* (cap) and *join* (cup) of two elements of a partial Boolean algebra are defined only if the two elements are compatible. It follows that they are unique, that is, independent of the particular Boolean algebra chosen to define the meet and join.

Any partial Boolean algebra becomes an orthocomplemented lattice with the following definitions:

Define the *partial ordering* of the lattice as follows: for two elements *a,b* of the *pBa*, $a \leqslant b$ if there is a Boolean algebra of the family to which both belong, and in which $a \leqslant b$.

Define orthocomplementation for the lattice as follows: the orthocomplement of any element of the *pBa* is the (unique) complement with respect to the *pBa*.

Define the *join* (*meet*) of two elements of the lattice as follows:

1. If the elements are compatible, the lattice joint (meet) is the same as the *pBa* join (meet);

2. If the elements are not compatible, the lattice join is the ϕ element, the lattice meet is the *1* element.

Convex Set

A set C of vectors of a linear vector space over the real number field F is called *convex* if, for any two vectors *a,b* in C and any number α in F such that $0 \leqslant \alpha \leqslant 1$, the vector $\alpha a + (1 - \alpha)\, b$ also belongs to C. Geometrically, if a and b are points, the line segment joining them also belongs to the convex set.

Extremal Elements

An element of a convex set C is called *extremal,* if it cannot be written as a linear combination of any two elements of C unless one of the coefficients is zero. Geometrically, an extremal element is a point on the boundary of the convex set which cannot be expressed as a linear combination of two other boundary points

Wall

A convex subset W of a convex set C is called a *wall* when for any two elements a,b in C and any number α such that $0 \leqslant \alpha \leqslant 1$, if $\alpha a + (1 - \alpha)\, b$ is in W then a,b are in W. Geometrically, a wall is a maximal plane portion (of any dimension) of the boundary of a convex set. Extremal elements are zero-dimensional walls.

Lattice of Walls

The set of walls of a convex set is partially ordered by the relation of set-theoretical inclusion. If we include the empty set as ϕ element, and the entire convex set as *1* element, then this set of walls has a *lattice* structure: the cap of two walls is the set-theoretical intersection of the two walls (which must itself be a wall), and the cup of the two walls is the smallest wall containing them both.

APPENDIX 3: PROOF THAT THE LÜDERS RULE FOLLOWS FROM THE FEYNMAN RULES FOR PROBABILITY AMPLITUDES

The following discussion is simplified by use of the Dirac bra-ket notation, in which bra(c)kets represent probability amplitudes. A state is represented by some symbol such as s placed within either a bra symbol $|s\rangle$ or a (dual) ket symbol $\langle t|$. When a bra and a ket are placed together in that order they form a bracket $\langle t|s\rangle$ which represents a probability amplitude, a complex number of modulus less than or equal to one. When they are placed together in the opposite order, they represent an operator $|s\rangle\langle t|$. Such an operator may act on a bra or a ket to produce another bra or ket. For example, $|s\rangle\langle t|$ acting on the bra $|u\rangle$ produces $|s\rangle\langle t|u\rangle$: the bra $|s\rangle$ multiplied by the amplitude $\langle t|u\rangle$.

Now let us apply the Feynman rules to the following situation. After initial preparation of an ensemble in the state a, an incomplete or nonmaximal registration is performed, followed by a maximal registration.

What is the (conditional) probability, given b as the outcome of the first (nonmaximal) registration, that the result of the ensuing (maximal) registration will be c? Suppose that the nonmaximal registration result, if extended to a maximal one, would give one of the alternative results b_i, $i = 1,2, \ldots m$ ($m \leqslant n$, the dimension of Hilbert space representing the ensemble, and $m > 1$ since the registration is nonmaximal.[157] Then the distinct process leading from a to c via b is composed of m indistinguishable alternative subprocesses $a \rightarrow b_i \rightarrow c$. Each such subprocess consists of two steps $a \rightarrow b_i$ and $b_i \rightarrow c$. The amplitudes for $a \rightarrow b_i \rightarrow c$, which we write $\langle c|a\rangle_{b_i}$, is the product of the amplitudes for $a \rightarrow b_i$ and $b_i \rightarrow c$ (Feynman Rule 4). Writing these amplitudes as $\langle b_i|a\rangle$ and $\langle c|b_i\rangle$, respectively,

$$\langle c|a\rangle_{b_i} = \langle c|b_i\rangle\langle b_i|a\rangle.$$

By Rule 2 the amplitude for the entire process $a \rightarrow b \rightarrow c$, which we symbolize by $\langle c|a\rangle_b$, is given by

$$\langle c|a\rangle_b = \sum_{i=1}^{m} \langle c|a\rangle_{b_i} = \sum_{i=1}^{m} \langle c|b_i\rangle\langle b_i|a\rangle.$$

This expression may be considerably simplified by noting that $|b_i\rangle\langle b_i|$ is the projection operator P_{b_i}, which projects onto the ray in Hilbert space representing the maximal registration result b_i. Thus:

$$\sum_{i=1}^{m} \langle c|b_i\rangle\langle b_i|a\rangle = \sum_{i=1}^{m} \langle c|P_{b_i}|a\rangle = \langle c|\sum_{i=1}^{m} P_{b_i}|a\rangle,$$

where the last step follows from linearity of the operations involved. But

$$\sum_{i=1}^{m} P_{b_i} = P_b,$$

the projection operator onto the m-dimensional subspace of Hilbert space representing the nonmaximal b. So finally:

$$\langle c|a\rangle_b = \langle c|P_b|a\rangle.$$

The total probability for this process is then given by the modulus squared of this amplitude. To form the complex conjugate of the bracket in the Dirac notation, it is just rewritten in reverse order. So the propability $P_{a \rightarrow b \rightarrow c}$ for the entire process is given by[158]

$$P_{a \rightarrow b \rightarrow c} = \langle a|P_b|c\rangle\langle c|P_b|a\rangle.$$

Since $|c\rangle\langle c|$ is just P_c, the projection operator onto the ray representing c, we may write:

$$P_{a \rightarrow b \rightarrow c} = \langle a|P_b\, P_c\, P_b|a\rangle.$$

This still does not answer our original question: what is the (conditional) probability of registering c, given that the result of the first nonmaximal registration was b? To find this, we must apply Rule 5, since the steps from $a \rightarrow b$ and $b \rightarrow c$ are distinguishable. Accordingly,

$$P_{a \rightarrow b \rightarrow c} = P_{a \rightarrow b} P_{b \rightarrow c},$$

so

$$P_{b \rightarrow c} = P_{a \rightarrow b \rightarrow c} / P_{a \rightarrow b}.$$

Summing $P_{a \rightarrow b \rightarrow c}$ over *all* possible outcomes c gives $P_{a \rightarrow b}$ for the given situation (starting from a). Since all possible outcomes span the entire Hilbert space:

$$\sum_{\text{all } c} P_c = I,$$

where I is the identity operator on the Hilbert space. Thus,

$$\begin{aligned} P_{a \rightarrow b} &= \sum_{\text{all } c} P_{a \rightarrow b \rightarrow c} \\ &= \sum_{\text{all } c} \langle a|P_b\, P_c\, P_b|a\rangle \\ &= \langle a|P_b\, (\sum_{\text{all } c} P_c) P_b|a\rangle \\ &= \langle a|P_b\, P_b|a\rangle \\ &= \langle a|P_b|a\rangle, \end{aligned}$$

since projection operators are idempotent (i.e., $P^2 = P$). So finally:

$$P_{b \rightarrow c} = \langle a|P_b P_c P_b|a\rangle / \langle a|P_b|a\rangle,$$

which is the form of Lüders's, rule used in the papers of Bub, Friedman and Putnam, and others.

This result is equivalent to application of the projection postulate for nonmaximal registrations: $P_b|a\rangle$ is the unnormalized projection of the initial state a onto the subspace representing b and $\langle a|P_b|a\rangle$ is just the square of the normalization factor for $P_b|a\rangle$. So, if $|a\rangle_b$ is the normalized projection of $|a\rangle$ onto the b-subspace,

$$P_{b \rightarrow c} = {}_b\langle a|P_c|a\rangle_b.$$

In many cases a number of alternative maximal measurements could be performed as extensions of the nonmaximal measurement b. It does not matter which of these is used in computing the amplitudes $\langle c|a\rangle_b$, since the final result clearly depends only on b. In other cases, such as the triple-slit experiment mentioned in the text, it is hard to imagine more than one physically realizable alternative maximal extension.

NOTES

Research for this essay was supported in part by the National Science Foundation.

1. L. D. Landau, "The Theory of Superfluidity of Helium II," *J. Phys.* (USSR) 5 (1941): 71. For a recent review, see M. I. Kaganov and I. M. Lifshits, *Quasi-Particles* (Moscow: Mir Publishers, 1979).

2. V. L. Ginzburg, "The Quantum Theory of Radiation of an Electron Uniformly Moving in a Medium," *J. Phys.* (USSR) 2 (1940): 441. For a review see I. Brevik and B. Lautrup, "Quantum Electrodynamics in Material Media," *Mat.-fys. Medd. Dan. Vid. Selsk.* 38, no. 1 (1970).

3. A recent textbook taking this approach to logic is even entitled *Logische Sprachregeln* (*Logical Rules of Language*) by A. Sinowjew (Zinov'ev) and H. Wessel (Berlin; VEB Deutsche Verlag der Wissenschaften, 1975). They state in the Foreword:

> The title of the book, *Logical Rules of Language,* was not chosen haphazardly. We treat logical rules, that is, as rules for operating with linguistic forms (constructions, expressions) of a certain type. The definition of logic handed down from traditional logic, and still often encountered as the science of the laws of correct thought, is only justified insofar as correct thought is understood as the carrying out of definite operations with linguistic expressions in accord with logical rules. However, in this case the definition is practically meaningless because of its nature. If, on the contrary, one conceives of thought as the process taking place in the human brain, then the rules of logic have only the following relation to it: They are rules which a person may apply in using language, but in no way do they guide the processes taking place in a human brain. Even less are logical rules general laws of being. Many difficulties and misunderstandings in connection with the rules of logic arise only because an attempt is made to foist upon them a role as laws of the human brain or as general laws of the world, a role which does not at all befit them. (P. 15)

4. David Finkelstein remarks casually, "By a quantum I mean an object whose propositional calculus is isomorphic to the lattice of subspaces of a Hilbert space." He goes on to explain that he now accepts the suggestion of Feynman "that a reasonable model for the world is a computer, a giant digital computer." Then he notes: "The idea that the world is some kind of computer is not far removed, incidentally, from the idea that the world is some kind of brain; which is not far removed from the idea that the world is the mind of God; which stems back to the hermetic doctrines of the second century B.C. (I am indebted to Professor Jauch for this reference), so we are working in an ancient and honorable tradition." See, D. Finkelstein, "Space-Time Structure in High Energy Interactions," pp. 324–43 in T. Gudehus, G. Kaiser, and A. Perlmutter, eds., *Coral Gables Conference on Fundamental Interactions at High Energies* (New York: Gordon and Breach, 1969). None of my comments refer to the content of Finkelstein's scientific program, but only to the philosophical views he expresses.

5. David Finkelstein, "The Physics of Logic," in R. S. Cohen and M. W. Wartofsky, eds., *Boston Studies in the Philosophy of Science 5* (Dordrecht: Reidel, 1969).

6. Hilary Putnam, "How to Think Quantum-Logically," *Synthese* 29 (1974): 55–61. The quotation is from p. 56, and a misprint has been corrected.

7. Jeffrey Bub and William Demopoulos, "The Interpretation of Quantum Mechanics," in Cohen and Wartofsky, *Boston Studies,* vol. 13, pp. 92–122; The quotation is from Jeffrey Bub, *The Interpretation of Quantum Mechanics* (Dordrecht: Reidel, 1974), p. viii. Here is the most explicit discussion of the meaning he gives to the word "logic" and the significance of quantum logic that I could find in Bub's book:

> In the previous chapters, I have represented classical and quantum mechanics as principle theories of logical structure, since they introduce constraints on the way in

> which the properties of a physical system are structured. The logical structure of a physical system is understood as imposing the most general kind of constraint on the occurrence and non-occurrence of events. I have argued that the transition from classical to quantum mechanics is to be understood as a generalization of the Boolean propositional structures of classical mechanics to a particular class of non-Boolean structures. There are two aspects to this thesis: firstly, the significance of a realist interpretation of logical structure analogous to Einstein's realist interpretation of geometric structure involved in the transition from classical to relativistic physics; secondly, the resolution of problems of interpretation by relating the peculiar statistical relations of quantum mechanics to the specific character of the underlying logical structure. Insofar as problems of interpretation remain, they are either problems about logic, or problems about the category of algebraic structures in relation to the Boolean structures of classical mechanics.

8. Karl Popper, in "Why Are the Calculi of Logic and Arithmetic Applicable to Reality," in *Conjectures and Refutations* (New York: Basic Books, 1962), pp. 207–08, has noted: "any theory which does not allow for the radical difference between the status of a physical truism (such as 'All rocks are heavy') and a logical truism (such as 'All rocks are rocks' or perhaps 'Either all rocks are heavy or some rocks are not heavy') must be unsatisfactory. We feel that such a logically true proposition is not true because it describes the behavior of all possible facts but simply because it does not take the risk of being falsified by any fact; it does not exclude any possible fact and it therefore does not assert anything whatsoever of any fact at all."

9. See app. 1, on panlogism.

10. Although the word "panlogism" is post-Kantian, one of the most interesting critiques of Leibniz's panlogism is due to Kant (see, especially, the "Note to the Amphiboly of Concepts of Reflection" in the *Critique of Pure Reason*). Commenting on this criticism, Koŕner notes that "it is perhaps unlikely that the thesis that judgments about perceived particulars 'really' assert relations between concepts will ever lose its attraction, which seems to be based on some obscure desire that reality should be fully rational" (S. Körner, *Kant* [Harmondsworth, England: Penguin Books, 1955]).

11. Putnam, "How to Think Quantum-Logically,"

12. This analogy is developed in Putnam, "How to Think Quantum-Logically"; see my earlier comments on this "proportion" in J. Stachel, "The Logic of Quantum Logic," in R. S. Cohen et al., eds., *PSA 1974* (Dordrecht: Reidel, 1976), pp. 515–26.

13. For an interesting discussion, see F. Waisman, "Are There Alternative Logics," in F. Waismann, *How I See Philosophy* (London: MacMillan, 1968). The article was originally published in 1945–46.

14. Waismann, *How I See Philosophy,* p. 70.

15. Arthur Fine, "Some Conceptual Problems of Quantum Theory," in R. G. Colodny, eds., *Paradigms and Paradoxes* (Pittsburgh: University of Pittsburgh Press, 1972).

16. Saul Kripke has expressed related views in a widely known but still unpublished talk on "The Question of Logic." Excerpts from it appeared in the 1978 Ph.D. thesis of Alan Stairs: "Quantum Mechanics, Logic and Reality" (University of Western Ontario). I quote Kripke from chap. 1:

> Why not say this? There aren't different logics, there is only logic. There are different formal systems. We use logic to reason about them. If a new formal system has an interesting informal interpretation, then it may embody sound principles of logic. But we can't adopt it. We can only try and prove that a given formal system embodies either (a) sound principles or (b) perhaps more ambitiously, complete principles for some domain of connectives. These may be new connectives which we hadn't thought about before and new laws which apply to them, and this may be called introducing a new logic but only in a loose sense. It is really that we have found out what holds in logic. We may also even, in a rare day . . . , discover that

something we thought for centuries as a sound principle of logic was actually based on, a fallacy. This is not because we are adopting a new logic but because we look at the old formal system and see that it wasn't really sound with respect to its informal interpretation and that the proof we thought we had that it was sound was fallacious. This can happen . . . just as our reasoning can sometimes embody fallacies, so our proof that a given formal system is sound with respect to a given interpretation can contain a fallacy, and occasionally perhaps it does. So all of these things can happen. But if instead the game is supposed to be: Logic is going to be up for grabs, we will choose which logic we are going to adopt, maybe this can mean something, but I don't think a sense has been given to it yet, and in spite of the common use of this kind of phrase in the literature, I don't know what it means. And I would await a further literature to see what it might mean.

17. For criticisms of these theses, see e.g., Max Black, "Carnap on Semantics and Logic," in his *Problems of Analysis* (Ithaca: Cornell University Press, 1954), pp. 255–90; Willard van Orman Quine, "Two Dogmas of Empiricism," in his *From a Logical Point of View* (Cambridge, Mass.: Harvard University Press, 1961), pp. 20–46.

18. K. Marx, "Critique of Hegelian Dialectic and Philosophy in General," quoted from L. D. Easton and K. H. Guddat, *Writings of the Young Marx on Philosophy and Society* (Garden City: Anchor Books, 1967), p. 320.

19. See the anthology of papers tracing the early development of quantum logic: C. A. Hooker, ed., *The Logico-Algebraic Approach to Quantum Mechanics 1: Historical Evolution* (Dordrecht: Reidel, 1975). This includes both the Birkhoff and von Neumann paper, "The Logic of Quantum Mechanics" (1936), and an early Strauss paper, "Mathematics as Logical Syntax—A Method to Formalize the Language of a Physical Theory" (1937).

See also the papers on quantum logic included in Martin Strauss, *Modern Physics and its Philosophy*, (Dordrecht: Reidel, 1972), and his "Logics for Quantum Mechanics," *Foundations of Physics* 3 (1973): 265. There might be more clarity in the discussion of quantum logic if the papers of Strauss were more widely known.

For a historical survey of early work on quantum logic, see "Quantum Logic," in Max Jammer, *The Philosophy of Quantum Mechanics* (New York: John Wiley & Sons, 1974), pp. 304–416.

My use of the phrase "the logic *of* quantum mechanics" as a shorthand characterization of the Finkelstein, Putnam, Bub-Demopoulos outlook should not be taken to imply that Birkhoff and/or von Neumann accepted this interpretation. Historically, the trend arose from the rediscovery of the Birkhoff–von Neumann paper by Finkelstein in the 1960s. But, of course, so did other trends within quantum logic, such as the Jauch-Piron approach, which might claim a closer affiliation with von Neumann's original aims.

20. M. Gardner, "Is Quantum Logic Really a Logic," *Philosophy of Science* 38 (1971): 508 was kind enough to cite my oral comment on this question: "John Stachel of Boston University remarked that quantum theory is full of misnomers: 'observables' which are not observable; 'the uncertainty principle' which gives quite certain information about dispersions; 'wave functions' which have nothing to do with waves; and now 'quantum logic' which is not a logic but an algebraic structure bearing certain formal similarities to the standard propositional calculus."

21. I claim no expertise in color theory, and even simplify somewhat the little I know, so I beg the indulgence of experts in the field. I am also not an expert in lattice theory and cut some corners in my exposition. Similar indulgence is craved from the mathematically sophisticated.

Color mixing may be either additive or subtractive, but I shall only discuss additive color mixing. For an anthology of original papers on color theory, including contributions by Newton, Grassmann, Maxwell, Helmholtz, and Schrödinger, see David L. MacAdam, ed., *Sources of Color Science* (Cambridge, Mass.: MIT Press, 1970). Especially valuable for the treatment of the mathematical structure of the theory are the articles of Schrödinger in the

Annalen der Physik 63 (1920): 397–456, 481–520, partially translated in MacAdam; and Schrödinger's article "Die Gesichtsempfindungen," in O. Lummer, ed., *Müller-Pouillets Lehrbuch der Physik II. Auflage Zweiter Band: Lehre von der Strahlenden Energie (Optik)* (Braunschweig: Vieweg, 1926), pp. 465–560. A recent work in this tradition is Joseph Weinberg's "The Geometry of Color," in *GRG Journal-General Relativity and Gravitation* 7 (1976): 135–69.

For a detailed treatment of orthomodular lattices, with full references to earlier literature and some discussion of their application to quantum logic, see Samuel S. Holland, "The Current Interest in Orthomodular Lattices," in J. C. Abbott, ed., *Trends in Lattice Theory* (New York: Van Nostrand Reinhold, 1970), pp. 41–126. A brief outline of some useful definitions and theorems will be found in app. 2 to this article. The standard treatise on mathematical aspects of quantum logic is V. Varadärajan, *Geometry of Quantum Theory* (Princeton: Van Nostrand 1968–70).

22. The reader familiar with lattice theory will realize that I have avoided discussion of the difference between (nonunique) complements and the (unique) orthocomplement. See app. 2.

23. Complex spaces, rather than real ones, are used in quantum mechanics. However, the difference is irrelevant for the present purposes.

24. Hooker made the point well in a discussion of quantum logic in his Pittsburgh talk:

> The quantum-logical rules are modeled on the formal relationships among the subspaces of a Hilbert space. Thus, one, in effect, rules out any awkward quantum-mechanical statements in advance as contravening the laws of (quantum) logic. To put the proposal in an uncharitable way, one legislates away all difficulty from the theory [or rather thinks one has, I would say, J. S.], and then proclaims that, in discovering that the laws of logic are different from what was thought heretofore, we have also discovered that quantum theory gives a perfectly sound and clearly complete (*logically* clear and complete!) account of physical reality. . . . The chief criticism of the quantum logic approach is not, of course, that its practitioners have succumbed to this crude seduction–one would not for a moment imagine so. . . . No, the criticism is that quantum logic, pursued as *normative* rather than simply as an important algebraic structure, diverts our attention from the *physical situations with which we are confronted* in the micro-domain and the *present inability of our conceptual structures to comprehend them adequately*. By seeking to rectify our difficulties merely by legislating them away in the logical rules, one undermines that sensitivity to physical situations vis-à-vis conceptual apparatus that physics so badly needs.

This quotation is from C. A. Hooker, "The Nature of Quantum Mechanical Reality: Einstein Versus Bohr," in Colodny, *Paradigms and Paradoxes,* pp. 180–81.

25. It may be thought that, in my zeal to wisecrack, I am setting up a straw man. Let me cite the paper by Bub and Demopoulos, "Interpretation of Quantum Mechanics." On p. 99, they state: "Finally, it is necessary to consider the objection that the concept of logical structure introduced here involves an unjustifiable extension of 'logic.' Insofar as this is not a completely verbal issue, it overlooks several considerations." I quote their second consideration *in full:* "The phase space structures with which we are concerned are Boolean algebras or generalizations of Boolean algebras. From a mathematical point of view, classical propositional logic is essentially a Boolean algebra when equivalent sentences are 'identified.' There is also the well-known equivalence of the representation theory of Boolean algebras with the metatheory of classical logic."

26. W. Stegmüller, *Hauptströmungen der Gegenwartsphilosophie, Band II* (Stuttgart: Alfred Kröner, 1960), pp. 208–09 (my translation).

27. Ibid. p. 209 (my translation).

28. In actual fact we should use equivalence classes of such statements since a number of

beams of light of differing spectral composition gives rise to the same color sensation. This of course makes the analogy with the filter interpretation of quantum logic even closer.

29. H. Putnam, "Is Logic Empirical?" in Cohen and Wartofsky, *Boston Studies,* vol. 5, reprinted in Hooker, *Logico-Algebraic Approach to Quantum Mechanics,* vol. 2. References are to the reprinted version.

30. Putnam, "Is Logic Empirical?" p. 198. He adds, however, "My own point of view . . . is that we simply do not possess a notion of 'change of meaning' refined enough to handle this issue."

31. See W. Kneale and M. Kneale, *The Development of Logic* (Oxford: Clarendon Press, 1962), chap. 3, sec. 3. "The Debate on the Nature of Conditionals," for early discussions of this question. Hans Reichenbach discusses some aspects of this problem in *Nomological Statements and Admissible Operations* (Amsterdam: North-Holland, 1954), as does E. W. Adams, *The Logic of Conditionals* (Dordrecht: Reidel, 1975). D. Goldstick, "The Truth-conditions of Counterfactual Conditional Sentences," in *Mind* 87 (1978): 1, is a recent article on the subject with the additional references. For a discussion in the context of an application to quantum logic, see Patrick Heelan, "Quantum and Classical Logic, Their Respective Roles," in Cohen and Wartofsky, *Boston Studies,* vol. 13.

32. Stegmüller, *Hauptströmugen.* Blau's work is discussed on pp. 187–90.

33. Strauss, in his "Foundations of Quantum Mechanics" (chap. 19 of *Modern Physics*), discusses "reactive properties"; and in his "Intertheory Relations (III)" (chap. 22) uses the phrase "stochastic modes of reaction." Paul Feyerabend, in his "Problems of Microphysics" (in R. G. Colodny, ed., *Frontiers of Science and Philosophy* (Pittsburgh: University of Pittsburgh Press, 1962) uses the term "relational properties." Carnap, in "Testability and Meaning," *Philosophy of Science* 3 (1936): 419 speaks of "disposition-terms." "Dispositional properties" is another related term often used in the philosophical literature. See, e.g., A. D. Smith, "Dispositional Properties," *Mind* 86 (1977): 439, which contains many references to earlier work. Smith advocates the use of a nonmaterial conditional to explicate the meaning of a dispositional property, summing up his discussion of fragility in the formula: "x is fragile at t (in environment E) = x is such (and E is such) that it follows by law that if x is (suitably) knocked at t, then x breaks at t as a direct result of being so knocked." David Hockney uses the term "dispositional properties" in his critique of the Bub-Demopoulos viewpoint: "The Significance of a Hidden Variable Proof and the Logical Interpretation of Quantum Mechanics," *Int'l J. Theor. Phys.* 17 (1978): 1253. Howard Stein uses the term "eventualities" for such predicates in the quantum-mechanical context. See his important discussion in "On the Conceptual Structure of Quantum Mechanics," in Colodny, *Paradigms and Paradoxes.* David Bohm, perhaps following Heisenberg, speaks of "Quantum Properties of Matter as Potentialities," in his *Quantum Theory* (Englewood Cliffs: Prentice-Hall, 1951), pp. 132–33 and passim. See also W. Heisenberg, "Planck's Discovery and the Philosophical Problems of Atomic Theory," in his *Across Frontiers* (New York: Harper and Row, 1974), pp. 8–29, esp. pp. 16–17. I prefer to use the more general term "conditional property," also mentioned by Stein. Closer investigation might well show that several types of conditional properties should be distinguished. For example, it might prove useful to distinguish between dispositional and relational properties.

34. Hooker has used the same example in "The Nature of Quantum Mechanical Reality." A small terminological point: We usually talk about qualities such as hardness, viscosity, etc., as properties of the system; but I shall be more specific and refer to a particular numerical value of the hardness or viscosity as a simple property of the system; so that two different values for the hardness would correspond to two *different* simple properties of the system. A quality such as hardness is, thus, in my terminology, a *compound* property corresponding to the disjunction of all the simple properties corresponding to possible values of the hardness. (Since I confine myself to idealized examples, in which only a finite number of such values are possible, it will not be necessary to consider existential quantification.)

The symbols h, h_1, h_2, refer to particular numerical values of the hardness, and similarly, for the other qualities discussed.

35. Strauss, "Foundations of Quantum Mechanics."

36. Carnap, "Testability and Meaning."

37. Let a be "The material has hardness h"; b be "The material is in the solid state"; c be "The material scratched m_1 and not m_2." Then, $(a \cdot b) \equiv c$ reads "The material has hardness h and the material is in the solid state if and only if the material scratches m_1 and not m_2" while $a \equiv (b \rightarrow c)$ reads "The material has hardness h if and only if: if the material is in the solid state, then it scratches m_1 and not m_2." The corresponding truth tables are:

a	b	c	$b \rightarrow c$	$a \equiv (b \rightarrow c)$	$a \cdot b$	$(a \cdot b) \equiv c$	$(a \equiv b \rightarrow c) \equiv ((a \cdot b) \equiv c)$
T	T	T	T	T	T	T	T
T	T	F	F	F	T	F	T
F	T	T	T	F	F	F	T
F	T	F	F	T	F	T	T
T	F	T	T	T	F	F	F
T	F	F	T	T	F	T	T
F	F	T	T	F	F	F	T
F	F	F	T	F	F	T	F

We see that they agree when b is true, but disagree in half the cases when b is false.

38. I shall not discuss the possibility that all properties are ultimately conditional. Even if they are, a set of properties with mutually compatible conditions may be treated without any mention of the conditions for many purposes.

39. See, e.g., Waismann, "Are There Alternative Logics"; B. C. van Fraassen, "The Labyrinth of Quantum Logics," in Cohen and Wartofsky, *Boston Studies,* vol. 5; and A. A. Zinov'ev, *Foundations of the Logical Theory of Scientific Knowledge* (Dordrecht; Reidel, 1973).

40. Van Fraassen, "Labyrinth."

41. Zinov'ev, "Foundations of the Logical Theory."

42. Ibid., p. 92.

43. Ibid., p. 95.

44. See especially the closing section of Strauss, "Logics for Quantum Mechanics."

45. I shall not consider such anomalous cases as glass, which has a viscosity but normally appears to be in the solid state.

46. Finkelstein, "Physics of Logic." Finkelstein uses the diagram in discussing the polarization of photons. I shall discuss this subject in sec. 4.

47. Note that there is a complete dualism between these two operations, so that the opposite identification would be just as good. It is merely a question of "reading" the lattice from bottom to top versus "reading" it from top to bottom.

48. Such considerations must always be based on some theory of the nature of things. In a moment, I shall distinguish between the system (S) and a theoretical model (M) of the system.

49. Of course, this does not exclude a discussion of two logics carried out within some formalized metalanguage rich enough to carry out the formalization of both. But then the overriding logical necessities would be those of the logic of the metalanguage.

50. K. Gödel, "Die Vollständigkeit der Axiome des logischen Funktionkalküls," *Mh. Math. Phys.* 37 (1930).

51. G. Birkhoff, *Lattice Theory,* 2d rev. ed. (New York: American Mathematical Society, 1948), p. 189.

52. Putnam, "Is Logic Empirical," and "How to Think Quantum-Logically."

53. B. Mielnik, "Generalized Quantum Mechanics," *Commun. Math. Phys.* 37 (1974): 221,

summarizes the convex set approach to quantum mechanics, with references to earlier literature, and discusses the analogy between quantum mechanics and geometry first proposed in B. Mielnik, "Theory of Filters," *Commun. Math. Phys.* 15 (1969): 1.

54. D. Finkelstein, "The Physics of Logic," in Colodny, *Paradigms and Paradoxes,* pp. 53–54.

55. Ibid., p. 54

56. Putnam, "How to Think Quantum Logically." See sec. 1 for a fuller quotation, and the reference.

57. See references cited in note 33.

58. See discussion in note 33.

59. Of course, the most profound understanding of light so far achieved in physics comes from its treatment as a quantum field. For reasons to be discussed shortly, I shall not consider polarization from this viewpoint. Discussion of light actually takes us beyond nonrelativistic quantum mechanics, but those aspects of polarization to be considered do not involve any essentially nonrelativistic considerations. For a quantum field-theoretical treatment of light, including polarization, see e.g., A. I. Akhiezer and V. B. Berestetskii, *Quantum Electrodynamics* (New York: Interscience Publishers, 1965) or V.B. Berestetskii, E. M. Lifshitz, and L. P. Pitaevskii, *Relativistic Quantum Theory,* part 1 (Oxford: Pergamon Press, 1971).

60. Putnam, "How to Think Quantum Logically," p. 56.

61. Werner Heisenberg, *Die Physikalischen Prinzipien der Quantentheorie,* 4, Auflage (Leipzig: S. Hirzel Verlag, 1944), pp. 107–08, 110. The corresponding passages in the English edition, *The Physical Principles of the Quantum Theory* (Chicago: University of Chicago Press, 1930), are on pp. 177–78, 181, but my translation has been made directly from the German text.

62. M. Born, W. Heisenberg, and P. Jordan, "Zur Quantenmechanik II," *Z. Phys.* 35 (1925): 557.

63. P. Jordan, *Anschauliche Quantentheorie* (Berlin: Springer-Verlag, 1936), p. 231. His comments refer specifically to bosons.

64. Strauss, *Modern Physics,* p. 246.

65. Finklestein discusses Bose-Einstein and Fermi-Dirac statistics in the course of the physical studies that he carries out under the rubric "quantum logic." See, e.g., D. Finklestein, "Space-Time Code. 2," *Physical Review* D 5 (1972): 320. However, neither in this paper, nor the earlier discussion which it supercedes, have I found a discussion of the implications of these statistical considerations for his concept of the quantum-logical properties of quantum systems. For a discussion of the classical limit of boson and fermion fields, see H. J. Lipkin, "Feynman Diagrams, Propagators and Fields," chap. 12 of *Quantum Mechanics: New Approaches to Selected Topics,* (New York: American Elsevier, 1973), pp. 370–76. For a discussion of some problems connected with the classical limit of quantum electrodynamics, see I. Bialynicki-Birula, "Classical Limit of Quantum Electrodynamics," *Acta Physica Austriaca,* suppl. (1977): 111.

66. I follow a tradition going back to Dirac. See P. A. M. Dirac, *Principles of Quantum Mechanics* (Oxford: Oxford University Press, 1930), and subsequent editions. For a recent treatment, see H. J. Lipkin, "Polarized Photons for Pedestrians," in *Quantum Mechanics.* For a treatment of polarization theory including fuller discussions of many points which I shall mention, such as the Poincaré sphere, see G. N. Ramachandran and S. Ramaseshan, "Crystal Optics. A. Polarization of Light," in S. Fluegge, ed., *Encyclopedia of Physics,* vol. 25, 1, *Crystal Optics. Diffraction* (Berlin: Springer-Verlag, 1961).

67. See e.g., E. Mach, "Polarization," in *The Principles of Physical Optics* (London: Methuen, 1926), chap. 10.

68. E. G., see the article by Ramachandran and Ramaseshan, "Crystal Optics," or R. W. Wood, *Physical Optics* (New York: Dover, 1967).

69. A Einstein, "Über eine die Erzeugung und Verwandlung des Lichtes betreffenden heuristischen Gesichtspunkt," *Ann.. d. Physik* 17 (1905): 132. Ironically, the photoelectric effect is not one of the phenomena that demands a particulate structure of light for its explanation. See W. E. Lamb and M. O. Scully, "The Photoelectric Effect Without Photons," in Societé Française de Physique, *Polarisation Matière et Rayonnement.* (Paris: P.U.F., 1969). I am indebted to Dudley Shapere for bringing this paper to my attention.

70. Y. Freundlich, "Copenhagenism and Popperism," *Brit. J. Phil. Sci.* 29 (1978): 145. See also his "In Defense of Copenhagenism," *Stud. Hist. Phil. Sci.* 9 (1978): 151.

71. Lipkin, "Polarized Photons," p. 12.

72. See H. Poincaré, *Théorie Mathématique de la Lumière,* vol. 2 (Paris: Gauthier-Villars, 1892).

73. I shall not even sketch the simple geometrical proof. See Ramachandran and Ramaseshan, "Crystal Optics."

74. The study of such birefringent crystals first led to the discovery of polarization. See, e.g., Mach, "Polarization."

75. For details of this and other constructions, see Ramachandran and Ramaseshan, "Crystal Optics."

76. Polarization has one advantage. A beam of light incident on a screen with two slits will have to lose intensity in passing through the slits, which introduces additional complications into a complete analysis of the double slit arrangement. Alternatively, one can regard the source and screen with double slit as forming the preparation of the beam, but then the only "manipulation" of the beam before its registration is free propagation through space. A birefringent crystal, in contrast, can accept the entire incident beam.

77. P. Davies, *Other Worlds* (New York: Simon and Schuster, 1980), p. 126.

78. N. Bohr, "Unity of Knowledge," in *Atomic Physics and Human Knowledge* (New York: Science Editions, 1961), p. 71.

79. My distinction between coherent and incoherent manipulations is similar to one made by M. Strauss in "Intertheory Relations (III)," chap. 22 of *Modern Physics.* Strauss first introduces the concept of the state of a quantum system, and then distinguishes between nonstochastic and stochastic state changers. He subdivides nonstochastic state changers (which correspond to my coherent manipulations of ensembles) into: evolution in time, scatterers, and diffractors. The effect of the first two on the state vector is represented by unitary operators; the effect of diffractors is represented by projection operators. I shall discuss this distinction later, as well as the various types of incoherent manipulations corresponding to Strauss's stochastic analyzers and separators.

80. M. Strauss emphasized this. He refers to transitions from one state of a quantum system to another (in the Copenhagenist sense of the word "state") as stochastic modes of reaction: "In fact QM [Quantum Mechanics] is a sort of *black box* theory in that the stochastic reactions (induced state transitions) are not explained (i.e., reduced to other processes) but taken as irreducible; what the theory does explain are the relative frequencies, or rather: the relative probabilities, with which the various possible reactions (induced state transitions) occur. Thus, *'reaction of a quantum system to macroscopic systems' is a built-in feature of QM* and, hence a PI [Physical Interpretation] in terms of internal properties only is impossible or, at best, a metaphysical construct (with ghosts that never appear)" (ibid., p. 251). I shall return in the last section to the question of ascribing properties to quantum systems.

81. The contrast between principle and constructive theories is due to Einstein: see "What Is the Theory of Relativity," in A. Einstein, *Ideas and Opinions* (New York: Bonanza Books, 1954), p. 228. For further discussion of this point in the context of special relativity, see my paper "Special Relativity from Measuring Rods," in R. S. Cohen and L. Laudan, eds., *Physics, Philosophy and Psychoanalysis: Essay in Honor of Adolf Grünbaum* (Dordrecht: Reidel, 1983), pp. 255–72.

82. For discussion of two possible approaches to this problem, see: (for the quantum ergodic approach) A. Danieri, A. Loinger, and G. M. Prosperi, "Quantum Theory of Measurement and Ergodicity Conditions," *Nuclear Physics* 33, (1962); 297; and L. Rosenfeld, "The Measuring Process in Quantum Mechanics," *Suppl. Prog. Theor. Phys.* 222. (1965), reprinted in R. S. Cohen and J. Stachel, eds., *Selected Papers of Leon Rosenfeld* (Dordrecht: Reidel, 1979); (for the kinetic approach) L. Rosenfeld, C. George, and I. Prigogine, "The Macroscopic Level of Quantum Mechanics," *Mat.-fys. Medd. Dan Vid. Selsk.* 38, no. 12 (1972), and *Nature* 240 (1972): 25; both reprinted (the first only in part) in Cohen and Stachel, *Selected Papers of Leon Rosenfeld.* See also K. Hepp. "Quantum Theory of Measurements and Macroscopic Observables," *Helv. Phys. Acta* 45 (1972): 27.

83. For example, to the nuclear reactors thought to have spontaneously formed over two billion years ago at the Oklo uranium deposits in Gabon. See M. Maurette, "Fossil Nuclear Reactors," *Ann. Rev. Nucl. Sci.* 26 (1976): 319.

84. See M. Strauss, "Logics for Quantum Mechanics," *Foundations of Physics* 3 (1973): 265. He distinguishes two concepts of probability in these two cases, and shows how a nonclassical probability calculus may be used for conditional probabilities involving incompatible properties.

85. See e.g., H. J. Groenewold, "Skeptical Quantum Theory," *Physics Reports* 11 (1974): 327, and references therein to his earlier work; and L. E. Ballantine, "The Statistical Interpretation of Quantum Mechanics," *Rev. Mod. Phys.* 42 (1970): 358.

86. Freundlich, "Copenhagenism Verses Popperianism." This definition of state, for the example of polarization of photons, has been discussed above. See also Freundlich, "In Defense of Copenhagenism."

87. Freundlich, "Copenhagenism verses Popperianism."

88. Preparation is usually just the selection of part of the output of a registration device. Of course, the act of preparation may be outside human control. For example, sunlight provides a source of unpolarized light. So "preparation" need not have an anthropomorphic connotation, any more than "measurement," as the word is often used in quantum mechanics, need involve an actual measurement by human agency, let alone the reading of the record of such a measurement by a human observer.

89. See H. J. Groenewold, "Information in Quantum Measurements," *Proc. Koninkl. Nederl. Akad. van Wetenschappen B* 55 (1952): 219. See also Y. Aharonov, P. G. Bergmann, and J. L. Lebowitz, "Time Symmetry in the Quantum Process of Measurement," *Phys. Rev.* 134 (1964): 1410; and P. G. Bergmann, "The Quantum State Vector and Physical Reality," in M. Bunge, ed., *Quantum Theory and Reality* (Berlin: Springer-Verlag, 1967).

90. See R. Schiller, *Phys. Rev.* 125 (1962): 1116, for a discussion of the relation between solutions to the Hamilton-Jacobi and Schrödinger equations.

91. Using the concept of state, I can be more specific about the difference between coherent and incoherent manipulation of quantum ensembles. The entropy of an ensemble in a given state may be defined. Pure states are then characterized by the fact that their entropy is zero. Then coherent manipulation will not change the entropy, while (nondestructive) incoherent manipulation will increase it. For a definition of the entropy associated with the state of an ensemble, see J. von Neumann, *Mathematische Grundlagen der Quantenmechanik* (Berlin: Springer-Verlag, 1932), chap. 5. The English translation is *Mathematical Foundations of Quantum Mechanics* (Princeton: Princeton University Press, 1955).

If we regard destruction or removal of a part of an ensemble as increasing the entropy of that part (total loss of information), we may drop the restriction and say that an incoherent manipulation *always* increases the entropy of an ensemble.

92. D. Finkelstein, J. M. Jauch, and D. Speiser, "Notes on Quaternion Quantum Mechanics," in Hooker, *The Logico-Algebraic Approach,* vol. 2, discuss this possibility.

93. B. Mielnik, "Generalized Quantum Mechanics," *Commun. Math. Phys.* 37 (1974): 221, reviews work on the convex set approach to quantum mechanics, with references to earlier

papers, including his own. B. Mielnik, "Convex Sets: Non-Linear Quantization," in *Proceedings of the International Symposium on Mathematical Physics,* Mexico City, January 5–8, 1976, (n.p.), vol. 1, p. 355, gives a brief account.

94. Mielnik, "Generalized Quantum Mechanics."

95. A nonmaximal registration is often associated with destruction (or dropping from further consideration) of part of the initial ensemble. The double-slit experiment, for example, may be looked upon as a nonmaximal registration, with the part of the ensemble not registered beyond the slits dropped from consideration. If one were to make the double slits in a mirror (perfect reflector), for example, and place photographic plates on *both* sides of the mirror, one would have a nonmaximal registration for the *full* incident ensemble.

96. This version incorporates the so-called Lüders rule for incomplete measurements, to be discussed in the next section.

97. H. Hertz, *Die Prinzipien der Mechanik* (Leipzig: Barth, 1894), pp. 2–3 (my translation).

98. J. C. Maxwell, "The Diagram of Colors," *Trans. R. Soc. Edinburgh* 21 (1857): 275; partially reprinted in MacAdam, *Sources of Color Science.*

99. I have discussed some issues in quantum logic in "The 'Logic' of Quantum Logic"; in "Comments on 'The Formal Representation of Physical Quantities'," in *Logical and Epistemological Studies in Contemporary Physics,* and in "Comments on 'Some Logical Problems Suggested by Empirical Theories' by Professor Dalla Chiara," in R. S. Cohen and M. W. Wartofsky, eds., *Language, Logic and Method* (Dordecht: Reidel, 1982).

100. I have in mind particularly the implications of the nonseparability of quantum systems. For a discussion of this issue see, e.g., B. D'Espagnat, *Conceptual Foundations of Quantum Mechanics,* 2d ed. (Menlo Park: W. A. Benjamin, 1976).

101. Note that any attempt to treat quantum systems "in themselves" inevitably leads to the need to define a "wave function of the universe." Since quantum systems that have interacted in the past must be treated as parts of a single system thereafter (quantum nonseparability), the assumption of an intrinsic wave function for any limited part of the universe leads to the need to assume one for the entire universe. This in turn raises the problem of how to interpret the quantum-mechanical formalism when applied to the universe. The only internally consistent interpretation seems to be the Everitt "many-universe" interpretation, which creates more problems than it solves, in my opinion. I have discussed some of these problems in "The Rise and Fall of Geometrodynamics," in K. F. Schaffner and R. S. Cohen, eds., *PSA 1972* (Dordrecht: D. Reidel 1974), pp. 31–54; see esp. pp. 40–43. Since no quantum logician has proposed that quantum theory be restricted in application to the universe as a whole, nor adopted the "many-universe" interpretation, I shall not comment further on this issue.

102. Heisenberg, *The Physical Principles of the Quantum Theory.*

103. N. Bohr and L. Rosenfeld, "Zur Frage der Messbarkeit der elektromagnetischen Feldgrössen," *Mat.-fys. Medd. Dan. Vid. Selsk.* 12, no. 8 (1933). An English translation is in Cohen and Stachel, *Selected Papers of Leon Rosenfeld.*

104. Blokhintsev refers to registration devices as spectral analyzers and notes:

> These spectral analysers should not be thought of as necessarily having the form of laboratory apparatus. On the contrary, an experimenter or technician who chooses a particular apparatus merely makes a certain combination of what already exists in Nature, and it would be absurd to think that if there were no "observer" quantum ensembles would no longer be meaningful. As soon as a situation arises in Nature where spectral resolution of the original ensemble occurs, there is a formation of new ensembles defined by new characteristics, i.e., what is usually called "the intervention of measurement." This process may or may not be observed by an observer; the objective phenomenon is unaffected.

D. I. Blokhintsev, *Quantum Mechanics* (Dordrecht: Reidel, 1964), p. 59. I will add that I believe I am following the Bohr tradition (as contrasted with the von Neumann tradition), in which a measurement is completed once an irreversible change has occurred in a piece of macroscopic apparatus—whether or not any human (or other living) eye is ever cast upon that mark. My discussion is, of course, quite independent of whether this particular belief of mine is correct or not.

105. By the spectral theorem (see, e.g., M. Reed and B. Simon, *Methods of Modern Mathematical Physics 1: Functional Analysis*, chap. 7, "Spectral Theorem," pp. 221–48), any well-behaved self-adjoint operator may be resolved into an integral over a maximal set of projection operators with coefficients equal to the eigenvalues of the self-adjoint operator. From the point of view I have presented, one may think of the operator corresponding to a physical quantity as built up from the eigenvalues and corresponding projection operators. It follows that if one operator is a function of another, then the same registration device will serve for registration of values of both the corresponding physical quantities. A registration of position coordinates, for example, is also a registration of the cubes of position coordinates.

106. Of course, it may turn out that quantities must be introduced for the description of a quantum system which have no classical analogue.

107. In discussing this example, I have neglected all mathematical complications stemming from the continuous eigenvalue spectra of position and momentum as essentially irrelevant to the point I am making.

108. Heisenberg, *Physical Principles*.

109. There is a well-known limitation on localization of a photon, because of its zero rest-mass. For a plane wave of wavelength λ, traveling in the z-direction, the z-component of its position could not be localized more precisely than the value of λ. It might be better to use a nonrelativistic particle in the example, since no such limitation would then be present. However, such complications are irrelevant to the discussion of *x-y* plane localization. See A. I. Akhiezer and V. B. Berestetskii, *Quantum Electrodynamics*, chap. 1, "Quantum Mechanics of the Photon," for a discussion of the localization of photons.

110. In terms of the Hilbert space model, the first device corresponds to projections onto two orthogonal rays of the Hilbert space; adding the signals corresponding to each ray would correspond to a set-theoretical union of the two rays. The second device corresponds to projection onto the *two-dimensional subspace* of Hilbert space spanned by the two rays—quite different from their set-theoretical union!

111. As mentioned earlier, such *Gedankenexperimente* merely illustrate or exemplify what the formalism excludes. Their logical function is similar to the role in special relativity of *Gedankenexperimente* seeking to establish the absolute simultaneity of distant events.

112. See Mielnik, "Theory of Filters," and "Quantum Logic: Is It Necessarily Orthocomplemented?" in M. Flato, et al., eds., *Quantum Mechanics, Determinism, Causality, and Particles* (Dordrecht: Reidel, 1976), for an extensive critique of the quantum-mechanical concept of filter and proposals for its generalization.

113. By D. Finkelstein. See his "Physics of Logic."

114. By J. M. Jauch, for example. See his "An Axiomatic Foundation of Quantum Mechanics," in B. D'Espagnat, ed., *Foundations of Quantum Mechanics. Proceedings of the International School of Physics "Enrico Fermi." Course 49* (New York: Academic Press, 1971).

115. By G. Ludwig, for example. See his "The Measuring Process and an Axiomatic Foundation of Quantum Mechanics," in D'Espagnat, ed., *Foundations of Quantum Mechanics.*

116. While there is no real harm in ambiguity of terminology employed by *different* persons, there is danger if the *same* person uses a term now to mean one thing, now another.

117. See Mielnik, "Quantum Logic: Is It Necessarily Orthocomplemented?"

118. See Finkelstein, "Physics of Logic."

119. See J. M. Jauch, *Foundations of Quantum Mechanics* (Reading: Addison-Wesley, 1973).

120. Mielnik "Theory of Filters" and "Quantum Logic."

121. J. von Neumann, comments on Bohr's paper, "The Causality Problem in Atomic Physics," *New Theories in Physics* (International Institute of Intellectual Cooperation, 1939), p. 38. Similar ideas are expressed in a manuscript from about 1937, "Quantum Logics (Strict and Probability Logics), reviewed by A. H. Taub" in J. von Neumann *Collected Works* (New York: Macmillan, 1962), vol. 4, pp. 195–7. He there considers "The system L of logics" which consists of "the set of all statements *a,b,c,* . . . which can be made concerning (the physical system) *S*." Such a statement is always one concerning the outcome of a certain measurement, which is to be performed on *S*. He then sets up a lattice structure on *L* by means of the implication relation and "the operation of 'negation': $-a$." He postulates the existence of a greatest lower bound of any two elements *a* and *b*, which he identifies with the logical operation of conjunction and symbolizes by *a b* and of a least upper bound, which he identifies with disjunction and symbolized by $a + b$. Thus far, he has done no more than briefly recapitulate the ideas of the earlier Birkhoff and von Neumann paper, "Logic of Quantum Mechanics." But then he continues:

> So far the only structure *L* possesses is defined by the "primitive notions" $a,b,-a,$ and the "derived notions" $a + b$, *a b*. To this extent we will call *L* the system of "*strict logics*." In application of *L* to actual physical reality, however, a further structure of *L* appears, which can only be expressed in terms of "probability." In other words: For any well defined state of our knowledge concerning the mathematical description of physical reality, that is for any reasonable mathematical model *S*, a probability function exists. So we have in *L:*
>
> (VI) The (*real number-valued*) functions called "probability function": $P(a,b)$.
>
> $P(a,b) = \Theta$ (Θ a real number) means this: If a measurement of *a* on *S* has shown *a* to be true, then the probability of an immediately subsequent measurement of *b* on *S* showing *b* to be true, is equal to Θ . . . .
>
> We prefer to disclaim any intention to interpret the relations $P(a, b) = \Theta (0 < \Theta < 1)$ in terms of strict logics. In other words, we admit: "Probability logics cannot be reduced to strict logics, but constitute an essentially wider system than the latter, and statements of the form $P(\text{a, b}) = \Theta$ $(0 < \Theta < 1)$ are perfectly new and *sui generis* aspects of physical reality."
>
> So probability logics appear as an essential extension of strict logics. This view, the so-called "logical theory of probability" is the foundation of J. M. Keynes's work on this subject.

122. See A Landé, *Quantum Mechanics in a New Key* (New York: Exposition Press, 1973), esp. chaps. 2 and 3.

123. Strauss, "Logics for Quantum Mechanics." See also his "Foundations of Quantum Mechanics" and "Intertheory Relations (3)," in *Modern Physics.*

124. See Mielnik, "Generalized Quantum Mechanics," for a survey with other references.

125. See J. M. Jauch and C. Piron, "What is Quantum Logic?" in P. Freund, C. Goebel, and Y. Nambu, eds., *Quanta* (Chicago: University of Chicago Press, 1970).

126. See Strauss, *Modern Physics* and "Logics for Quantum Mechanics."

127. See P. Mittelstaedt, *Quantum Logic* (Dordrecht: Reidel, 1978).

128. See van Fraassen, "The Labyrinth of Quantum Logics" and "Semantic Analysis of Quantum Logic," in C. Hooker, ed., *Contemporary Research in the Foundations and Philosophy of Quantum Mechanics* (Dordrecht: Reidel, 1973), pp. 80–113.

129. M. L. Dalla Chiara, "Some Logical Problems Suggested by Empirical Theories," in Cohen and Wartofsky, *Language, Logic and Method*; and "Quantum Logic and Physical Modalities," *J. Phil. Logic* 6 (1977): 391.

130. See K. Popper, "Quantum Mechanics Without 'the Observer'," in M. Bunge, ed.,

Quantum Theory and Reality (New York: Springer-Verlag, 1967); and "Birkhoff and von Neumann's interpretation of Quantum Mechanics," *Nature* 219 (1968): 682.

131. See Putnam, "Is Logic Empirical?"

132. See Stairs, "Quantum Mechanics." I quote from this. Kripke considers the case in which it is asserted that "A = 1 [or] A = 2" and "B = 1 [or] B = 2" are both true "in their usual interpretation" before any measurements are made. Nevertheless "one . . . wishes to maintain" that "A = 1 B = 1," "A = 1 B = 2," "A = 2 B = 1," and "A = 2 B = 2" are all false. Kripke continues:

> It would seem that . . . if one of the statements "A = 1" and "A = 2" is true and one of the statements "B = 1" and "B = 2" was already true before a measurement was made . . . then there are four cases. If A = 1, there are two cases for B . . . But he says that (A = 1 B = 1) and (A = 2 B = 2) are both false. And if A equals 2 then there are two cases for B. But he says that both . . . are both false. Well Putnam says, we should adopt a nonstandard logic, which he takes over from Birkhoff and von Neumann, called quantum logic, in which the distributive law, essentially proof by cases of this kind, fails. . . .
>
> How is one going to refute this paper? . . . I should like if possible to have refuted it as follows. I should say, "Well, let's say that A has been measured and is really equal to 1. Then you say that B is one of the two values 1 and 2. So consider by cases. Suppose B was 1. But then that contradicts the assertion that "A = 1 B = 1" is false. Suppose that B was 2. That contradicts the assertion that "A = 1 B = 2" was false. . . . That is what I would like to say and that is the most conclusive proof I think that could possibly be given that this paper is wrong. . . . I don't think any extra premises besides "A = 1" and "B = 1 [or] B = 2" are required. And to say that there is some extra premise which you call the distributive law is already to say that my reasoning was fallacious or at least was a suppressed enthymeme. . . . But I would say that there is no extra premise.

133. See Putnam, "Is Logic Empirical?"

134. M. Gardner, in "Is Quantum Logic Really Logic?" *Phil. Sci.* 38 (1971): 508, has claimed that the distributive law does not break down in the double-slit experiment. It is clear from his illustrations that he considers only the case where D is small; he does note that points of the plate not in the diffracted image of the two slits are redundant.

135. M. Friedman and H. Putnam, "Quantum Logic, Conditional Probability, and Interference," *Dialectica* 32 (1978): 305.

136. See S. Kochen and E. Specker, "Logical Structures Arising in Quantum Theory," in J. W. Addison, L. Henkin, and A. Tarski, eds., *The Theory of Models* Amsterdam: North-Holland, 1965), p. 177. The proof is on p. 184.

137. Putnam, "Is Logic Empirical?" p. 192.

138. Y. Freundlich, "Copenhagensim and Popperism," *Brit. J. Phil. Sci.* 29 (1978): 145.

139. H. J. Groenewold, "Information in Quantum Measurements," *Proc. Koninkl. Nederl. Akad. van Wetenschappen B* 55 (1952): 219.

140. Y. Aharonov, P. G. Bergmann, and J. Lebowitz, "Time Symmetry in the Quantum Process of Measurement," *Phys. Rev.* 134 (1964): B 1410. See also P. G. Bergmann, "The Quantum State Vector and Physical Reality," in Bunge, *Quantum Theory and Reality*.

141. This is reminiscent on von Neumann's final reflections on quantum logic, except that here no conditionalizing of the probabilities seems to be involved. See note 121.

142. J. Bub, "Some Reflections on Quantum Logic and Schrödinger's Cat," *Brit. J. Phil. Sci.* 30 (1979): 27.

143. K. Popper, "Particle Annihilation and the Argument of Einstein, Podolsky and Rosen," in W. Yourgrau and A. van der Merwe, eds., *Perspectives in Quantum Theory* (Cambridge, Mass.: MIT Press, 1971), p. 182. See esp. p. 187.

144. M. Friedman and H. Putnam, "Quantum Logic, Conditional Probability, and Interference."

145. G. Lüders, "Über die Zustandsänderung durch den Messprozess," *Ann. d. Physik* 8 (1951): 322. An exposition of Lüders's work, including other relevant material is in W. Furry, "Some Aspects of the Quantum Theory of Measurement," in W. Britten, ed., *Lectures in Theoretical Physics,* vol. 8A, *Statistical Physics and Solid State Physics* (Boulder: University of Colorado Press, 1966). Bub brought Lüders rule to the attention of the quantum logicians. See J. Bub, "Von Neumann's Projection Postulate as a Probability Conditionalization Rule in Quantum Mechanics," *J. Phil. Logic* 6 (1977): 381. Friedman and Putnam do not mention Lüders rule by name, but it is the content of their formula (1).

146. R. P. Feynman, *Rev. Mod. Phys.* 20 (1948): 307. Feynman calls the rule for the addition of probability amplitudes "a typical representation of the wave nature of matter" (p. 369). Heisenberg practically stated the Feynman rules around 1929 in *The Physical Principles of Quantum Mechanics,* pp. 59–62, esp., p. 61. See also R. P. Feynman and A. R. Hibbs, *Quantum Mechanics and Path Integrals,* (New York: McGraw-Hill, 1965).

147. Feynman and Hibbs, *Quantum Mechanics,* p. 14.

148. S. Kochen and E. P. Specker, "The Problem of Hidden Variables in Quantum Theory." *Journal of Mathematics and Mechanics* 17 (1967): 59.

149. See, e.g., G. M. Hardgree, "The Conditional in Abstract and Concrete Quantum Logic," in Hooker, *Logico-Algebraic Approach,* vol. 2.

See. G. Hellman, "Quantum Logic and the Projection Postulate," *Phil. Sci.* 48 (1981): 469. The Sasaki hook of two elements a and b of an orthomodular lattice is defined as the element $a' \vee (a \wedge b)$. It then follows that $a \sim b = (a \rightarrow b) \wedge (b \rightarrow a)$.

150. See Hardgree, "Abstract and Concrete Quantum Logic," for a discussion of the properties of the Sasaki hook. Hardgree shows that it may be interpreted as a Stalnaker conditional, a type of counterfactual conditional.

151. J. Bub, "Von Neumann's Projection Postulate," p. 389.

152. S. Kochen, "The Interpretation of Quantum Mechanics." I thank Dr. Kochen for a preprint of this paper.

153. See Marx, "Critique of Hegelian Dialectic."

154. J. Stachel, "The 'Logic' of Quantum Logic."

155. Ibid., p. 517.

156. Aron Gurwitsch, *Leibniz: Philosophie des Panlogismus* (Berlin: Walter de Gruyter, 1974).

157. Of course, the formalism includes the maximal case if $m = 1$.

158. Since projection operators are Hermitean, there is no need to worry about Hermitean conjugation in reversing the order of terms in the bracket.

JOHN STACHEL
Boston University

Einstein and the Quantum: Fifty Years of Struggle

> Science is the attempt to make the chaotic diversity of our sense experience correspond to a logically uniform system of thought.
>
> —Albert Einstein
> *The Fundaments of Theoretical Physics*

In the correspondence of his last years, Einstein returns a number of times to the theme of his half-century-long struggle with the quantum. On December 12, 1951, for example, he wrote to his old friend and patent office colleague Michele Besso: "The whole fifty years of conscious brooding [*Grübelei*] have not brought me nearer to the answer to the question 'What are light quanta?' Nowadays every scalawag [*Lump*] believes that he knows what they are, but he deceives himself."[1]

He wrote to Max von Laue, the only "Aryan" German colleague from Einstein's Berlin years to whom he still felt close, on January 17, 1952, commenting on the latest edition of von Laue's textbook on special relativity. After pointing out that he was aware of the limits of validity of Maxwell's theory before 1905, he added: "But unfortunately the fifty years that have passed since then have not brought us closer to understanding the atomistic structure of radiation. On the contrary! The De Broglie waves are obviously the counterpart of the electromagnetic waves."[2]

Again writing to Besso on December 10, 1952, Einstein said: "at any rate we are just as far from a really rational theory (of the dual nature [*Doppelnatur*] of light quanta and particles) as fifty years ago! P.S. A really rational theory would have to *deduce* the elementary structures (electrons, etc.) not *posit* them from the outset" (*Besso Correspondance*, pp. 482–3).

I shall discuss a few of the things Einstein accomplished in the course of those fifty years of struggle with the quantum; what he was trying to accomplish but felt he had failed to do; the methods that guided him both in his successes and his failures; and why he regarded what he—and others—had done as bringing us no nearer to a really rational understanding of the quantum.

Fortunately, I feel no obligation to give an exhaustive survey of Einstein's work on quantum theory, since there are a number of excellent recent studies. These include the work of Martin Klein, conveniently summarized with references to his earlier papers in his paper at the IAS Einstein Centennial;[3] an article and book by Abraham Pais, as well as his IAS Centennial talk;[4] and Max Jammer's book and papers at the Berlin and Jerusalem Einstein Centennials.[5] References in these sources lead into further literature on this topic, so I can proceed in good conscience with my unsystematic survey, which mixes detailed discussion of a few points with a broad, but only sketchily substantiated, characterization of Einstein's work on quantum problems.

I thank the Hebrew University of Jerusalem, which holds the copyright, for allowing me to quote from both published and unpublished letters and papers by Einstein.

1. EINSTEIN AND THE BOHR MODEL

I shall start this paper with a minor historical mystery and end with some major physical ones. Einstein described his first published work on the quantum problem, in a letter to his friend Konrad Habicht, as "very revolutionary." This paper, received March 17, 1905, by the *Annalen der Physik,* was entitled "On a Heuristic Viewpoint Concerning the Generation and Transformation of Light." Later, I shall discuss the nature of Einstein's approach to the quantum in this and his other early papers, but I want to draw your attention to an almost unnoticed remark by Michele Besso. Besso was Einstein's sounding board for his ideas about relativity: he is the only person thanked in the 1905 relativity paper. Writing to Einstein on January 17, 1928, Besso said: "For my part, in the years 1904 and '05 I was your public. If, in connection with the drafting of your papers on the quantum problem, I deprived you of a part of your fame, yet in return I secured a friend for you in Planck."[6] No reply has been found. I shall leave aside the question of why Besso claims to have gained Planck's friendship for Einstein. The rest of the quote suggests that, at the time of Einstein's early work on the quantum theory, Besso held Einstein back from publishing some idea that would have brought Einstein even more fame. Considering the fame Einstein had achieved by 1928, Besso could have been referring to no small item! Is there any evidence that Einstein held back some potentially important speculation around 1905? Yes, there is.

It is sometimes thought that Einstein was not originally interested in atomic structure and did not concern himself with this problem in his early years. However, there are several contemporary items of evidence to the contrary. Einstein wrote to Habicht in 1905, probably between June and September: "there is not always a fully developed subject for my musings. At least none that appeals. There would of course be the subject of the spectral lines; but I believe that a simple connection of these phenomena with others already studied does not exist at all, so that the subject for the moment seems to promise very little."[7]

A letter to Philip Lenard a few months later makes it clear that Einstein had closely followed Lenard's studies of atomic spectra, and was ready to express some of his "musings":

> The experiments known to me do not exclude that possibility that the emission or absorption of each individual spectral line is connected with a definite state of the emitting or absorbing center (atom), which [state] is characteristic for it [i.e., the emission or absorption]. . . . According to the indicated conception the absorption of a series by a (cold) vapor should be interpreted thus, that the absorption of light of the line ν_1 makes the absorbing center in question receptive for light of the line ν_2, etc. Then an absorption of ν_2 by the vapor would only be possible with simultaneous absorption of ν_1.[8]

This was written *after* the 1905 paper in which Einstein had proposed that emission and absorption of light of frequency ν be conceived as taking place in quanta of energy $h\nu$ (actually, Einstein at that time did not use Planck's h, but the equivalent $R/N\beta$). Perhaps it is not too far fetched to speculate, based on this letter, that around the end of 1905 Einstein had the idea of discrete energy states (levels, we would say today) of the atom in which it did not radiate. If we multiply ν_1 and ν_2 by h, we might try to interpret Einstein's remarks in the letter according to figure 1.[9]

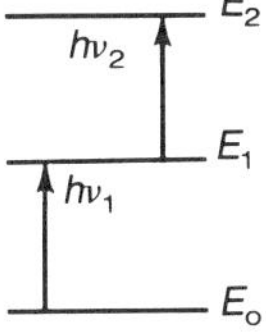

Figure 1

At the time, Lenard did not reply to this letter; when he did write Einstein, in 1909, he mentioned that it was still on his writing desk.

Is there any further evidence to support this wild speculation? Yes, but only a little more. In September 1913, Georg von Hevesy, a radiochemist then working at Rutherford's laboratory, attended the Vienna meeting of the German Society of Natural Scientists and Physicians (the

German AAAS) at which Einstein delivered a review of his work on general relativity up to that date. Hevesy met Einstein, and they discussed Bohr's theory of the hydrogen atom, which had been published that spring. Hevesy gave accounts of the conversation in letters to Rutherford and Bohr. Hevesy to Rutherford:

> Speaking with Einstein on different topics we came to speak of Bohr's theory, he told me that he had once similar ideas but he did not dare to publish them. "Should Bohr's theory be right, it is of the greatest importance." When I told him about the Fowler spectrum the big eyes of Einstein looked still bigger and he told me "Then it is one of the greatest discoveries."
>
> I felt very happy hearing Einstein saying so.[10]

Hevesy to Bohr:

> Then I asked him about his view on your theorie. He told me, it is a very interesting one, important one if it is right and so on and he had very similar ideas many years ago but had no pluck to develop it; I told him then that it is established now with certenety that the Pickering-Fowler spectrum belongs to He. When he heard this he was extremely astonished and told me: "Than the frequency of the light does not depand at all on the frequency of the electron"—(I understood him so??) And this is an *enormous acheiwement.* The theory of Bohr must be then wright. I can hardly tell you how pleased I have been and indeed hardly anything else could make me such a pleasure then this spontaneus judgement of Einstein.[11]

Why Einstein was so impressed by the fact that the frequency of the spectral lines was distinct from any mechanical frequency of the Bohr orbits can be gathered from a comment he made much later: "The important fact that, according to Bohr's theory, the frequency of the emitted radiation is not determined by electrical masses that undergo periodic processes of *the same* frequency, can only strengthen us in this doubt of the independent reality of the wave field [i.e., the classical Maxwell field]."[12]

There is also a reference to the relation of Bohr's work to Einstein's early work in a letter from Einstein to Besso in 1948. I cannot resist including the unrelated but delightful parenthetical remark of Einstein which immediately follows: "It is also not correct that the Bohr conditions for electron orbits follow from my quantum papers. One can only conclude that, if the mechanical picture of the atom is retained, such rules must exist. (You know quite well too, that in spite of its great practical successes, I do not consider the present statistical quantum theory a good approach. It's the same with me as the Jews with the 'Messiah')" (*Besso Correspondance,* p. 391).

There are two more very slight bits of evidence, perhaps more psychohistory than history of science, which may interest you if you have been at all swayed by my argument. The conversation with Hevesy is not the only time Einstein's reaction to Bohr's theory was recorded. In a conversation with Pais in the fifties, A. D. Fokker reminisced about working with Einstein in Zürich: "In 1913–14 he [Fokker] studied with Einstein in Zurich and there he gave the first colloquium on Bohr's theory of the hydrogen atom. Einstein, Laue and Stern were among the audience. Einstein did not react immediately, but kept a meditative silence."[13] If he was learning that some of his own unpublished ideas of almost a decade earlier had been developed and carried to such a successful conclusion by someone else, he might well have meditated.

Finally, I quote from Einstein's 1949 "Autobiographical Notes," in many ways the summing up of his scientific career. He never makes any claim to have anticipated Bohr in this document—or anywhere else, as far as I know—but there is a curious transition at one point. In talking about the early years of this century, he says:

> All of this was quite clear to me shortly after the appearance of Planck's fundamental work; so that, without having a substitute for classical mechanics, I could nevertheless see to what kind of consequences this law of temperature-radiation leads for the photo-electric effect and for other related phenomena of the transformation of radiation-energy, as well as for the specific heat of (especially) solid bodies. All my attempts, however, to adapt the theoretical foundation of physics to this [new type of] knowledge failed completely. It was as if the ground had been pulled out from under one, with no firm foundation to be seen anywhere, upon which one could have built. That this insecure and contradictory foundation was sufficient to enable a man of Bohr's unique instinct and sensitivity to discover the principle laws of the spectral lines and of the electron-shells of the atoms together with their significance for chemistry appeared to me as a miracle—and appears to me as a miracle even today. This is the highest form of musicality in the sphere of thought. (P. 45, translation corrected from new edition)[14]

The reference to Bohr's work—the only one in the "Autobiographical Notes"— is completely out of the chronological order that Einstein by and large follows in this essay. One may speculate that there was an association of ideas at this point: recalling his own early attempts to understand atomic spectra could have led him to pay tribute at this point to the man who later so successfully developed similar ideas.

This is all the evidence I have for my conjecture. If Besso's advice influenced Einstein to drop a train of ideas that might have led to the Bohr atom in 1905, then Besso's negative influence on the history of modern physics is at least as great as his positive influence through encouragement of Einstein's work in relativity. The lack of a response from Lenard might also have discouraged Einstein from going on with this work.

Of course, it might be argued that it was primarily Einstein's basic approach to physics which led him away from this problem. Just after the lines quoted above, Einstein continues: "My own interest in those years (after the turn of the century) was less concerned with the detailed consequences of Planck's results, however important these might be. My main question was: What general conclusions can be drawn from the radiation-formula [i.e., Planck's] concerning the structure of radiation and even more generally concerning the electromagnetic foundations of physics?" ("Autobiographical Notes," p. 47).

2. EINSTEIN AND THE QUANTUM HYPOTHESIS

The tone of respect—perhaps awe would not be too strong a word—which Hevesy used in recounting Einstein's views to Rutherford and Bohr indicates how highly regarded Einstein was around 1913 (before general relativity was completed) in the physics community. Some of his renown at that time no doubt came from his work on special relativity, but probably more of it was due to his work on quantum theory. It did not come, however, from his espousal of what came to be called the photon concept. The idea that electromagnetic radiation possesses particulate characteristics, put forward by Einstein in his first quantum paper of 1905, was generally looked upon as a quirk of Einstein's, to be tolerated rather than taken seriously. In a formal recommendation of Einstein for the Berlin post he assumed in 1914, Planck, Nernst, Rubens, and Warburg—leading lights of the Berlin physics community and all deeply interested in quantum theory—said: "That he may sometimes have missed the target in his speculations, as, for example, in his theory of light quanta, cannot really be held against him." What work of Einstein's did they single out for praise? After mentioning his work on special relativity, they went on:

> Far more important for practical physics is his penetration of other questions on which, for the moment, interest is focused. Thus he was the first man to show the importance of the quantum theory for the energy of atomic and molecular movements, and from this he produced a formula for the specific heat of solids which, although not yet entirely proved in detail, has become a basis for further development of the newer atomic kinetics. He has also linked the quantum hypothesis with photoelectric and photochemical effects by the discovery of interesting new relationships capable of being checked by measurement, and he was one of the first to point out the close relationship between the constant of elasticity and those in the optical vibrations of crystals.[15]

It was this work of Einstein, starting with his 1907 paper (written late in 1906) "Planck's Theory of Radiation and the Theory of Specific Heat," that actually put quantum theory in the mainstream of physics. The theory of black body radiation was considered a rather esoteric specialty by all but a handful of physicists—but a handful that included Planck, Einstein, Lorentz, and Ehrenfest, of course. Among that handful, only Einstein and Ehrenfest accepted the idea that Maxwell's classical theory of the electromagnetic field would have to be profoundly modified to account for quantum effects. The prevailing opinion was that some modification of the theory of matter and/or the interaction between matter and radiation was all that was needed. I shall return to this question shortly, but I want to emphaize that it was largely because of Einstein's work, applying the quantum hypothesis to the most varied topics in the structure of matter and the interaction between matter and radiation, that quantum theory really caught on and became a "mainstream" topic of experimental and theoretical work in the physics community. Kuhn has called attention to this point in his recent book.[16]

Among the theoreticians, Einstein was primarily responsible for this major shift of attention. Peter Debye reminisced in 1964:

> The whole thing started with a kind of interpolation formula by Planck. Nobody wanted to accept it then because it did not appear logical. In Planck's radiation formula half of the argument was continuous and the other half was based on the concept of quanta of energy, which he set equal to $h\nu$ in order to get Wien's Radiation Law. There was much trouble at the beginning. The only man who appeared sensible was Einstein. He had the feeling that if there was anything to Planck's idea then it should also appear in other parts of physics. Well, at that time, he talked about the photoeffect, specific heats, and so forth. Then I tried to formulate the theory of specific heats in a more general way.[17]

What distinguished Einstein's approach from the outset was his firm conviction that neither classical mechanics nor classical electrodynamics—nor a combination of the two—provided a "firm foundation," as he put it in the "Autobiographical Notes," on which to build theoretical physics. Thus, he rejected not only the mechanical world view inherited from the nineteenth-century masters of physics; but the new electromagnetic world view, as well as a combination of the two approaches such as Lorentz's theory of the electron. He rejected them, of course, not as useful approximations with limited domains of applicability, but as "firm foundations" for the edifice of theoretical physics. His later recollection of this early rejection was not an artifice of Einstein's memory. In 1907 he wrote.

> We do not have for the moment a complete world-picture [*Weltbild*] consistent with the relativity principle. . . . In earlier papers I have shown that our present electromechanical world-picture is not suited to explain the entropy properties of radiation nor the laws of emission and absorption of radiation and those of specific heat. Rather, it is necessary to assume, in my opinion, that the state of every periodic process is such that a transfer of energy can only take place in definite quanta of finite magnitude (light quanta); therefore, that the manifold of really possible processes is a smaller one than the manifold of possible processes in the sense of our current theoretical outlook.[18]

He went on to explain why it was nevertheless possible to use current theories under certain conditions.

For a person of Einstein's temperament, to whom the scientific *Weltbild* meant so much both intellectually and emotionally, this was indeed a situation of crisis. Einstein's often-cited 1918 Planck birthday celebration talk, entitled "Principles of Research," gives some idea of the emotional significance of the *Weltbild* for him:

Man tries to make for himself in the fashion that suits him best a simplified and intelligible picture of the world; he then tries to some extent to substitute this cosmos of his for the world of experience, and thus to overcome it. This is what the painter, the poet, the speculative philosopher, and the natural scientist do, each in his own fashion. Each makes this cosmos and its construction the pivot of his emotional life, in order to find in this way the peace and security which he cannot find in the narrow whirlpool of personal experience.[19]

Perhaps even more revealing is an address to the students of UCLA given in February 1932:

Science as something already in existence, already completed, is the most objective, impersonal thing that we humans know. Science as something coming into being, as a goal, however, is just as subjectively, psychologically conditioned as all other human endeavors. This is so much the case that the question of the goal and meaning of science will receive quite different answers at different times and from different personalities.

To be sure, all are agreed that science must establish a connection between facts of experience such that on the basis of experienced facts we are enabled to predict other such facts. According to the view of many positivists the most complete possible solution of this task is the only goal of science.

I do not believe, however, that so primitive an ideal would really permit the kindling to a high degree of that researcher's passion from which really great accomplishments arise. A stronger, but also more obscure drive lies behind the tireless exertions tied to such achievements: one wants to comprehend being, reality. . . . At the basis of all such attempts lies the belief that being is completely harmonious in its structure. Today we have less ground than ever before to allow ourselves to stray from this wondrous belief. (Item 2-110).[20]

The large cathexis of emotion attached to his views on quantum mechanics can also be seen in a letter to the physicist Tatiana Ehrenfest, widow of Paul Ehrenfest: "I find the idea that there should not be laws for being ['*das Seiende*'], but only laws for probabilities, simply monstrous [*Scheusslich*] (a nauseatingly indirect description)"(Einstein to T. Ehrenfest, October 12, 1938, Item 10-296).[21]

Perhaps one can appreciate now the full significance of the words, quoted earlier, in which Einstein recalled his feelings at the turn of the century: "It was as if the ground had been pulled from under one, with no firm foundation to be seen anywhere, upon which one could have built" ("Autobiographical Notes," p. 45). A little later in the "Autobiographical Notes" he states: "Reflections of this type made it clear to me as long ago as shortly after 1900, i.e., shortly after Planck's trail-blazing work, that neither mechanics nor electrodynamics could (except in limiting cases) claim exact validity. By and by I despaired of the possibility of discovering the true laws by means of constructive efforts based on known facts. The longer and more despairingly I tried, the more I came to the conviction that only the discovery of a universal formal principle could lead us to assured results" (pp. 51–53).

Note the emotion-laden content of Einstein's wording: I do not think it is rhetorical exaggeration, but takes us to the deep well-springs of Einstein's extraordinary creativity. It would require an insightful psychobiographer to follow up these clues. Lacking such skills, I shall confine myself to explaining what Einstein meant by the statement: "I became convinced that only the discovery of a universal formal principal could lead us to assured results." For this will lead to an understanding of how Einstein approached the quantum riddle and why he felt that fifty years of pondering [*Grübelei*] had not led him closer to an answer.

3. CONSTRUCTIVE AND PRINCIPLE THEORIES

We can begin by recalling the contrast between two types of theory Einstein drew in a 1919 article for the *Times* of London:

> We can distinguish various kinds of theories in physics. Most of them are constructive. They attempt to build up a picture of the more complex phenomena out of the materials of a relatively simple formal scheme from which they start out. Thus the kinetic theory of gases seeks to reduce mechanical, thermal, and diffusional processes to movements of molecules—i.e., to build them up out of the hypothesis of molecular motion. When we say that we have succeeded in *understanding* a group of natural processes, we invariably mean that a constuctive theory has been found which covers the processes in question.
>
> Along with this most important class of theories there exists a second, which I will call "principle-theories." These employ the analytic, not the synthetic, method. The elements which form their basis and starting-point are not hypothetically constructed but empirically discovered ones, general characteristics of natural processes, principles that give rise to mathematically formulated criteria which the separate processes or the theoretical representations of them have to satisfy. Thus the science of thermodynamics seeks by analytical means to deduce necessary conditions, which separate events have to satisfy, from the universally experienced fact that perpetual motion is impossible.
>
> The advantages of the constructive theory are completeness, adaptability, and clearness, those of the principle theory are logical perfection and security of the foundations.
>
> The theory of relativity belongs to the latter class. (*Ideas and Opinions,* p. 228)

I maintain that Einstein adopted a variant of this "principle-theory" approach in all of the quantum work for which he is so justly famous: development of the photon concept; quantum theory of specific heats; wave-particle duality; the thermodynamic derivation of Planck's law; the derivation based on transition probabilities; and numerous other contributions I shall not even mention. Yet he always maintained: "When we say that we have succeeded in *understanding* a group of natural processes, we *invariably* mean that a *constructive* theory has been found which covers the processes in question" (emphasis added); so "A really rational theory would have to *deduce* the elementary structures (electrons, etc.) not *posit* them from the outset." Never satisfied with the extraordinary successes of his "principle-theory" approach to the quantum, he continued to search for a constructive theory which would give him the *understanding* which, as indicated above, meant so much to him both intellectually and emotionally. He sought it through various unified field theories which he attempted to construct (note the significance of this word!) from 1909 (*before* general relativity) until the end of his life.

These preliminary remarks must serve to motivate my division of Einstein's efforts to come to terms with the quantum into four constituents or strands:

1. Statistical studies of radiation and matter, which led him to the firm conviction that one would never understand Planck's radiation law, matter-radiation interactions, specific heats at low temperatures, or many other properties of solid bodies without radical revision of both classical mechanics and classical electromagnetic theory.

2. The attempt to use trustworthy general principles, such as those of thermodynamics and statistical reasoning,[22] combined with the empirically warranted "quantum hypothesis" as Einstein called it, to derive reliable consequences about the behavior of radiation and matter, alone and in interaction. This procedure was not to be confused with an *understanding* of the quantum since the "quantum hypothesis" has been posited without explanation.

In his report to the first Solvay Congress in 1911, Einstein discusses what general principles can be relied upon in deriving the consequences of the quantum hypothesis. He says:

> We are all agreed that the so-called quantum theory of today is indeed a useful tool, but no theory in the ordinary meaning of the word, at any rate not a theory which could now be developed in a coherent manner. On the other hand, it has been proved that classical mechanics, as expressed in Lagrange's and Hamilton's equations, no longer can be regarded as a usable system for the theoretical representation of all physical phenomena. . . .
>
> So the question arises, on the validity of which general principles of physics we may hope to rely in the field of concern to us [i.e., quantum phenomena]. In the first place we are all agreed that we should retain the energy principle.
>
> A second principle to the validity of which, in my opinion, we absolutely have to adhere is Boltzmann's definition of entropy by means of probability. The weak glimmer of theoretical light that we see today over equilibrium states of processes of an escillatory nature we owe to this principle.[23]

Einstein goes on to discuss how to attach meaning to Boltzmann's principle which is independent of theoretical assumptions about the nature of the system to which it is applied. But I want to emphasize Einstein's approach: Take universal formal principles on which we can rely (based on their wide range of empirical success, simplicity, etc.) and apply them to empirically validated results for some quantum system, such as Planck's law for black body radiation, in order to see what can be learned about the nature and/or structure of the system in question, such as black body radiation. Other formal principles in which Einstein placed great confidence were the relativity principle[24] and the laws of thermodynamics. The latter can be looked upon as applications of the energy principle and the Boltzmann principle, but Einstein thought it important to demonstrate that certain results could be deduced from the quantum principle by purely thermodynamic arguments.

3. The search for a constructive theory of matter and radiation which would yield an understanding of the quantum. Such a theory should be based upon "hypothetically constructed" "elements" fitted together into a "relatively simple formal scheme" from which a "picture of the more complex phenomena" could be deduced.

In his 1909 survey paper "On the Present State of the Problem of Radiation," Einstein reported on his efforts to set up a nonlinear electromagnetic field theory: "I have not yet succeeded in finding a system of equations satisfying these conditions, from which I could see that it was suited to the construction of the electrical elementary quantum [i.e., the electron] and the light quantum. The manifold of possibilities does not seem to be so great that one need be scared away from the task."[25] Thus, at least as early as 1909 Einstein was searching for a nonlinear, "unified" field theory from which electrons and light quanta could be derived. Within a few years he had given up this ambitious program:

> I have also come to the opinion, as a result of many fruitless attempts, that by purely constructive efforts [*blosses Konstruieren*] one cannot put radiation theory back on its feet. Therefore, I have attempted to arrive at a new formulation of the question purely thermodynamically, without making use of any model [*Bild*]. . . .
>
> One cannot really seriously believe in the existence of countable quanta, since the interference properties of light emitted in various directions from a luminous point are really not compatible with them. In spite of this the "honorable" quantum theory is still more preferable to me than the compromises found *up to now* as a substitute. (Albert Einstein to W. Wien, May 17, 1912, Item 23-558)

In later years, after he had developed the general theory of relativity, Einstein came to doubt whether synthetic-constructive efforts starting from concepts closely linked to empirical evidence could ever lead to such a unified theory. He shifted to the search for a formal scheme which, starting from a small number of highly abstract concepts, would lead after a long chain of deductions to an explanation of quantum effects:

> I am firmly convinced that every attempt to arrive at a rational theory by *synthetic* construction will have an unsatisfactory result. Only a new basis for all of physics, from which all possible processes can be deduced with logical necessity (as for

example is the case with thermodynamics) can bring a convincing solution. (Albert Einstein to M. Renninger, June 11, 1953, Item 20-032)

Now you will understand why I lapsed into my apparently Don Quixotic attempts to generalize the gravitational equations. If one cannot trust Maxwell's equations, and a representation [*Darstellung*] by means of field and differential equations is indicated, on account of the principle of general relativity, and one has come to despair of arriving at deeper basis [*Tieferlegung*] of the theory by intuitive [*anschaulich*]—constructive means; then no other sort of effort seems open. (Einstein to von Laue, January 17, 1952, Item 16–167)

I do not believe in micro- and macro- laws, but only in (structure) laws of general rigorous validity. And I believe that these laws are *logically simple,* and that reliance on this logical simplicity is our best guide. Thus, it would not be necessary to start with more than a relatively small number of empirical facts. If nature is not arranged in correspondence with this belief, then we have altogether very little hope of understanding it more deeply. . . .

This is not an attempt to convince you in any way. I just wanted to show you how I came to my attitude. I was especially strongly impressed with the realization that, by using a semi-empirical method, one would never have arrived at the gravitational equations of empty space. (Einstein to David Bohm, November 24, 1954, Item 8-055)

The line he had earlier draw between a theory of principle and a constructive theory here becomes blurred; but the goal of *deducing* quantum phenomena from some unified theory, rather than assuming their existence from the outset, remains

The three constituents or strands so far mentioned are not to be thought of as chronologically related. They overlap and even sometimes intertwine. For example, Einstein continued to examine and test classical theories from new angles to demonstrate their insufficiency to his colleagues. And clues provided by the "principle theory" approach to the quantum hypothesis suggested leads for a constructive theory.

4. When, in the mid-1920s, matrix mechanics was developed by Heisenberg, Born, and Jordan and wave mechanics by de Broglie and Schrödinger, a fourth constituent or strand enters the picture. At first Einstein participated eagerly, if critically, in the exploration of the new ideas. After he became convinced that neither approach—nor the unification of the two in the new quantum mechanics—was "the real thing," "the true Jacob," as he sometimes expressed it, he began a critical exploration of quantum mechanics aimed at demonstrating that it could not be regarded as a complete description of physical reality. This critical examination should always be viewed, however, in the context of his search for a complete, explanatory theory.

There is no space for even a brief survey of Einstein's work in each of these areas. I shall discuss a few characteristic highlights of Einstein's approach and a few points I feel are treated inadequately in the literature, including the nature of Einstein's critique of quantum mechanics which is still sometimes inadequately understood.

4. THE WAVE-PARTICLE DUALITY

Einstein's discovery of what is now called the wave-particle duality, made in late 1908 or early 1909, has been extensively and well treated (notably by Klein) so I can be brief. But I cannot forbear saying a few words about it, since it is so characteristic of Einstein's approach and was so pregnant with future significance. In his early papers on statistical mechanics, Einstein had developed general methods for analyzing the fluctuations of any mechanical system, methods which he applied first to testing the atomic hypothesis by considering observable fluctuations of small particles suspended in a fluid (Brownian motion), In his first quantum paper of 1905 Einstein boldly applied this approach to a nonmechanical system: black body radiation. He showed that, in the high-frequency limit where Wien's law holds, this radiation behaved (with respect to its energy fluctuations, at least) like a gas of independent particles. In 1908–1909 he applied his fluctuation approach to Planck's law which holds for radiation at all frequencies. His strategy was to forget about the origin of Planck's law for the moment (he had much to say about this both earlier and later). Taking Planck's law as a valid empirical description of the radiation spectrum, what can we learn about the structure of that radiation by calculating the average fluctuation of the energy in some given frequency range in a small volume of the radiation?

To answer this question, he applied his general formula for energy fluctuations to Planck's law. The result consists of two terms: one was linear in the energy density, the other was quadratic. The linear term, as he knew from his work on Wien's law, corresponds to the fluctuations of a gas of independent particles. The second term corresponds to the fluctuations resulting from superposition of random standing electromagnetic waves in a cavity, the energy fluctuations in a small volume arising from interference. (Such a classical wave model of black body radiation leads to the Rayleigh-Jeans law, of course.) He showed that there is an analogous two-term result for the momentum fluctuations of black body radiation.

Since Einstein's early views on light quanta are still sometimes misunderstood, let me quote a letter to Lorentz: "I am not at all of the opinion that one should think of light as composed of quanta which are independent of each other, and localized in relatively small regions, For the explanation of the Wien limit of the radiation formula this would indeed be the easiest way. But even the splitting of a light ray at the surface of a refracting medium completely forbids this approach. A light ray divides, but a light quantum indeed cannot divide without an alteration of frequency" (Einstein to Lorentz, May 23, 1909, Item 16-419).

During the next few years, Einstein was not always certain that the concept of light quanta was the best way to approach the study of radiation. At one particularly low moment, after the failure of several attempts at a constructive quantum theory, he wrote: "Right now I am trying to derive the law of heat conduction in rigid dielectrics from the quantum hypothesis. Whether these quanta really exist, I don't ask any more. I am also not trying any more to construct them, because I know now that my brain is not capable of it. But I am searching as diligently as possible to learn the domain of applicability of this concept [*Vorstellung*]" (Albert Einstein to M. Besso, May 13, 1911, *Besso Correspondance,* pp. 19–20). The success of his well-known 1916–1917 derivation of the Planck law, showing that individual acts of emission and absorption of radiation by atoms must be directed processes involving momentum recoil, finally reassured him about the existence and particulate nature of light quanta.[26]

Einstein wrote to Besso: "The quantum paper sent to you has again led me back to the view of the spatial corpuscularity [*Quantenhaftigkeit*] of radiation energy. But I feel that the real jest that the eternal riddle-poser has set before us is still absolutely not understood. Shall we live to see the saving idea?" (Einstein to Besso, March 7, 1917, *Besso Correspondance,* p. 103).

His 1909 work convinced Einstein that radiation could be understood only through some sort of synthesis of wave and particle concepts. He placed great emphasis on his fluctuations results, later making the ability to explain fluctuation phenomena a challenge with which to test the new quantum mechanics. The famous "*Dreimännerarbeit*" of Heisenberg, Born, and Jordan which appeared in 1926[27] claimed to resolve a paradox sharply emphasized by Ehrenfest the previous year:[28] while quantization of the proper vibrations in a cavity yielded the Planck formula for spectral energy distribution (as Debye showed in 1911), it produced the *wrong* formula for energy fluctuations in a small volume (as Einstein indicated in 1909 and Lorentz proved in detail a few years later). By introduction of the so-called zero-point energy of the proper vibrations into which the field was analyzed, Jordan—who wrote this section of the paper—claimed to show that the new matrix mechanics, when applied to fields, could produce not only the quadratic (classical wave) term but also the linear (classical particle) term. The latter term results from interference between the zero point and the other oscillations. Einstein, who had been given proofs of the article, was not impressed. In a postcard to Jordan he wrote: "The thing with the fluctuations is rotten [*faul*]. One can indeed calculate the average magnitude of fluctuations with the zero point term $\frac{1}{2}h\nu$, but not the probability of a very large fluctuation" (Einstein to Pascual Jordan, April 6, 1926, Item 13-479).

A discussion of fluctuations had been going on between Einstein and Jordan previously and continued for some time after. In one of his papers on field quantization[29] Jordan tried to answer Einstein's critique, but it is clear that Einstein was never satisfied with the quantum-mechanical explanation of fluctuations or the wave-particle duality.

A couple of quotations illustrate Einstein's attitude:

As far as 'fluctuations' are concerned, they are completely incompatible with the Maxwell energy tensor, even if the latter does explain a part (for weak fields an insignificant one) of the fluctuations. If one doesn't explain the fluctuations by a swindle (I mean the renunciation of the description of the individual system), then the radiation must be richer in structure than can be expressed by a wave theory in the conventional sense. (Einstein to Schrödinger, June 15, 1950, Item 22-168).

Encouraging a physicist critical of conventional quantum theory, Einstein wrote:

I find it quite reasonable that, by your Gedankenexperiment, you have again drawn people's attention to the fact that the wave-particle duality is a reality which one should not allow to be deluded away by metaphysical artifices. (Einstein to M. Renninger, February 27, 1954, Item 20-036)

Describing to Laue how a study of the fluctuations of a mirror in a cavity filled with black body radiation first convinced him of the inadequacy of Maxwell's theory, he adds:

The quantum theory doesn't help me at all here; it seems to me to be self-deception when one envelops the movement of a plate in a radiation field in the famous probability cloud, and the field as well. Nothing better will come from this. (Einstein to von Laue, January 17, 1952, Item 16-167)

To show that Einstein was perhaps not exaggerating in his comment about self-deception which opened this article, let me quote the interview with Peter Debye which I cited earlier.[30] Reminiscing about the Einstein fluctuation result, Debye says:

Einstein had calculated the fluctuations of energy in a space containing radiation. Now, there are two parts to it; if you say that it is all waves, you get a certain fluctuation. But, if you take the entropy, from this, with Planck's formula, you may calculate a second fluctuation which is independent of the fluctuations due to the waves. But that is not a real fluctuation; it is inherent in the quantum formulation.

Bauer: And this caused the trouble. Well, nowdays, we don't feel troubled by this quality.

Debye: Well, of course not. People have become accustomed to it. At that time, one had to try to answer whether an electron was a wave or a particle. Of course, it is both.

Corson: Should we change the subject?

I do not quote this to single out Debye, but just because his comment is so typical of many discussions of the wave-particle duality. Einstein comments on such "explanations": "We know that light has certain characteristics which we designate for short, respectively, as undulatory and corpuscular. It has no meaning to say, it *is* a wave and it *is* a corpuscle. Up to now we just have no reasonable theory which explains all its characteristics. However there is no *contradiction,* any more that it signifies a contradiction that a man feels *and* has weight" (Albert Einstein to Robert Federn, July 27, 1954, Item 59-654).

Even if one fully accepts quantum mechanics, the fluctuation story does not end so simply. A 1931 paper by Heisenberg shows that the calculation in the *Dreimännerarbeit* actually *diverges* if one does it right—and not because of a divergent zero-point energy, either. I shall not go into the further history of this interesting problem, the resolution of which from the quantum-mechanical standpoint, I believe, hinges on a specification of the conditions of measurement of the fluctuations.[31]

Recently, the topic of fluctuations has become of central research interest in general relativity, through the study of the Hawking process.[32] Without discussing this fascinating topic, I want to emphasize its significance. For the first time, there appears to be an area in which quantum mechanics and general relativity may be fruitfully applied jointly. Just for this reason I feel there is a need for extreme caution in accepting all the claims that have been made, and for critical analysis of the calculations, and especially of the justification for the application, in this context, of the physical concepts on which these calculations are based.

5. EINSTEIN AND THE BORN INTERPRETATION

Although it is fairly well known that Einstein first introduced transition probabilities into quantum theory, it is not generally realized how large a role he played in the development of the so-called Born or statistical interpretation of the wave function. In his work of 1916–1917, mentioned earlier, Einstein postulated the existence of transition probabilities between the discrete energy levels of an atomic system. Transition probabilities for spontaneous and induced emission and (induced) absorption were assumed to exist by analogy with similar processes in the classical theory of a charged oscillator developed by Planck (by a correspondence principle argument, one would say today) and with the treatment of radioactivity decay by Rutherford. Einstein noted this analogy between radioactivity and absorption and emission as early as 1911 in a letter to Besso (*Besso Correspondance,* pp. 26–27). Although he was very pleased with the resulting derivation of the Planck distribution formula, Einstein never regarded his introduction of transition probabilities as anything but a temporary expedient: "The weakness of the theory lies on the one hand in the fact that it does not get us any closer to making the connection with the wave theory [i.e., overcoming the wave-particle duality]; on the other, that it leaves the duration and direction of the elementary processes to 'chance.' Nevertheless I am fully confident that the approach chosen here is a reliable one" (Van der Waerden, *Sources*, p. 76).

In January 1920, he wrote to Born: "That business about causality causes me a lot of trouble, too. Can the quantum absorption and emission of light ever be understood in the sense of the complete causality requirement, or would a statistical residue remain? I must admit that there I lack the courage of my convictions. But I would be very unhappy to renounce *complete* causality. . . . (The question whether strict causality exists or not has a definite meaning, even though there can probably never be a definite answer to it.)" (*Born-Einstein Letters*, p. 23).[33]

His doubts about introducing probabilistic assumptions as a matter of principle did not prevent him from discussing a probabilistic interpretation of Maxwell's equations to link the wave and particle concepts. Although these speculations were well known to a number of physicists during the early twenties, Einstein never published anything about his concept of the "ghost-field" ["*Gespensterfeld*"].[34] Lorentz wrote a long letter to Einstein in 1921 discussing this idea and gave a very similar account in his 1922 lectures at the California Institute of Technology, later published as "Problems of Modern Physics."[35] Since the book is readily available, I shall quote only a couple of sentences from it:

> The hypothesis of light-quanta, however, is in contradiction with the phenomena of interference. Can the two views be reconciled? I should like to put forward some considerations about this question, but I must first say that Einstein is to be given credit for whatever in them may be sound. As I know his ideas concerning the points to be discussed only by verbal communication, however, and even by hearsay, I have to take the responsibility for all that remains unsatisfactory. (*Problems of Modern Physics*, pp. 156–57)

Now an excerpt from the letter:

> Basic ideas. In emission of light two things are radiated. There is namely:
>
> 1. An interference radiation, which occurs according to the ordinary laws of optics, but still carries no energy. One can, for example, imagine that this radiation consists of ordinary electromagnetic waves, but with vanishingly small amplitudes. As a consequence they cannot themselves be observed; they serve only to prepare the way for the radiation of energy. It is like a dead pattern, that is first brought to life by the energy radiation. [In the book, he says; "On a screen you will have something like an undeveloped photographic image."]
>
> 2. The energy radiation. This consists of indivisible quanta of magnitude $h\nu$. Their path is prescribed by the (vanishingly small) energy flux in the interference radiation, and they can never reach places where this flux is zero (dark interference bands).
>
> In an individual act of radiation the full interference radiation arises, but only a single quantum is radiated, which therefore can only reach one place on a screen placed in the radiation. However, this elementary act is repeated innumerably many times, with as good as identical interference radiation (the same pattern). The different quanta now distribute themselves statistically over the pattern, in the sense that the average number of them at each point of the screen

is proportional to the intensity of the interference radiation reaching that point. In this way the observed interference phenomena arise, corresponding to the classical results. (Lorentz to Einstein, November 13, 1921, Item 16-544)

Lorentz explicitly attributes the ideas quoted to Einstein.

If these ideas reminded you of the Born interpretation of the wave function, this is not a coincidence. In the second, more detailed, of two papers entitled "The Quantum Mechanics of Collision Processes," received July 21, 1926, Born contrasts Heisenberg's interpretation of matrix mechanics, according to which "an exact representation [*Darstellung*] of processes in space and time is impossible in general" with Schrödinger's interpretation of wave mechanics, according to which a reality similar to that of classical light waves is attributed to de Broglie waves.[36] He continues: "Neither of these two conceptions seems satisfactory to me. I shall attempt to give a third interpretation and test its usefulness for collision processes. Here I connect with an observation of Einstein on the relation of wave field and light quantum; he said more or less, that the waves are only there in order to show the path to the corpuscular quanta, and he spoke in this sense of a ghost-field [*Gespensterfeld*]. This determines the probability for a light quantum, the carrier of energy and momentum, to take a particular path; however, no energy or momentum belongs to the field itself."

Leaving aside the puzzling question why no such acknowledgment appears in Born's first paper on the subject, written about a month earlier, note that Born made similar acknowledgments on a number of later occasions. The first seems to have been in a letter to Einstein that is not included in the published *Born-Einstein Letters.* On November 30, 1926, Born wrote: "To report about myself, I am quite satisfied as far as physics goes, since my idea to conceive of the Schrödinger wave field as a ghost field in your sense, is constantly proving to be better" (Item 8-179).

This might seem to be the whole story: Einstein invented the "ghost-field" interpretation of the Maxwell field, and Born later applied it to the de Broglie-Schrödinger waves, in particular to collision phenomena. So the story appears in Jammer's book, for example, but two footnotes in Heisenberg's first paper on the uncertainty relations suggest there is more to the story. Early in the paper is a footnote which reads: "The present paper arose out of efforts and desires to which others gave clear expression much earlier, before the origin of quantum mechanics. I recall here especially Bohr's works on the fundamental postulate of quantum theory and Einstein's discussion of the connection between wave field and light quantum" (pp. 173–74).[37]

This note makes it clear that Heisenberg knew about Einstein's *Gespensterfeld* idea.[38] Perhaps he had even discussed it with Einstein, although Heisenberg's later reminiscences of his conversations with Einstein do not report such a discussion. The really remarkable footnote, however, comes later: "The statistical interpretation of the de Broglie waves was first formulated by Einstein. (*Sitzungsber. d. preuss. Akad. d. Wiss. 1925*, p. 3)" (p. 176). After citing a paper of Born and Heisenberg, and one by Jordan, in which "this statistical element in quantum mechanics plays an essential role," he continues: "It was mathematically analyzed in a fundamental paper of M. Born (*Zeitschr. f. Physik* 38, 803, 1926) and used for the interpretation of collision phenomena."

Since Heisenberg had previously acknowledged Einstein's work on "the connection between wave field and light quantum" his citation of Einstein as originator of the statistical interpretation of *de Broglie waves* must be taken seriously. The paper he cites is Einstein's famous second paper on "Quantum Theory of the Monatomic Ideal Gas," submitted January 8, 1925, in which he applied what we now call Bose-Einstein statistics to such a gas.[39] Sections 8 and 9 of the paper are of relevance to our discussion. In the first, entitled "The Fluctuation Properties of the Ideal Gas," he uses his ubiquitous fluctuation technique to calculate the average fluctuation in the number of molecules within a small volume of the gas. As in the radiation case, he arrives at two terms: one corresponds to a classical particle fluctuation term, as might be expected; the other corresponds to a fluctuation term arising from interference of classical waves. The wave-particle duality has now reappeared for a corpuscular system.

Einstein states that this result should be taken seriously "because I believe it is a question of more than a mere analogy." He goes on to cite de Broglie's work, known to him from a copy of de Broglie's 1924 thesis sent to him by Paul Langevin, Einstein's old friend and de Broglie's thesis adviser. This discussion of de Broglie's work is well known; de Broglie acknowledged the role it played in getting a hearing for his ideas:

> In November 1924 I had defended before the Faculty of Science of Paris a Doctoral Thesis in which I had summarized my new ideas on wave mechanics. Paul Langevin had communicated the manuscript of my work to *M.* Einstein who had immediately perceived its interest. Shortly afterwards, in January 1925 the illustrious scientist presented a note to the Berlin Academy of Sciences in which he stressed the importance of the concepts expounded in my Thesis and deducing numerous consequences from them. This memoir of *M.* Einstein drew attention to my work, little noticed up to then and for that reason I have always felt that I owed him a great personal debt for the precious encouragement he thus brought me.[40]

But the following (ninth) section of the paper does not seem to be as well known, perhaps because its title, "Remark on the Viscosity of the Gas at Low Temperatures," does not seem to promise much of general interest. I shall therefore quote from it at length:

According to the considerations of the previous paragraph, it appears that an undulatory field is connected with every motion [*Bewegungsvorgang*], just as the optical undulatory field is connected with the motion of light quanta. This undulatory field, whose physical nature is for the moment still unclear, must in principle permit its existence to be demonstrated by the corresponding phenomena of its motion [*Bewegungserscheinungen*].

He goes on to discuss the diffraction of a stream of gas molecules by a slit, presumably following de Broglie, who had given a similar discussion in his thesis. Einstein concludes that, at ordinary speeds, the wavelength is too small to allow observation of such diffraction by a slit. He goes on:

At low temperatures, λ [the de Broglie wavelength] is of the order of magnitude of σ [the molecular diameter] for the gases hydrogen and helium, and it seems indeed that the influence on the frictional coefficient to be expected from the theory makes itself felt.

If the stream of molecules moving with the speed v strikes another molecule, which for convenience we represent as immobile, then this is comparable with the case of a wave train of a certain wavelength λ striking a foil of diameter 2σ. A (*Fraunhofer*) diffraction effect occurs which is analogous to that yielded by an aperture of the same magnitude. Large diffraction angles occur if λ is of the same order of magnitude as σ or larger. Thus, in addition to deflections due to collision in accord with [classical] mechanics, mechanically inexplicable deflections of the molecules also will occur with frequency comparable to the former, which will diminish the mean free path.

Einstein applies this idea to the interpretation of data on the variation of the viscosity of a gas with its temperature. Note how close Einstein came here to the statistical interpretation of de Broglie waves scattering from a hard sphere. Born was certainly following Einstein's work on gas degeneracy, as his letter of July 15, 1925, to Einstein shows (*Born-Einstein Letters,* p. 83) and also read de Broglie's work "At Einstein's instigation," as he puts it (*Born-Einstein Letters,* p. 86). Born's reference, in his letter to Einstein of November 30, 1926 (previously quoted), to "my idea, to conceive of the Schrödinger wave field as a ghost field in your sense" may be an overstatement of his claim to originality.

This example also illustrates how deeply Einstein was involved in the development of quantum mechanics, both directly and indirectly. I could cite other examples, some well known, like his correspondence with Schrödinger during the development of wave mechanics;[41] others, not so well known, like his correspondence with Heisenberg and Jordan during the development of matrix mechanics.[42] However, I shall only comment on two points that I feel are often misunderstood.

Einstein's relationship with quantum mechanics is sometimes presented as if Einstein were repelled by matrix mechanics from the start while initially rather attracted to wave mechanics. Actually, his attitude to both was similar: initially intrigued by each, he immersed himself in the details of each approach. He came to feel that—although each contained "a piece of the truth" as he sometimes put it—neither was ultimately satisfactory. This happened *before* the discovery that the two theories were mathematically identical, made independently by Schrödinger and Pauli in mid-1926. Rather than document this process in detail, I shall only quote a remarkable anecdote:

Hartmut Kallman recalled one of the famed Berlin colloquia at which the relative merits of Heisenberg's and Schrödinger's points of view were being discussed:

"When a Colloquium on this theme was held at the Berlin University, not 20 but 200 physicists were present. People were packed into the room as lectures on Heisenberg's and Schrödinger's theories were given. At the end of these reports Einstein stood up and said: 'Now just listen! Up until now we had no exact quantum theory, and now we suddenly have two. You will agree with me,' he continued, 'that these two exclude each other. Which theory is correct? Perhaps neither is correct.' At that moment—I shall never forget it—[Walter] Gordon stood up and said: 'I have just returned from Zurich. Pauli has proved that both theories are identical'."[43]

6. OBSERVABLE QUANTITIES

The other comment is on the Heisenberg-Einstein discussion, reported in Heisenberg's book, *Physics and Beyond,*[44] over whether quantum theory should be based exclusively on "quantities observable in principle" [*prinzipiell beobachtbare Grössen*]. I am not certain when this phrase was first used in the physics literature, but Minkowski used a related expression. In his 1908 paper on relativistic electrodynamics[45] ("Die Grundgleichungen für die elektromagnetischen Vorgänge in bewegten Körpern," he speaks of "*lauter beobachbaren Grössen*" (purely observable quantities). Born, of course, was working with Minkowski at the time, and it is possible that this expression influenced his later thinking about quantum theory. It is more likely, however, that it was Einstein who introduced the phrase "quantities observable in principle" to the relevant circle of physicists. A long tradition associates this concept with the special theory of relativity. The closest he gets to using it, as far as I have been able to determine, is at the beginning of the 1905 paper on the special theory where he contrasts "observable phenomena," which only depend on the relative motion of a magnet and a conductor, with the accepted interpretation of Maxwell's theory which "leads to asymmetries, which do not seem to be attached to the phenomena." This clearly suggests a criterion for theories which Einstein made explicit a decade later in a letter to Lorentz, justifying his *general* relativity theory:

In the description of the relative motion (of an arbitrary type) of two coordinate systems K_1 and K_2, it is immaterial whether I relate K_2 to K_1 or inversely K_1 to K_2. If in spite of this, K_1 is distinguished because, relative to K_1, the general laws of nature are supposed to be simpler than relative to K_2, then this preferential status is a fact without physical cause. Of two things K_1 and K_2, equivalent according to their definition, one is distinguished without physical (accessible to observation in principle) basis. My trust in the consistency of natural processes resists this most forcefully. (Einstein to H. A. Lorentz, January 23, 1915, Item 16-436)

Here he uses the phrase "in principle accessible to observation" (*der Beobachtung prinzipiell zugänglich*). A few years earlier he used the expression "quantities observable in principle" (the first use I have found) in comments at the first (1911) Solvay Congress. Discussing his interpretation of Boltzmann's principle, Einstein starts by saying: "If an externally isolated physical system of fixed energy is given, then the system can still assume the most varied states, which are characterized by a number of quantities observable in principle (e.g., volumes, concentrations, the energies of parts of the system, etc.)."[46] After discussing how to define the probability *W* of each such state Einstein continues: "If one defines *W* in the indicated way as temporal frequency, then Boltzmann's equation directly contains a physical assertion. It contains a relation between quantities observable in principle."[47]

It was in justification of the principle of general covariance that he again employed this concept. As I have discussed in detail elsewhere,[48] his dicovery of the flaw in his "hole" argument against general covariance—an argument that helped to delay formulation of the final gravitational field equations for over two years—led him to emphasize the concept of observability in principle. In a letter to Besso he stated: "Nothing is physically real but the totality of spatio-temporal coincidences. If, for example, physical events were to be built exclusively from the motions of material points, then meetings of the points, i.e., the intersections of their world lines would be the only reality, i.e., observable in principle" (*Besso Correspondance,* p. 64, letter of January 3, 1916).

The first use of the phrase I have located by someone central to the development of the new quantum theory is again in a relativistic context. It occurs in a paper by Pauli on Weyl's unified field theory: "One would like to keep to the introduction into physics of only quantities observable in principle."[49]

By 1923 at the latest, Pauli was applying this criterion to questions of atomic theory, as a letter to Eddington shows.[50] At the end of 1924, he wrote to Bohr: "I believe that the values of the energy and momentum of the stationary states are something much more real than the 'orbits.' The (still unattained) goal must be to deduce these and all other physically real, observable properties of the stationary states from the (whole) quantum numbers and the quantum-theoretical laws" (*Pauli Briefwechsel,* p. 189).

Near the beginning of their paper "On the Quantum Theory of Aperiodic Processes," which just preceded Heisenberg's first paper on what was later called matrix mechanics, Born and Jordan remark: "A fundamental proposition of great significance and fruitfulness asserts that in the true laws of nature only such quantities enter as are observable, determinable in principle."[51]

They add in a footnote: "Thus, the theory of relativity arose from the circumstance that Einstein recognized the impossibility in principle of determining absolute simultaneity of two events taking place at different places."

Even Born's first papers on the statistical interpretation of the wave function do not deviate from this approach. In a 1926 a lecture on "Physical Aspects of Quantum Mechanics," he still rejected positions as unobservable:

> Formal quantum mechanics obviously provides no means for the determination of the positions of particles in space and time . . . The quantum theoretical description of the system contains certain declarations about the energy, the momenta, the angular momenta of the system; but does not answer, or at least only answers in the limiting case of classical mechanics, the question of where a certain particle is at a given time. In this respect the quantum theory is in agreement with the experimentalists, for whom microscopic coordinates are also out of reach.[52]

Thus, originally Born did not envisage the interpretation of the amplitude squared of the wave function as proportional to the probability of a position measurement, as Linda Wessels and Mara Beller have pointed out.[53]

By the time of Heisenberg's discussions with Einstein there was thus a considerable tradition citing the authority of Einstein to assert that only quantities observable in principle should be introduced in a physical theory. Heisenberg claimed to base his approach to quantum theory on this principle.[54] In his first paper (July 1925) he states:

> It is well known that the formal rules which are used in quantum theory for calculating observable quantities, such as the energy of the hydrogen atom, may be seriously criticized on the grounds that they contain, as a basic element, relations between quantities that are apparently unobservable in principle, e.g., position and period of revolution of the electron. Thus, these rules lack an evident physical foundation, unless one still wants to retain the hope that the hitherto unobservable quantities may later come within the realm of experimental determination.

After giving reasons why such a hope seems unfounded, he continues:

> In this situation it seems sensible to discard all hope of observing hitherto unobservable quantities, such as the position and period of the electron, and to concede that the partial agreement of the quantum rules with experience is more or less fortuitous. Instead it seems more reasonable to try to establish a theoretical quantum mechanics, analogous to classical mechanics, but in which only relations between observable quantities occur.[55]

Heisenberg was in correspondence with Einstein soon after this. Only Heisenberg's replies are available; it appears from these that, as a result of Einstein's objections, he was anxious for a discussion of his approach.

A talk in Berlin gave Heisenberg the occasion for this discussion. In a chapter entitled "Quantum Mechanics and a Talk with Einstein," he gives an admittedly Thucydidean reconstruction of that conversation. Heisenberg reports that he stated:

> We cannot observe electron orbits inside the atom but the radiation which an atom emits during discharges enables us to deduce the frequencies and corresponding amplitudes of its electrons. After all, even in the older physics wave numbers and amplitudes could be considered substitutes for electron orbits. Now, since a good theory must be based on directly observable magnitudes, I thought it more fitting to restrict myself to these, treating them, as it were, as representatives of the electron orbits.
>
> But you don't seriously believe, Einstein protested, that none but observable magnitudes must go into a physical theory?
>
> Isn't that precisely what you have done with relativity? I asked in some surprise. After all, you did stress the fact that it is impermissible to speak of absolute time, simply because absolute time cannot be observed; that only clock readings, be it in the moving reference system or the system at rest, are relevant to the determination of time.
>
> Possibly I did use this kind of reasoning, Einstein admitted, but it is nonsense all the same. Perhaps I could put it more diplomatically by saying that it may be heuristically useful to keep in mind what one has actually observed. But on principle, it is quite wrong to try founding a theory on observable magnitudes alone. In reality the very opposite happens. It is the theory which decides what we can observe. (*Physics and Beyond,* p. 63)

This discussion is often cited as the quintessential clash between the positivism of Heisenberg and the realistic approach of Einstein. But that is *not* the reason Heisenberg recounts it. The denouement comes only in the next chapter, in which Heisenberg recalls the discovery of the uncertainty relations in February 1927. Bohr and Heisenberg were then beset by doubts about how to reconcile the quantum-mechanical formalism with the circumstance that:

> The path of the electron through the cloud chamber obviously existed; one could easily observe it. The mathematical framework of quantum mechanics existed as well, and was much too convincing to allow for any changes. Hence it ought to be possible to establish a connection between the two, hard though it appeared to be.
>
> It must have been one evening after midnight when I suddenly remembered my conversation with Einstein and particularly his statement, "It is the theory which decides what we can observe." I was immediately convinced that the key to the gate that had been closed for so long must be sought right here. . . .
>
> We had always said so glibly that the path of the electron in the cloud chamber could be observed. But perhaps what we really observed was something much less. Perhaps we merely saw a series of discrete and ill-defined spots through which the electron had passed. In fact, all we do see in the cloud chamber are individual water droplets which must certainly be much larger that the electron. The right question should therefore be: Can quantum mechanics represent the fact that an electron finds itself approximately in a given place and that it moves approximately with a given velocity, and can we make these approximations so close that they do not cause experimental difficulties? (*Physics and Beyond* pp. 77–78)

Heisenberg thus says that a *retreat* from his original extreme positivist or operationalist viewpoint, motivated by Einstein's critique, helped him to find the uncertainty relations. I shall not retell Bohr's story of how Einstein originally tried to circumvent the uncertainty relations by a series of ingenious thought experiments.[56] But I must emphasize that he finally came to accept them fully as a limit on possible experimental determination of the properties of a system. Indeed, he often stressed this point in his later discussions of quantum mechanics, as we shall see.

I shall only mention Einstein's contributions to the experimental study of the question of the nature of light. As we have seen, Einstein never believed in what he called the "naïve conception" of light quanta as totally independent point particles. After he became convinced, in 1916–1917, that light quanta had the particlelike property of momentum as well as energy, he tried to design a crucial experiment, which would distinguish between any reasonable sort of particle picture of light and the classical undulatory picture. If there ever was a time when Einstein was an "insider" in the physics community it was during the twenties in Berlin. So his two suggestions were quickly carried out by experimentalists. But in both cases it turned out, upon more careful theoretical analysis of the situation, that the proposed experiment was actually unable to distinguish between the wave and particle pictures of light.[57] The failure of these "crucial" experiments designed by Einstein probably played a role in the evolution of Bohr's views on light. Bohr held firmly to the undulatory picture for a long time. As late as the early 1920s, he regarded Einstein's light quantum concept as an aberration. Bohr was willing to give up conservation of energy for individual interactions between light and matter to preserve an exclusively undulatory model of light (the Bohr-Kramers-Slater theory of 1923).[58] When the results of the Compton-Simon and Bothe-Geiger experiments showed this approach to be untenable, Bohr was forced to take the particle picture more seriously; but he still would not give up the undulatory picture. I believe the demonstration that both pictures led to the same prediction for the outcome of Einstein's experiments was one of the elements that contributed to shaping Bohr's concept of complementarity of wave and particle descriptions.[59] Bohr continued to regard the particle aspect of light, however, as playing a secondary, more formal role compared to the undulatory aspect, the only one to manifest itself in the classical limit. In Bohr's view, the wave aspects of the electron played a similar formal role. Although Einstein never accepted complementarity (saying he had failed to make sense of Bohr's concept of complementarity in spite of repeated attempts to do so), he came to a rather similar conclusion: "I do not believe that the light-quanta have reality in the same immediate sense as the corpuscles of electricity. Likewise I do not believe that the particle-waves have reality in the same sense as the particles themselves. The wave-character of particles and the particle-character of light will—in my opinion—be understood in a more indirect way, not as immediate physical reality." (Einstein to Paul Bonofield, September 18, 1939, Item 6-118.1).

7. EINSTEIN'S CRITIQUE OF QUANTUM MECHANICS

In concluding this essay, I shall discuss the nature of Einstein's critique of quantum mechanics. This question was closely tied up with his hopes for—and doubts about—the future course of physics. As mentioned previously, Einstein was troubled by the inherently probabilistic nature of the quantum-mechanical description of reality. He indicated that he could more easily conceive of a completely chaotic universe than one governed by probabilistic *laws:* "I still do not believe that the Lord God plays dice. If he had wanted to do this, then he would have done it quite thoroughly and not stopped with a plan for his gambling: In for a penny, in for a pound [*Wenn schon, den schon*]. Then we wouldn't have to search for laws at all." (Einstein to F. Reiche and wife, August 15, 1942, Item 20-019).

He drafted a reply to Born's essay on "Einstein's Statistical Theories": "This article is a moving hymn to a beloved friend, who in his old age (shall we say) unfortunately has succumbed to occultism, one who will not believe in spite of all the evidence, that God plays dice. In one point however, Born does me an injustice, namely, when he thinks that I have been untrue to myself in this respect since earlier I often availed myself of statistical methods. In truth, I never believed that the foundations [*Grundlage*] of physics could consist of laws of a statistical nature" (unpublished draft comment on Born's Essay for *Albert Einstein: Philosopher-Scientist,* Item 2-027).

But this issue was not the deepest source of his dissatisfaction with the prevailing interpretation of quantum mechanics: "The sore point [*Der wunde Punkt*] lies less in the renunciation of causality than in the renunciation of the representation of a reality thought of as independent of observation" (Einstein to Georg Jaffe, January 19, 1954, Item 13-405).

It is important to emphasize that he did not see this as a defect of quantum mechanics as such, but as a defect of the prevailing interpretation of the theory as the most complete possible description of an individual system:

> In what relation does the "state" ("*Quantum state*") described by a ψ function stand to a definite real situation [*Sachverhalt*] (let us call it "real state")? Does the quantum state characterize a real state (1) completely or (2) only incompletely?
>
> The question cannot be answered at once [*ohne Weiteres*] because indeed every measurement signifies an uncontrollable real intervention [*Eingriff*] in the system (Heisenberg). The real state is thus not immediately accessible to experience and its judgement always remains hypothetical (comparable with the concept of force in classical mechanics, if we do not set up the laws of motion a priori). Assumptions (1) and (2) are therefore both possible in principle. A decision

between them can only be reached by investigation and comparison of their consequences.

I reject (1) because this outlook necessitates the assumption that a rigid coupling exists between parts of a system spatially arbitrarily far apart from each other (instantaneous action at a distance, which does not decrease with increasing distance). (*Besso Correspondance,* p. 487, letter of October 8, 1952)

The last sentence is an allusion to the Einstein-Podolsky-Rosen (EPR) argument. Since this argument is well known, I shall only quote a short account by Einstein:

It is not my opinion that there is a logical inconsistency in the quantum-theory itself and the "paradoxon" does not try to show it. The intention is to show that statistical quantum theory is not compatible with certain principles the convincing-power of which is independent of the present quantum theory.

There is the question: Does it make sense to say that two parts *A* and *B* of a system do exist independently of each other if they are (in ordinary language) located in different parts of space at a certain time, if there are no considerable interactions between those parts (expressed in terms of potential energy) at the considered time? . . . I mean by "independent of each other" that an action on *A* has no immediate influence on the part *B*. In this sense I express a principle a)

a) independent existence of the spacially separated. This has to be considered with the other thesis b)

b) the ψ-function is the complete description of an individual physical situation.

My thesis is that a) and b) can not be true together, for if they would hold together the special kind of measurement concerning *A* could not influence the resulting ψ-function for *B*, (after measurement of *A*).

The majority of quantum theorists discard a) tacitly to be able to conserve b). I, however, have strong confidence in a) so that I feel compelled to relinquish b).

(Einstein to L. Cooper, October 31, 1949, Item 8-411)

Einstein held that if one adopted the ensemble or statistical interpretation of the ψ-function, which he identified with the Born interpretation, there was no problem. After a concise exposition of the EPR paper, Einstein continues:

It became clear to me, meanwhile, how one has to choose the interpretation of the formalism [*des Schemas*] of quantum mechanics, in order that the concept of the real state of an (isolated) system not give rise to any sort of paradox; it is none other than the Born interpretation, about which I do not know for certain if it is represented by Born himself with complete consistency.

It is . . . . to be expected that behind quantum mechanics there lies a lawfulness and a description which refer to the individual system. That it is not attainable within the bounds of concepts [taken] from classical mechanics is clear; the latter however is in any case outmoded as the foundation [*Fundament*] of physics. (Einstein to Gregory Breit, August 2, 1935, Item 6-173)

Elsewhere, he elaborates on the inadequacy of the classical-mechanical starting point of quantum mechanics:

I do not at all doubt that the contemporary quantum theory (more exactly "quantum mechanics") is the most complete theory compatible with experience, as long as one bases the description on the concepts of material point and potential energy as fundamental concepts. [The difficulties of quantum mechanics] are connected with the fact that one retains the classical concepts of force or potential energy and only replaces the laws of motion by something entirely new. The completeness of the mathematical mechanism of the theory and its significant success turn attention away from the difficulty of the sacrifice that has been made.

To me it seems, however, that one will finally recognize that something must take the place of forces acting or potential energy (or in the Compton effect, of wave fields), something which has an atomistic structure in the same sense as the electron itself. "Weak fields" or forces as active causes will then not occur at all, just as little as mixed states.[60]

In one of his last letters he again touched on this point, also noting a source of resistance to such objections.

I believe however that the renunciation of the objective description of "reality" is based upon the fact that one operates with fundamental concepts which are untenable in the long run (like f.i. classical thermodynamics). Quite understandably most physicists resist the idea that we are still very far from a deeper insight into the structure of reality. (Einstein to Andre Lamouche, March 20, 1955, Item 15-044)

Did Einstein believe in some sort of hidden variable program to underpin quantum mechanics? Bernard d'Espagnat, after a clear exposition of Einstein's approach to the nonseparability problem in quantum mechanics, maintains that: "Einstein's conclusion was that hidden parameters must exist. These parameters he seems to have conceived as attached to each system, as in the classical case, or as we would say now, as 'local variables': for in such a way his principle of separability would not be violated."[61]

J. S. Bell expresses similar views in a note, "Einstein and Hidden Variables," defending his position against criticism by Max Jammer: "I had for long thought it quite conventional and uncontroversial to regard Einstein as a proponent of hidden variables and indeed . . . as the most profound advocate of hidden variables."[62] A recent review article on the topic by Clauser and Shimony repeats this view.[63]

While there are statements by Einstein that might seem to allow such an interpretation, I believe that this view misses the basic thrust of Einstein's search for a unified field theory:

In a consistent field theory there is no real definition of the field. . . . *A priori* no bridge to the empirical is given. There is not, for example, any "particle" in the strict sense of the word, since it does not fit into the program of representing reality by everywhere continuous, indeed even analytic functions. . . . The upshot is that a comparison with the empirically known can only be expected to come from finding exact solutions of the system of equations in which empirically "known" structures and their interactions are "reflected." Since this is immensely difficult, the sceptical attitude of contemporary physicists is quite understandable. (*Besso Correspondance,* pp. 438–39, letter of April 15, 1950)

Our situation is this. We stand before a closed box, that we cannot open, and try to discuss what is inside and what is not. The similarity of the theory to Maxwell's is only external, so that we cannot transfer the concept of "force" from this theory to the asymmetric field theory. If this theory is at all useful, then one cannot assume any separation between particles and field of interaction. In addition, there is no concept at all of the *motion* of something more or less rigid. The question here is exclusively: are there singularity-free solutions? Is their energy in particular localized in such a way as demanded by our knowledge of the atomic and quantum character of reality [*Realität*]? The answer to this question is really not attainable with contemporary mathematical methods. Thus I do not see how one can guess whether any sort of action-at-a-distance or any type of object, insofar as we have attained a semi-empirical knowledge of them, can be represented by the theory. (Einstein to John Moffat, August 24, 1953, Item 17-394)

If the concept of "hidden variable theory," not to speak of "*local* hidden variable theory," is given a precise meaning—and not used as shorthand for *any* non-quantum-mechanical approach—then what Einstein has in mind is not a hidden variable approach. Fuller discussions of this question are found in the book and articles by Max Jammer cited at the beginning of this paper.

Einstein never denied the great explanatory power of quantum mechanics, yet he did not feel this success required acceptance of its conceptual structure as the basis for further progress in theoretical physics. He wrote to Schrödinger: "The wonder [about quantum mechanics] is only that one can represent so much with it, although the most important theoretical source of knowledge, group invariance, finds such incomplete application there . . . . It is the case that a logically coherent theory that is connected appropriately to the real state of affairs usually has great extrapolatory power, even if it is little related to the deeper truth [*der Wahrheit in der Tiefe*]" (Einstein to Schrödinger, July 16, 1946, Item 22-109).

8. GENERAL COVARIANCE

The mention of group invariance brings up one of the most profound sources of his skepticism about the ultimate significance of quantum mechanics. Einstein attached primary importance to the principle of general convariance: "You consider the transition to special relativity as the most essential thought of relativity, not the transition to general relativity. I consider the reverse to be correct. I see the most essential thing in the overcoming of the inertial system, a thing that acts upon all processes, but undergoes no reaction. This concept is in principle no better than that of the center of the universe in Aristotelian physics" (Einstein to Georg Jaffe, January 19, 1954, Item 13-405). Even relativistic quantum mechanics is constructed on a nongenerally covariant basis, but:

Contemporary physicists do not see that it is hopeless to take a theory that is based on an independent rigid space (Lorentz-invariance) and later hope to make it general-relativistic (in some natural way). (Einstein to von Laue, September, 1950, Item 16-147).

I have not really studied quantum field theory. This is because I cannot believe that special relativity theory suffices as the basis for a theory of matter, and that one can afterwards make a non generally relativistic theory into a generally relativistic one. But I am well aware of the possibility that this opinion may be erroneous. (Einstein to K. Roberts, September 6, 1954, Item 20-049)

Einstein felt it important to explore the possibility that a field theory based upon a continuous manifold, the principle of general covariance and partial differential equations, could provide an explanation of quantum phenomena:

I am firmly convinced, that the [EPR] dilemma depicted can only be overcome through a quite different outlook on the situation, and indeed by a description that lies much closer to the "classical" than we now hold probable or even conceivable. One must always bear in mind that up to now we know absolutely nothing about the laws of motion of material points from the standpoint of "classical field theory." For the mastery of this problem, however, no special physical hypothesis is needed, but "only" the solution of certain mathematical problems. (Einstein to Ernst Cassirer, March 16, 1937, Item 8-394)

To Schrödinger he wrote explaining in more detail why exact solutions were needed:

The quantum facts seem . . . to show that arbitrarily weak interactions can produce discrete changes (quantum jumps). Since this seems contradictory for the individual event, one is forced to the statistical interpretation: small forces produce small changes not in the individual things, but in the probability of their states. In reality however it must be something like this, that interactions have just as atomistic a character as the structures on which they act. Then it is all over with the quasi-static interpretation of interactions! Therefore I don't believe that one can advance further with field theory by approximative considerations (weak fields). (Einstein to Schrödinger, July 25, 1951, Item 22-180)

Sometimes, he indicated that a fundamental length might be needed to explain the existence of stable structures in such a field theory:

If one does *not* want to introduce rods and clocks as independent objects into the theory, then one must have a structural theory in which a fundamental length enters, which then leads to the existence of a solution in which this length occurs, so that there no longer exists a continuous sequence of "similar" solutions. This is indeed the case in the present quantum theory, but has nothing to do with its basic characteristics. Any theory which has a universal length in its foundations and, on the basis of this circumstance, qualitatively distinguished solutions of definite extent would offer the same thing with respect to the question envisioned here. (Einstein to W. F. G. Swann, January 24, 1942, Item 20-624)

At other times he proposed another idea: the overdetermination of a system of equations to restrict the manifold of possible initial conditions.

In "Does Field Theory Offer the Possibility for a Solution of the Quantum Problem," he wrote:

According to our theories up to now the initial state of a system can be freely chosen; the differential equations then give the temporal development. From our knowledge of quantum states . . . this trait of the theory does not correspond to reality. The initial state of an electron moving around the nucleus of a hydrogen atom cannot be freely chosen, but the choice must correspond to the quantum conditions. More generally; not only the temporal development but also the initial conditions obey laws.

Can one do justice to this knowledge about natural processes, to which indeed we must ascribe general significance, in a theory based upon partial differential equations? Quite certainly: we need only "overdetermine" the field variables by the equations. That is, the number of differential equations must be greater than the number of field variables determined by them.[64]

Einstein carried out a decades-long search for a classical field theory whose singularity-free solutions could reproduce the atomistic, quantum structure of matter and radiation. In spite of these efforts, he acknowledged the possibility that *any* theory based upon continuum concepts might be inadequate and that totally new mathematical concepts might be needed to explain quantum effects. As early as November 1916, he wrote to a former student:

But you have correctly grasped the drawback that the continuum brings. If the molecular view of matter is the correct (appropriate) one, i.e., if a part of the universe is to be represented by a finite number of moving points, then the continuum of the present theory contains too great a manifold of possibilities. I also believe that this too great is responsible for the fact that our present means of description miscarry with the quantum theory. The problem seems to me how one can formulate statements about a discontinuum without calling upon a continuum (space-time) as an aid; the latter should be banned from the theory as a supplementary construction not justified by the essence of the problem, which corresponds to nothing "real." But we still lack the mathematical structure unfortunately. How much have I already plagued myself in this way!

Yet I see difficulties of principle here too. The electrons (as points) would be the ultimate entities in such a system (building blocks). *Are there* indeed such building blocks? Why are they all of equal magnitude? It is satisfactory to say?: God in his wisdom made them all equally big, each like every other, because he wanted it that way; if it had pleased him, he could also have created them different. With the continuum viewpoint one is better off in this respect, because one doesn't have to prescribe elementary building blocks from the beginning. Further, the old question of the vacuum! But these considerations must pale beside the overwhelming fact: The continuum is more ample than the things to be described. (Einstein to Walter Dällenbach, November 1916, Item 9-072)

He continued to discuss this possibility over the years. In 1935 he wrote to Paul Langevin:

In spite of all successes of quantum mechanics I do not believe that this method can offer a useable *foundation* [*Fundament*] of physics. I see in it something analogous to classical statistical mechanics, only with the difference that here we have not found the equations corresponding to those of classical mechanics.

In any case one does not have the right today to maintain that the foundation must consist in a *field theory* in the sense of Maxwell. The other possibility, however, leads in my opinion to a renunciation of the time-space continuum and to a purely algebraic physics. Logically this is quite possible (the system is completely described by a number of integers; "time" is only a possible standpoint [*Gesichtspunkt*], from which the other "observables" can be considered—an observable logically coordinated to all the others). Such a theory doesn't have to be based upon the probability concept. For the present, however, instinct rebels against such a theory. (Einstein to Langevin, October 3, 1935, Item 15-408)

In his last years, he made a number of rather pessimistic comments about the prospects for a continuum theory:

In present-day physics there is manifested a kind of battle between the particle-concept and the field-concept for leadership, which will probably not be decided for a long time. It is even doubtful if one of the two rivals finally will be able to maintain itself as a fundamental concept. (Einstein to Herbert Kondo, August 11, 1952, Item 14-306)

I consider it as entirely possible that physics cannot be based upon the field concept, that is on continuous structures. Then *nothing* will remain of my whole castle in the air including the theory of gravitation, but also nothing of the rest of contemporary physics. (*Besso Correspondance,* p. 527, letter of August 10, 1954)

Your objections regarding the existence of singularity-free solutions which could represent the field together with the particles I find most justified. I also share this doubt. If it should finally turn out to be the case, then I doubt in general the existence of a rational physically useful *field* theory. But what then? Heine's classical line comes to mind "And a fool waits for the answer." (Einstein to André Lichnerowicz, February 25, 1954, Item 16-321)

I must confess that I was not able to find a way to explain the atomistic character of nature. My opinion is that if the objective description through the field as an elementary concept is not possible, then one has to find a possibility to avoid the continuum (together with space and time) altogether. But I have not the slightest idea what kind of elementary concepts could be used in such a theory. (Einstein to Bohm, October 28, 1954, Item 8-050)

At least once, he speculated in somewhat greater detail on what a non-continuum theory might involve:

The alternative continuum-discontinuum seems to me to be a real alternative; i.e., there is here no compromise. By discontinuum theory I understand one in which there are no differential quotients. In such a theory space and time cannot occur, but only numbers and number-fields and rules for the formation of such on the basis of algebraic rules with exclusion of limiting processes. Which way will prove itself, only success can teach us.

Physics up to now is naturally in its essence a continuum physics, in spite of the use of the material point, which looks like a discontinuous conceptual element, and has no more right of existence in field description. Its strength lies in the fact that it posits parts which exist quasi-independently, *beside* one another. Upon this rests the fact that there are reasonable laws, that is rules which can be formulated and tested for the individual parts. Its weakness lies in the fact that it has not been possible up to now to see how that atomistic aspect including quantum relations can result as consequences. On the other hand dimensionality (as four-dimensionality) lies at the foundation of the theory.

An algebraic theory of physics is affected with just the inverted advantages and weaknesses, aside from the fact that no one has been able to propose a possible logical schema for such a theory. It will be especially difficult to derive something like a spatio-temporal quasi-order from such a schema. I cannot imagine how the axiomatic framework of such a physics would appear, and don't like it, when one talks about it in dark apostrophes (*Anredungen*). But I hold it entirely possible that the development will lead there; for it seems that the state of any finite spatially limited system may be fully characterized by a finite number of numbers. This speaks against the continuum with its infinitely many degrees of freedom. This objection is not decisive only because one doesn't know, in the contemporary state of mathematics, in what way the demand for freedom from singularity (in the continuum theory) limits the manifold of solutions. (Einstein to H. S. Joachim, August 24, 1954, Item 13-453)

Even near the end of his life, he was on the lookout for new mathematical tools, which could turn such speculations—best kept private—into the basis for a real theory. The well-known mathematician Abraham Fraenkel reports:

In December 1951 I had the privilege of talking to Professor Einstein and describing the recent controversies between (neo-) intuitionists and their "formalistic" and "logicistic" antagonists; I pointed out that the first attitude would mean a kind of atomistic theory of functions, comparable to the atomistic structure of matter and energy. Einstein showed a lively interest in the subject and pointed out that to the physicist such a theory would seem by far preferable to the classical theory of continuity. I objected by stressing the main difficulty, namely, the fact that the procedures of mathematical analysis, e.g., of differential equations, are based on the assumption of mathematical continuity, while a modification sufficient to cover an intuitionistic-discrete medium cannot easily be imagined. Einstein did not share this pessimism and urged mathematicians to try to develop suitable new methods not based on continuity.[65]

9. CONCLUSION

One thing seems certain. The many attempts to reconcile general relativity and quantum theory have not yet brought us anything like a fully successful theory. Einstein expressed himself skeptically about all such attempts: "I do not believe that it will lead to the goal if one sets up a classical theory and then 'quantizes' it. This way was indeed successful in connection with the interpretation of classical mechanics and the interpretation of the quantum facts [*Quantentatsachen*] by modification of that theory on a fundamentally [*prinzipiell*] statistical basis. But I believe that, in attempts to transfer this method to field theories, one will hit upon steadily mounting complications and upon the necessity to multiply the independent assumptions enormously" (Einstein to John Moffat, June 4, 1953, Item 17-390).

The reconciliation of gravitation and quantization is still one of the most intractable problems in contemporary theoretical physics. The most profound comment one can make on it is that nobody has any really sound basis for believing that he or she knows how it will be solved—or bypassed—in the future. We can agree with Einstein: "One thing I have learnt in a long life: it is devilishly hard to get closer to 'Him', if one doesn't want to remain on the surface" (*Besso Correspondance,* p. 439, letter of April 15, 1950).

NOTES

1. P. Speziali, ed., *Albert Einstein–Michele Besso Correspondance, 1903–1955* (Paris: Hermann, 1972), p. 453 (hereafter referred to as *Besso Correspondance*). I shall give an example later of the sort of thing Einstein might have had in mind when he referred to self-deception.

2. A copy of the letter is in the Einstein Archive, Hebrew University of Jerusalem. I shall refer to such unpublished items by the Control Index number of the item in the duplicate of this Archive in Mudd Library, Princeton University (in this case, Item 16-167). Laue's was the first such treatise, published in 1911 after Einstein had turned down the publisher's offer to write one.

3. M. Klein, "No Firm Foundation: Einstein and the Early Quantum Theory," in H. Woolf, ed., *Some Strangeness in the Proportion* (Reading:Addison-Wesley, 1980), pp. 161–85.

4. A. Pais, "Einstein and the Quantum Theory," *Rev. Mod. Phys.* 51 (1979): 863. This article is included almost unchanged in A. Pais, "*Subtle is the Lord . . .* " *The Science and the Life of Albert Einstein* (Oxford: Clarendon Press, 1982). See also A. Pais, "Einstein on Particles, Fields and the Quantum Theory," in Woolf, *Some Strangeness in the Proportion,* pp. 197–251.

5. M. Jammer, *The Conceptual Development of Quantum Mechanics* (New York: McGraw-Hill, 1966); M. Jammer, *Albert Einstein und das Quantenproblem,* in H. Nelkowski et al., eds., *Einstein Symposion Berlin* (Berlin: Springer-Verlag, 1980); G. Holton and Y. Elkana, eds., *Albert Einstein/Historical and Cultural Perspectives* (Princeton: Princeton University Press, 1982), pp. 59–76.

6. *Besso Correspondance,* p. 238. There is a word following "fame" not transcribed in the printed version. It is illegible in the copy in the Einstein Archive and Professor Speziali kindly informed me that the original is no longer available to him.

7. The letter is quoted in Carl Seelig, *Albert Einstein und die Schweiz* (Zürich: Europa Verlag, 1952), pp. 77–78. The letter can be dated from its reference to the discovery of the mass-energy equivalence relation. I am indebted to Martin Klein for pointing out this quotation and for encouraging me to publish these speculations.

8. See A. Kleinert and C. Schoenbeck, "Lenard und Einstein," *Gesnerus* 35 (1978): 318, for the text of both letters cited here.

9. Of course, we now know that emission of a series of spectral lines is connected with transitions from a number of higher energy states to one lower state or level. I am not claiming that Einstein correctly anticipated the 1913 Bohr model in this letter.

10. This letter is printed in A. S. Eve, *Rutherford* (New York: Macmillan, 1930), pp. 224–26. The Pickering-Fowler series of spectral lines was first attributed to hydrogen. But the Bohr theory's prediction that it was actually due to singly ionized helium was experimentally confirmed in 1913 by Evans and, later, by Fowler and Paschen.

11. Letter of September 23, 1913, published in Ulrich Hoyer, ed., *Collected Works of Niels Bohr*, vol. 2, *Work on Atomic Physics, 1912–1917* (Amsterdam: North-Holland, 1981), p. 532. I am indebted to Dr. Erik Rüdinger for a copy of this letter before its publication.

12. *Über den Aether,* lecture of October 4, 1924, published in A. Einstein, *Verh d. Schweiz. Naturf. Ges.* 105 (1924): 85.

13. A. Pais, "Reminiscences from Post-war Years," in Rozental, *Niels Bohr* (New York: Interscience, 1964), p. 222.

14. A. Einstein, "Autobiographical Notes," in P. A. Schilpp, ed., *Albert Einstein: Philosopher-Scientist* (La Salle: Open Court, 1951), pp. 3–94. A new edition with corrected translation appeared in 1979 entitled *Albert Einstein: Autobiographical Notes.* The first edition will hereafter be cited as "Autobiographical Notes." In a draft statement he wrote in the early 1920s, proposing Bohr as a corresponding member of the Prussian Academy of Sciences, Einstein elaborated: "The empirically widely known structure of [atomic] spectra differed so essentially from what was to be expected, according to our old theories, that no one ⟨had the courage⟩ saw a possibility ⟨for saw for a compelling⟩ for a convincing theoretical interpretation of the observed regularities." (The words Einstein crossed out are in angle brackets; the lack of "courage" may well have been a self-reference.)

15. Cited in Ronald Clark, *Einstein: The Life and Times* (New York: World 1971), p. 169.

16. T. Kuhn, *Black Body Theory and the Quantum Discontinuity, 1894–1912* (Oxford: Oxford University Press, 1978).

17. "Peter J. W. Debye: An Interview," *Science* 145 (1964): 554.

18. A. Einstein, *Ann. Phys.* 23 (1907): 371.

19. A. Einstein, *Ideas and Opinions* (New York: Crown, 1954), p. 225. It will hereafter be cited as *Ideas and Opinions.*

20. The German text, as well as an English translation, appear in *Builders of the Universe* (Los Angeles: U. S. Library Association, 1932). I have retranslated the German version, pp. 94–96.

21. Einstein was clearly in a foul mood at the time this letter was written, shortly after Hitler's triumph at Munich. He added a P.S.: "What do you say to the way the fine democracies [*sauberen Demokratien*] are behaving? I say 'Pfui Teufel!'"

22. But, not classical statistical mechanics proper which led to disastrous consequences such as the equipartition theorem giving the wrong black body spectrum—the so-called Rayleigh-Jeans law (although Einstein was the first to write it down completely correctly)—and the wrong low temperature behavior of specific heats.

23. A. Einstein, "Zum gegenwärtigen Stande des Problems der spezifischen Wärme," in W. Nernst, ed., *Die Theorie der Strahlung und der Quanten* (Halle: Krapp Verlag, 1914), p. 353. This edition contains the original German text of Einstein's comments. The proceedings of the 1911 Solvay Congress were first published in French: P. Langevin and M. de Broglie, eds., *La Théorie du Rayonnement et les Quanta* (Paris: Gauther-Villars, 1912). Einstein had already considered and rejected the possibility of a statistical interpretation of the energy conservation principle. In a letter to Jacob Laub of Nov. 4, 1910, Einstein wrote: "Currently I have great hope of solving the radiation problem, and indeed *without light quanta.* . . . One must renounce the energy principle in its present form."

24. However, Einstein showed great reluctance to mixing relativistic considerations into his work on quantum theory: witness his over-a-decade-long delay between writing $E = h\nu$ and $p = h\nu/c$. Pais has discussed the reasons for this in detail.

25. A. Einstein, *Physik. Zeitschr.* 10 (1909): 185.

26. Einstein published his results three times: A. Einstein, *Verhandl. Deutsch. Phys. Ges.* 18 (1916): 318, *Mitt. Phys. Ges. Zürich* 16 (1916): 47, *Phys. Z.* 18 (1917): 121, the last two being the same paper, translated in B. L. van der Waerden, ed., *Sources of Quantum Mechanics* (Amsterdam: North-Holland, 1967), pp. 63–77. This book will be cited hereafter as van der Waerden, *Sources.*

27. M. Born, W. Heisenberg, and P. Jordan, *Z. Phys.* 35 (1926): 557, translated in van der Waerdan, *Sources,* pp. 312–85. Fluctuations are discussed in the final section of the paper.

28. P. Ehrenfest, *Z. Phys.* 34 (1925): 362.

29. P. Jordan, *Z. Phys.* 45 (1927): 766. See pp. 773–75.

30. See note 17.

31. See J. J. Gonzalez and H. Wergeland, *Kongel. Norske Videnskab. Selsk. Skr. 1973,* no. 4, "Einstein-Lorentz's Formula for the Fluctuations of Electromagnetic Energy," for a recent review, including reference to Heisenberg's paper. See also N. Bohr and L. Rosenfeld, *Mat.-fys. Medd. Dan. Vid. Selsk.* 12, no. 8 (1933), for the fundamental paper on the conditions of measurability of field quantities. This paper is translated in R. S. Cohen and J. J. Stachel, eds., *Selected Papers of Leon Rosenfeld* (Dordrecht: Reidel, 1979), pp. 357–400.

32. For recent surveys of this topic, see D. W. Sciama, "Black Holes and Fluctuations of Quantum Particles: An Einstein Synthesis," in M. Pantaleo and F. de Finis, eds., *Relativity, Quanta and Cosmology in the Development of the Scientific Thought of Albert Einstein* (New York: Johnson Reprint Corp., 1979), vol. 2, pp. 681–724; and J. D. Bekenstein, "Gravitation, the Quantum and Statistical Physics," in Y. Ne'eman, ed., *To Fulfill a Vision* (Reading: Addison-Wesley, 1981), pp. 42–59.

33. M. Born, ed., *The Born-Einstein Letters* (New York: Walker, 1971).

34. He did use the term at least once in his correspondence. See Einstein to Ehrenfest, January 11, 1922, Item 10-003.

35. H. A. Lorentz, *Problems of Modern Physics* (Boston: Ginn and Co., 1927; rpt. New York: Dover, 1967).

36. M. Born, *Zeitschr. f. Physik* 38 (1926): 803.

37. W. Heisenberg, *Zeitschr. f. Physik* 43 (1927):172.

38. A reference to "something which he [Einstein] calls a 'ghost' field of radiation," in H. A. Kramers and H. Holst, *The Atom and the Bohr Theory of Its Structure* (New York: Knopf, 1923) proves that Einstein's idea was well known in Copenhagen in the early twenties.

39. A. Einstein, *Sitzungsber. preuss. Akad. Wiss., physik.-math. Kl.* (1925): 3.

40. L. de Broglie, "Meeting with Einstein at the Solvay Council in 1927," in L. de Broglie, *New Perspectives in Physics* (New York: Basic Books, 1962), pp. 180–85.

41. Part of this correspondence has been published: K. Przibram, ed., *Schrödinger-Planck-Einstein-Lorentz: Briefe zur Wellenmechanik* (Vienna; Springer, 1963), translated as K. Przibram, ed., *Letters on Wave Mechanics: Schrödinger, Planck, Einstein, Lorentz* (New York: Philosophical Library, 1967).

42. Copies of Heisenberg's and Jordan's letters to Einstein are in the duplicate Einstein Archive. Einstein's letters to both, except fot one postcard to Jordan, appear to have been destroyed during the war.

43. Ulrich Benz, *Arnold Sommerfeld* (Stuttgart: Wissenschafliche Verlagsgesellschaft, 1975), pp. 152–53.

44. W. Heisenberg, *Der Teil und das Ganze* (Munich: DTV, 1973), translated as *Physics and Beyond* (New York: Harper & Row, 1971). This will be cited hereafter as *Physics and Beyond.*

45. H. Minkowski, *Goett, Nachr. Math-Phys. K1.* (1908): 54.

46. Einstein, "Zum gegenwärtigen Stande," p. 353. Einstein had used the phrase "variables observable in principle" in a similar context the previous year. See A. Einstein, *Ann. Phys.* 33 (1910): 1275.

47. Einstein, "Zum gegenwärtigen Stande," p. 355.

48. J. Stachel, "Einstein's Search for General Covariance, 1912–1915," talk given at the Ninth International Conference on General Relativity and Gravitation, July 17, 1980, to be published.

49. W. Pauli, *Verhandl. Deutsch. Physik. Ges.* 21 (1919): 742.

50. A. Hermann, K. V. Meyenn, and V. F. Weisskopf, eds., *Wolfgang Pauli/*

Wissenschaftliche Briefwechsel, vol. 1: *1919–1929* (Berlin: Springer Verlag, 1979). This will be referred to hereafter as *Pauli Briefwechsel.* See pp. 115–19 for the letter to Eddington.

51. M. Born and P. Jordan, *Zeitschr. f. Physik* 33 (1925): 479.

52. M. Born, *Nature* 119 (1927): 354, reprinted in M. Born, *Physics in My Generation* (New York: Springer Verlag, 1969), pp. 6–12.

53. L. Wessels, "What Was Born's Statistical Interpretation," in P. O. Asquith and R. N. Giesel, eds., *PSA 1980* (East Lansing; P. S. A., 1980), vol. 2; M. Beller, "The Myth of Born's Probabilistic Interpretation" (preprint). I thank Dr. Beller for a copy of this paper.

54. The historical accuracy of this claim has been challenged: see E. MacKinnon's paper, "Heisenberg, Models, and the Rise of Matrix Mechanics," in R. McCormmach and L. Pyenson, eds., *Historical Studies in the Physical Sciences,* vol. 8 (Baltimore: Johns Hopkins University Press, 1977).

55. W. Heisenberg, *Zeitschr. f. Physik.* 33 (1925): 879. Translation cited from Van der Waerden, *Sources,* pp. 261–62.

56. For Bohr's account, see N. Bohr, "Discussions with Einstein on Epistemological Problems in Atomic Physics," in Schilpp, *Albert Einstein* (1951), pp. 201–41.

57. See A. Einstein, *Sitzungsber. preuss. Akad. Wiss., physik.-math. K1.* (1921): 882, for the first proposed experiment and (1922): 18, for the reanalysis; A. Einstein, *Naturwiss.*14 (1926): 300, for the second proposed experiment, and A. Einstein, *Sitzungsber. preuss. Akad. Wiss., physik.-math. K1.* (1926): 234, for the reanalysis.

58. As mentioned earlier, Einstein had considered and rejected this idea over a decade before.

59. See Bohr's letter to Einstein of April 13, 1927 (Item 8-084).

60. A. Einstein, "Introductory Remarks on Basic Concepts," in *Louis de Broglie, Physicien et Penseur* (Paris: Albin Michel, 1953), pp. 4–14.

61. B. d'Espagnat, "Quantum Non-separability: How the Problem Evolved from Einstein's Time to Ours." Orsay preprint, December, 1976.

62. In *Frontier Problems in High Energy Physics* (Pisa: Scuola Normale Superiore, n.d.), pp. 41–45.

63. J. F. Clauser and A. Shimony, *Rep. Prog. Phys.* 41 (1978): 1881.

64. A. Einstein, *Sitzungsber. preuss. Akad. Wiss., physik.-math. K1.* (1923): 359.

65. A. Fraenkel, *Bull. Res. Coun. Israel* 3 (1954): 283.

Index of Names

Index of Topics